INFRARED THERMOGRAPHY

Edited by **Raghu V. Prakash**

Infrared Thermography
Edited by Raghu V. Prakash

Published by InTech
Janeza Trdine 9, 51000 Rijeka, Croatia

Publishing Process Manager Silvia Vlase

Technical Editor Teodora Smiljanic

Cover Designer InTech Design Team

Additional hard copies can be obtained from orders@intechopen.com

Infrared Thermography, Edited by Raghu V. Prakash
p. cm.
ISBN 978-953-51-0242-7

For EU product safety concerns:
IN TECH d.o.o., Prolaz Marije Krucifikse Kozulić 3, 51000 Rijeka, Croatia,
info@intechopen.com or visit our website at intechopen.com.

Meet the editor

Dr Raghu V. Prakash is an Associate Professor working with the Department of Mechanical Engineering, Indian Institute of Technology Madras, India, specializes in the areas of fatigue, fracture of materials (metals, composites, hybrids), structural integrity assessment and remaining life prediction of critical components used in aircraft, automotive, power-plant and process industries. He has more than twenty five years of professional experience in the field of fatigue and fracture and has hands-on experience in handling industrial problems relating to life prediction. Dr Prakash has won several prestigious awards and scholarships. His work experience includes Scientist at National Aerospace Laboratories, Bangalore, India (1993-98), Technical Director, BiSS Research, Bangalore, India (1998-2003) prior to joining academics. He has published more than 90 journal papers and conference papers on the areas of fatigue, fracture, materials characterization, of which many are with ASTM.

Contents

Preface

Infrared Thermography (IRT) is commonly used these days as a NDE tool to identify damages and provide remedial action. The fields of application are vast, that amongst others, includes materials science, life sciences and applied engineering. This book offers a collection of ten chapters with three major sections - relating to application of infrared thermography to study problems in materials science, agriculture, veterinary and sports fields as well as in engineering applications. The authors who have contributed to this book have a thorough knowledge in this select field. Aspects relating to both experimental application of IRT, as well as to mathematical modeling of temperature fields have been covered in this book. There is still a wide scope to make the technique as a quantitative measure of changes in any physical, biological system and it is our hope that more and more researchers in this field will contribute to the future compilations of research articles.

Dr. Raghu V. Prakash
Machine Design Section,
Department of Mechanical Engineering, IIT, Madras
India

Part 1

Materials Science

1

Application of IR Thermography for Studying Deformation and Fracture of Paper

Tatsuo Yamauchi
Kyoto University
Japan

1. Introduction

The deformation and fracture of paper have long been investigated from various points of view using many methods. Phenomena such as sound emission and thermal change occurring with the deformation have also been examined. Paper is a planar material and its thermal conductivity is low and further its specific heat is high; thus, the thermal change shown in surface temperature using IR thermography can be investigated related to the mechanical behavior. Advanced thermography with extensive computer memory and various imaging processes enables us to investigate the process of the deformation and fracture of paper. In this paper, studies on the fundamental thermodynamics of paper material are first critically introduced, discussing the effects of defects and unevenness of paper formation. Secondly, uneven thermal distribution and its time-sequential changes during the fracture toughness testing of paper are presented; that is to say, stress concentration and crack development at the notch tip which occurred during the deformation of paper with a notch could be visualized as serial images using IR thermography. The results helped to develop fracture toughness testing of paper and further gave an understanding of the fracture toughness of paper itself.

2. Application of thermography to investigate material deformation

The deforming and fracturing processes of solid material under stress are directly related to its mechanical strength, and thus are very important for many of its practical uses. A great number of studies relating to deforming and fracturing processes have been made from both experimental and theoretical points of view. Regarding paper material, a representative of two-dimensional planar material, its tensile load / elongation behavior has exclusively been investigated (Seth and Page, 1981; Chao and Sutton, 1988; Yamauchi et al., 1990). Among those studies thermodynamics of deformation is one of the basic subjects. Studying the fracture mechanisms of solid material from the viewpoint of thermodynamics has been proposed for a long time (Muller,1969). It is well known that the deformation process of solid material accompanies the generation and absorption of heat; however, it is not easy to measure the heat change or the change of the temperature of solid material under deformation and the related knowledge is still fragmentary even at the present time. In these circumstances, the measurement of infrared ray radiation from material is fairly popular as a non—contact method to measure temperature. This method started as a point measurement, progressed to a line measurement and further developed to measure a series of sequential two-dimensional temperature images for a material surface, that is, modern infrared thermography. Thus, the overall surface temperature was experimentally obtained without any difficulty; however, it did not assign the internal temperature of the three-dimensional material to study the thermodynamics. This gap between the surface and interior is one reason why the thermodynamics of solid deformation has been little studied. The surface temperature of two-dimensional planar material could be regarded as the real temperature of the material, as experimentally shown later (referring to the last paragraph of **6**. Effect of paper formation, see Fig. 12).; therefore, thermography was applied to observe the deformation of paper materials (Yamauchi et al., 1993).

From a technical point of view, the advancements of IR sensing, image processing with help of computer enabled a commercially available thermography measurement system equipped with high sensitive temperature detection of 0.1℃, image subtraction among serial images, histogram processing, mean value calculating functions and so on. This modern thermography measurement system was first commercialized for medical uses especially for breast cancer detection, and has since been used for non-destructive testing (NDT) of heating systems and quality inspection in commercial manufacturing. These are almost all static uses and dynamic use was limited, excluding the use of a special equipment combined with thermography to determine stress distribution within a solid material under strain (Koizumi, 1983). Thus, using this commercially available equipment named a stress analyser, imaginary calculation was conducted using temperature image changes caused by periodically given ultra-low distortion. The first trial of the dynamic use of thermography to display temperature images was probably conducted to detect fatigue-related heat emission in composite materials (Reifsnider and Williams, 1974).; however, first-stage thermography could not produce successive images to show the deformation process. A full—scale study of the deformation process was then conducted using advanced modern thermography for paper material (Yamauchi et al., 1993), partly because the thermal conductivity of which is low and thus suitable for consecutive observation of temperature images.

3. Temperature (heat) change during the tensile deformation of paper

Heat generation and absorption occurring during the tensile deformation process of paper was investigated with an in-house developed micro-calorimeter attached with straining device (Ebeling et al., 1974) by Ebeling. Figure 1 gives a representative result showing the heat absorption firstly occurred in the elastic tensile deformation region and heat generation caused by further straining in the plastic deformation region (Ebeling, 1976). He concluded that the heat absorption was caused by Kelvin's thermoelastic effect (heat generation in compression mode) and heat generation in the plastic deformation region was mainly due to irreversible deformation of fibers. At the same time, heat generation and adsorption occurring during the tensile deformation of paper were investigated with an infra-red line scanner (Dumbleton et al., 1973). This method also detected the initial cooling of paper specimens during elastic extension as well as the temperature rise due to energy dissipation during plastic deformation, as shown in Fig. 2. It was further found that energy change occurred throughout the specimen in a fairly uniform manner up to the moment of rupture, at which time the temperature increased instantaneously throughout the specimen and it was possible to account for from 40 to 80 % of the mechanical energy losses as increases in thermal energy.

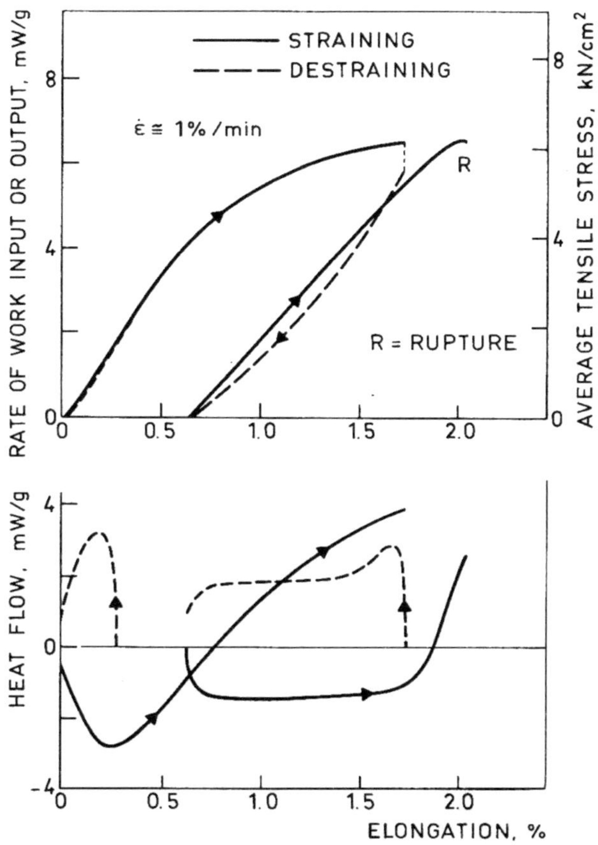

Fig. 1. Thermodynamic behavior of paper on tensile loading (Ebeling, 1976).

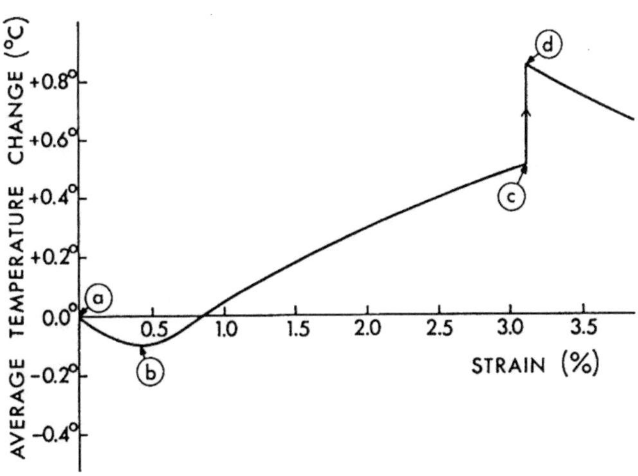

Fig. 2. Integrated average thermal response of paper on tensile loading (Dumbleton et al., 1973).

As to the temperature determination based on infrared radiation, the emissivity of paper is indispensable. Although it is considered to be about 0.9, 1.0 has been used for a long time as a perfect black body (Yamauchi et al., 1993), partly because of experimental difficulty in the determination. On the other hand, taking advantage of thermal emissivity difference in part, thermograph is applicable to measure moisture distribution and water absorption behavior of paper (Vickery et al., 1978; Tajima et al., 2011).

4. Application of thermography to study the tensile deforming process of paper

As shown in Fig. 3, the tensile behavior of paper was experimentally determined as the load-elongation relation and sequential surface temperature images of paper during tensile loading were spontaneously observed (Fig, 4) using modern thermography, and then the average temperature and histogram display of the temperature distribution were calculated (Fig. 5) (Yamauchi et al., 1993). Furthermore, a change in the average temperature during loading and the corresponding load—elongation curve were obtained, as shown in Fig. 6. Although the sensor sensitivity is 0.1℃ and the temperature change is very slight, averaging many temperature pixels caused a marked improvement of the S/N ratio using modern thermography, and thus it was measured with extremely high precision. The result is quite similar to those determined using the calorimeter and the infra-red line scanner, the thermal energy and mechanical rupture energy ratio being 57% within a predictive range (Ebeling, 1976; Dumbleton, 1973).

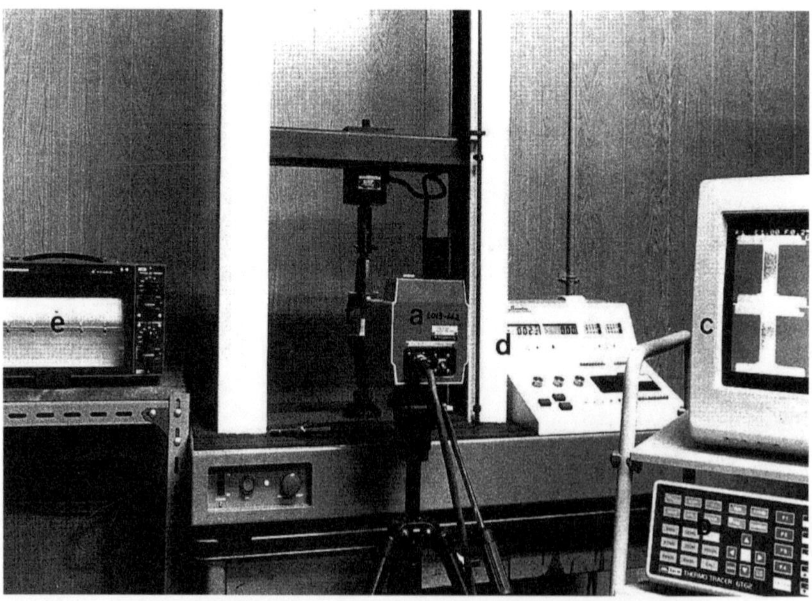

Fig. 3. Setup view of thermography for measuring the surface temperature of paper specimen under tensile strain: (a) infrared camera; (b) control unit; (c) monitor display; (d) Instron-type machine; (e) recorder; (f) paper specimen (Yamauchi and Tanaka, 1994).

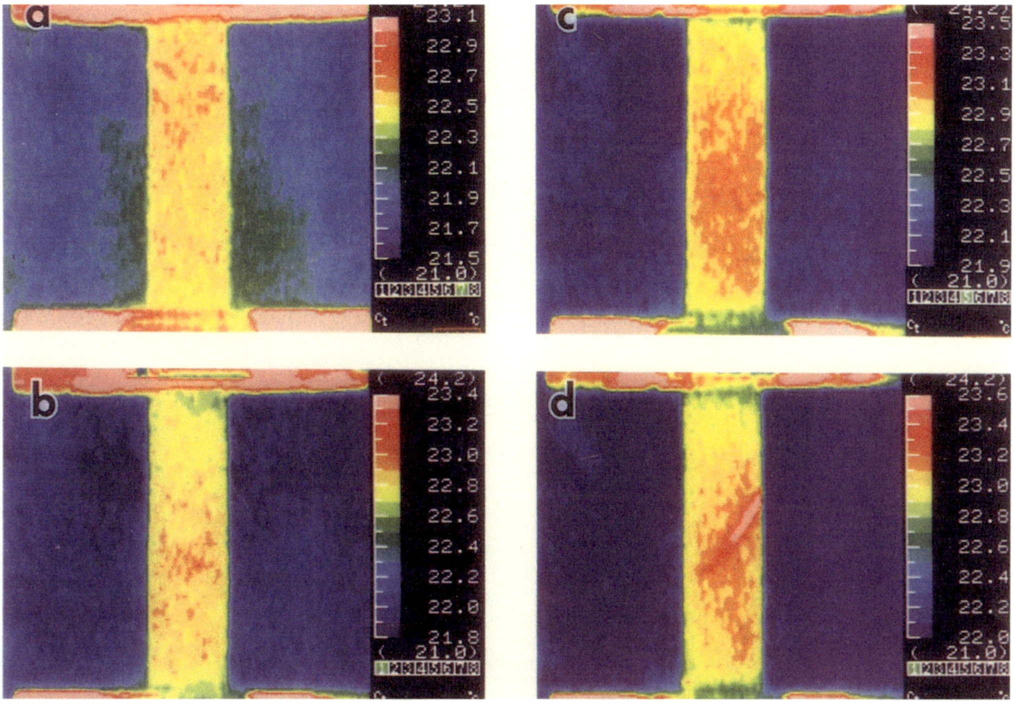

Fig. 4. A series of surface temperature images of paper under tensile strain. (a)-(d) correspond to the positions on the load-elongation curve in Fig. 6 (Yamauchi et al., 1993).

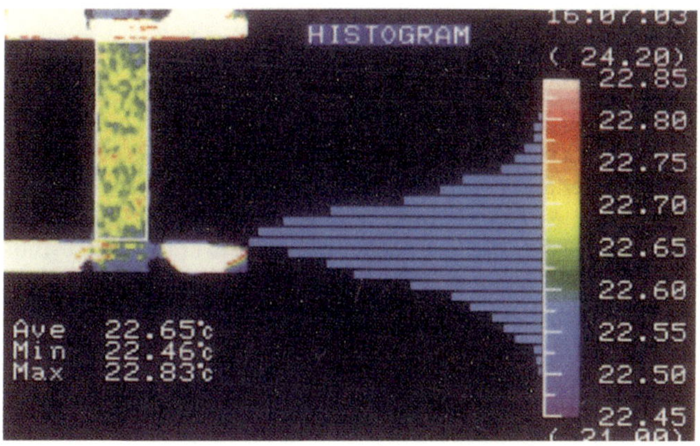

Fig. 5. Temperature image of paper under strain, histogram display of its temperature distribution, and calculated average temperature (Yamauchi et al., 1993).

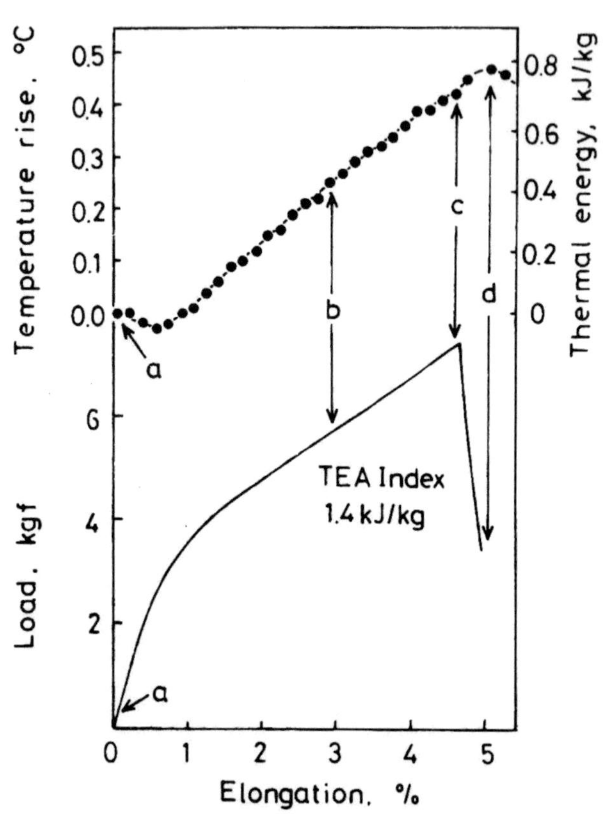

Fig. 6. Change of average temperature of paper under strain and the corresponding load-elongation curve (Yamauchi et al., 1993).

5. Thermal behavior of paper due to elastic elasticity and entropy elasticity

The initial tensile loading of paper accompanied cooling, which was in accordance with Kelvin's thermoelastic theory based on elastic elasticity. On the other hand, the heating of paper during plastic region elongation was interpreted as irreversible fiber deformation and the role of entropy elasticity was rejected, partly because of heat generation during unloading (Ebeling, 1976), although the mechanism of paper heating is still not clear. Subsequently, the thermal behavior of paper and rubber during unloading after highly tensile loading and ample successive stress-relaxation periods was examined in order to confirm this interpretation. Thus, a paper specimen was first strained up to about 80% of the breaking load, followed by being held at constant elongation for 1 h (stress-relaxation period) and then was destrained to zero load within 6 s (Yamauchi and Tanaka, 1994). The changes of load and the average temperature of the specimen during the above-mentioned process are shown in Fig. 7. Following the start of the tensile test, the temperature began to fall and reached a minimum at a point somewhat after the end point of the elastic deformation region. The temperature then rose almost linearly in the plastic deformation region. These behaviors are essentially the same as those described above. During the period of stress relaxation, the temperature becomes the surrounding temperature; that is to say, a period of 1h may be long enough to reach a temperature equilibrium. On destraining,

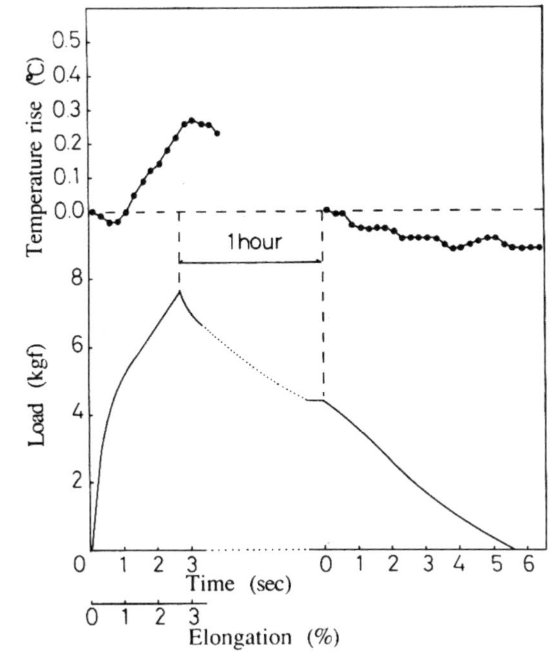

Fig. 7. Changes of load and temperature of paper during tensile straining, stress relaxation, and destraining periods (Yamauchi and Tanaka, 1994).

unexpectedly, the temperature gradually falls, although the extent of the temperature drop is smaller than that of the temperature rise during straining. For comparison, the changes of load and the average temperature of rubber during the same loading and unloading process are shown in Fig. 8. The temperature change indicates typical entropy elasticity, that is, the

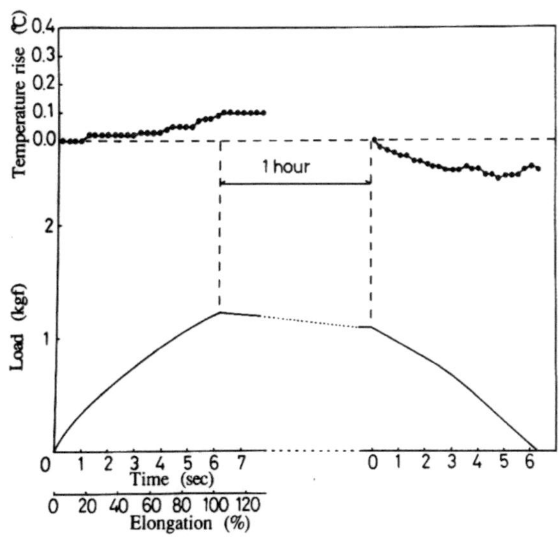

Fig. 8. Changes of load and temperature of rubber during tensile straining, stress relaxation, and destraining periods (Yamauchi and Tanaka, 1994).

temperature increase on straining from the start, and falls on destraining after the stress relaxation period. These results for paper suggest that elastic deformation based on energy elasticity occurred from the beginning of elongation and plastic elongation partly arose from deformation based on entropy elasticity. Further study is expected to clarify the role of entropy elasticity to plastic deformation for various types of paper.

6. Effect of paper formation on the deformation pattern shown as thermal images

Successively obtained thermal images of paper on loading (see Fig. 4) included a temperature irregularity arising from the measuring location condition. This is irrelevant in a thermodynamic study based on the change of average temperature; however a discussion of thermal distribution within an image needs a real thermal image and the thus a net thermal image was obtained by image subtraction between chronologically adjacent images, since the dimensional change of paper during that time was negligible (Yamauchi and Murakami, 1992). Paper is a fairly homogenous material and thermal change under loading occurred in fairly uniform manner throughout the specimen up to the moment of breakage (Dumbleton et al., 1973; Yamauchi and Murakami, 1992; 1993) .On the other hand, poorly formed paper, which has heterogeneous mass distribution, showed a marked uneven distribution in the higher temperature region (red part surrounded by yellow part) in the early stage of plastic deformation, but at a later stage of plastic deformation the temperature rise image became homogenous, as shown in Fig. 9 (Yamauchi and Murakami, 1993; 1994). Taking into consideration that paper often starts to break from the low mass region at the edge, more precise serial thermal images and soft X-ray images measured before elongation giving the mass distribution were required, as shown in Fig. 10 (Yamauchi and Murakami, 1994). These images allow us to follow the deformation process in more detail and to find a relationship between the temperature rise distribution and the mass distribution. The final

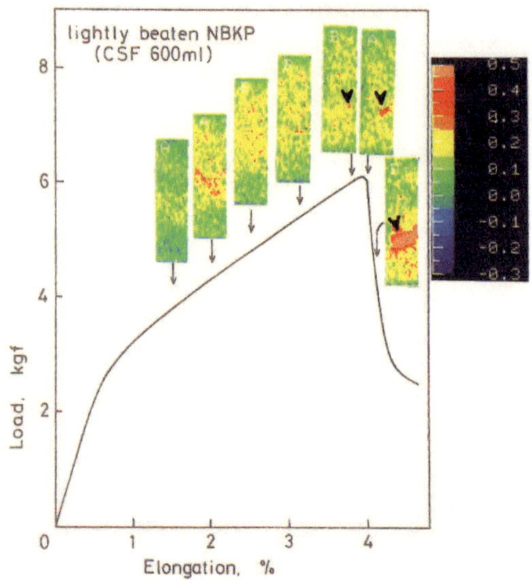

Fig. 9. Load/elongation relation and subtracted temperature rise distribution images during tensile straining of poorly —formed paper (Yamauchi and Murakami, 1994).

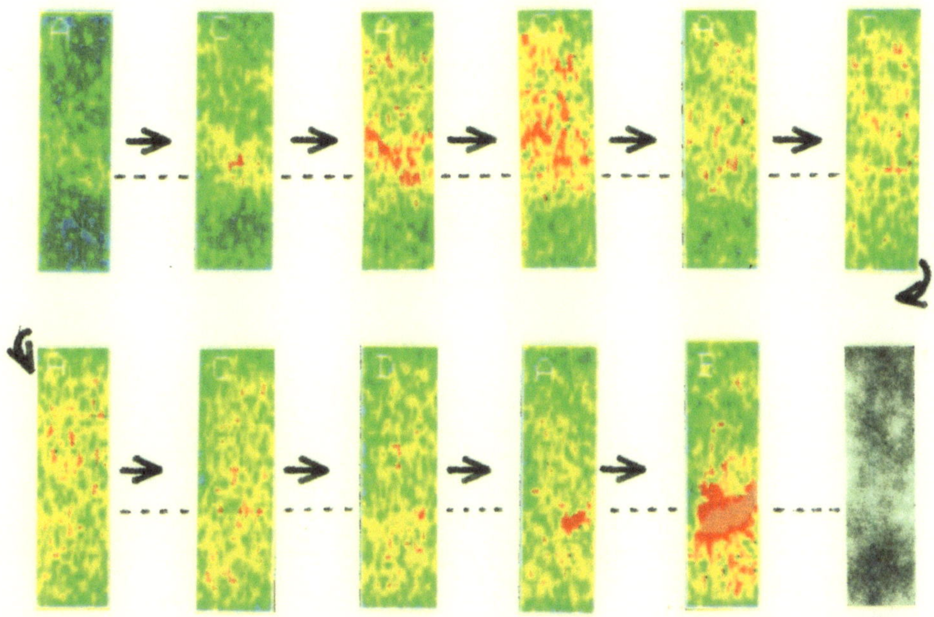

Fig. 10. Sequential subtracted temperature rise distribution images during straining and the soft X-ray mass distribution image of poorly—formed paper (Yamauchi and Murakami, 1994).

failure line marked by black arrows in Fig. 9 did not start from the edge but from a point within the specimen in this case.

Even within the same sample having the same poor formation (mass distribution), the relation between the deformation process, shown as sequential temperature rise distribution images, and breaking load/elongation, somewhat differs in each test. The results of four specimens from the same paper were compared in Fig. 11 with the breaking load, elongation and the soft X-ray image. Case 1 shows the shortened form of Fig. 10. In all these cases except case 4, the final failure line (pink part) started to run from one of the higher temperature regions (red and yellow part) in the early stage of plastic deformation, although these points often became indistinctive in the middle stage of plastic deformation. Thus, the failure-starting point can be predicted in the early stage of plastic deformation. The final failure is interestingly running through the one low mass (thin) part of the specimen, as shown on the soft X-ray image.

Temperature rise distribution images in case 4 were comparatively uniform, i.e., there is no distinctive sign showing the final failure line before breakage. Based on the fact that well—formed paper having uniform mass distribution showed a uniform deformation pattern, shown as uniform images of the temperature rise, and gave a higher breaking load (7.1 kgf), the higher breaking load in case 4 (6.8 kgf) suggested that the extent of uniformity in temperature rise distribution during straining, rather than that in mass distribution, was more directly related to the degree of stress concentration and thus to the breaking load.

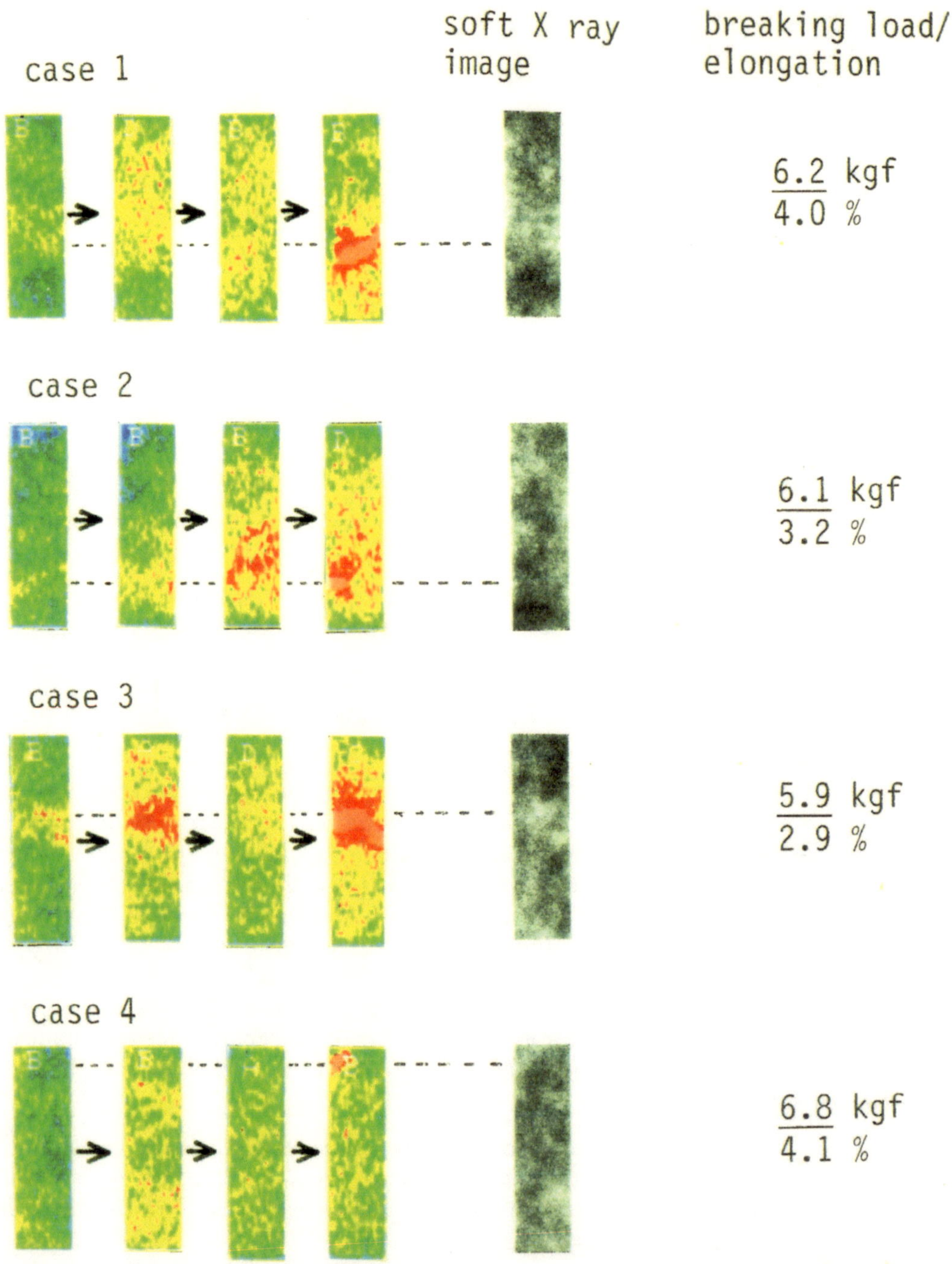

Fig. 11. Variation in the subtracted temperature rise distribution images, mass distribution image and breaking load/elongation of poorly—formed paper (Yamauchi and Murakami, 1994).

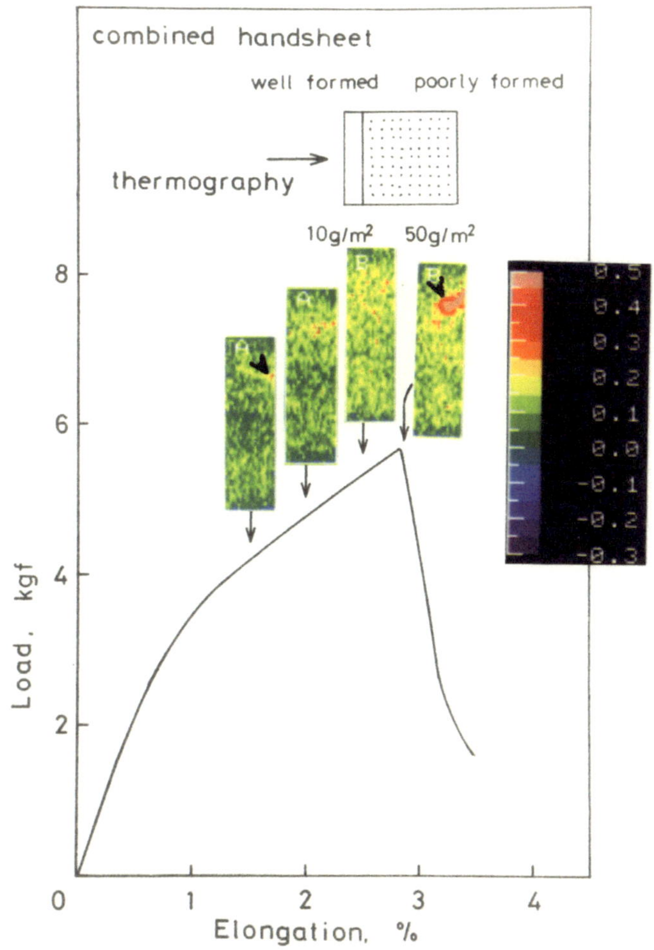

Fig. 12. Load/elongation relation and subtracted temperature rise distribution images during tensile straining of paper composed with well—and poorly—formed layers; the images were observed from the well—formed side.

In order to confirm the assumption that thermal behavior based on the surface temperature of paper is the real thermal behavior of paper itself using the difference derived from the formation in thermal behavior on loading, a special paper combined with well and poorly formed paper layers was prepared and its sequential temperature rise distribution images during tensile straining were observed from the well—formed layer side. Figure 12 shows the result of paper composed of a 10 g/m^2 well—formed layer and a 50g/m^2 poorly—formed layer. If the observed sequential temperature rise distribution image is not the thermal behavior of the paper itself but the surface thermal behavior, thermal change under loading should be uniform throughout the specimen up to the moment of breakage. However, uneven distribution of the higher temperature region, which finally developed to the breaking point, was observed at the early stage of plastic deformation. That is to say, the thermal behavior of paper observed as the surface temperature could be assigned to that of the paper itself.

7. Application of thermography for studying fracture toughness testing of paper

Relating to the simple elongation testing of paper to determine the tensile strength, fracture toughness testing was conducted using a paper specimen with notch (Niskanen, 1993). The ability of a paper to resist crack propagation from the notch tip under tensile loading is quite important in a number of end-use situations, as well as in papermaking processes. At the manufacturing, printing, and converting operations, stresses are concentrated around the defects in the plane of paper sheet. Thus, in-plane fracture toughness testing has been mainly investigated for paper specimens with center, single or double side notches. During testing, stress is concentrated on around the notch tip and resistance (work) to failure before the start of crack propagation was determined as crack tip opening displacement, J-integral and essential work of fracture (EWF) (Hirano and Yamauchi, 2000). The change in stress distribution, shown as sequential temperature rise distribution, and movement of the crack tip, shown as the shift of the highest temperature point during testing, could be visualized using thermography (Tanaka et al.,1997; Yamauchi and Hirano, 2000). The advantages of thermography are fully used to obtain the sequential changes of thermal images during testing to study the fracture toughness and the testing method itself.

In the measurement of the EWF for fracture toughness developed originally for ductile metals (Cotterell and Reddel, 1977), deep double-edge notched tension (DENT) specimens of various widths were employed as shown in Fig. 13. When a specimen completely yields

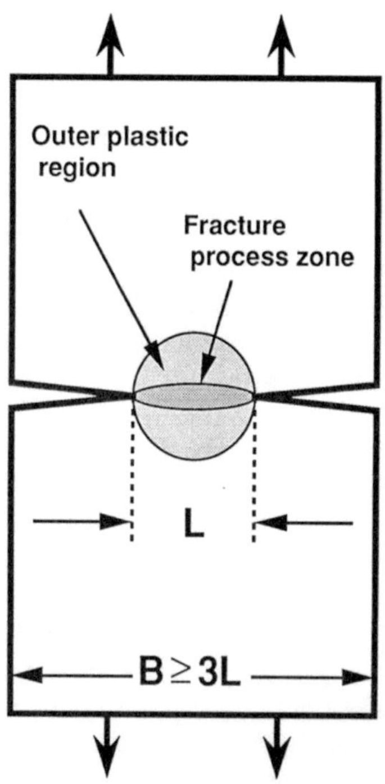

Fig. 13. Deep double-edge notched tension specimen for in-plane fracture toughness testing, showing the fracture process zone and the outer plastic deformation region (Cotterell and Reddel,1977) (B: specimen width, **L**.: ligament length).

before fracture, a plastic deformation zone is theoretically supposed to be a circular area centered on the ligament (between double notches, one third of the width, displayed in ***L***). Further, the work performed to fracture paper specimen can be separated into two components: (1) the essential work performed in the fracture process zone and (2) the non essential work performed in the plastic deformation zone, and a linear relationship is expected between ***L*** and (specific) work for fracture (w_f). In this method the appearance of the circular plastic deformation zone on local process zone before onset of stable crack growth should be a prerequisite for this measurement. Thus, detection of the fracture process zone was tried by both methods, silicone impregnation and thermography (Tanaka et al., 2003), and further development of the plastic deformation zone during testing was examined on sequential close-up temperature rise distribution images, as shown in Fig.14,15. At the moment of fracture during testing, stress is concentrated as expected at the tip of notches, shown as a rise of temperature around the notch tips (I) and no thermal change at the region far from notch tips(II) in Fig. 14. Sequential close-up temperature rise distribution images displayed in more detail the stress concentration and further the appearance of a circular plastic deformation zone from halfway through the plastic deformation region. In practice, the blue spot, whose temperature is higher than the surroundings, firstly appeared around the notch tips (Fig. 15; d,e). These spots spread out toward the inside, and then the enlarged spots joined together just after the maximum load point (Fig. 15; h). Finally, the whole shape of the amalgamated high-temperature zone became circular just before failure (Fig. 15; i). On the other hand, the notch tip has moved

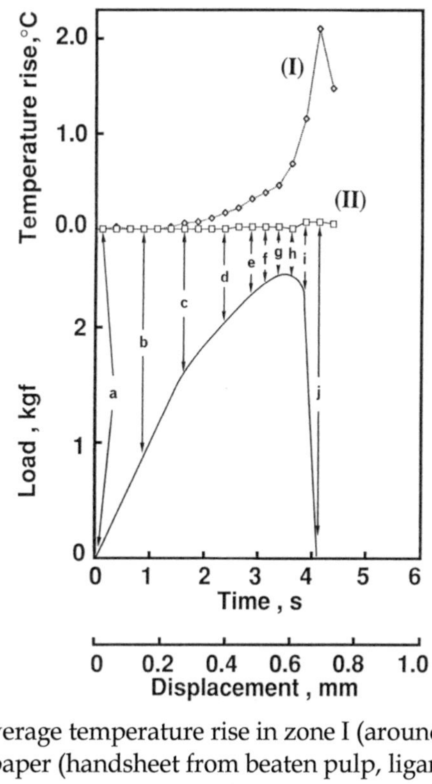

Fig. 14. Changes of the average temperature rise in zone I (around notch tip) and II (far from notch) during testing of paper (handsheet from beaten pulp, ligament: 5mm) and the corresponding load-displacement curve (Tanaka et al., 1997).

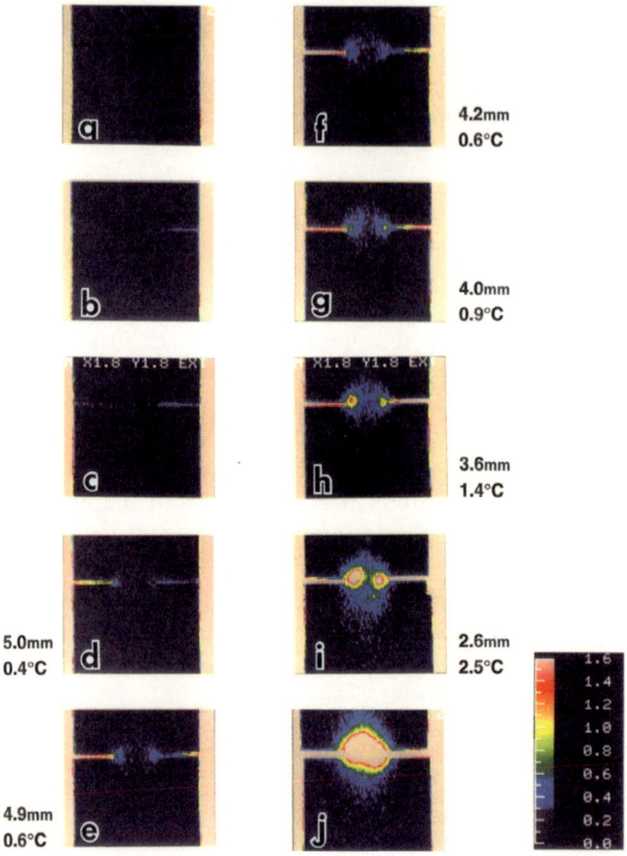

Fig. 15. Sequential close-up temperature rise distribution images of paper (handsheet from beaten pulp, ligament: 5mm); (a) – (j) correspond to the positions on the load-displacement curve in Fig. 8a. Beside each image, the distances between the maximum temperature points around both notch tips are shown with the temperature rise (Tanaka et al., 1997).

toward the inside even before the stage of maximum load, as shown by a decrease in the distance between the maximum temperature points (Fig. 15; e,f,g) This movement means stable crack growth, and further onset of crack growth before maximum load point to sheet failure was confirmed by means of direct observation of the notch tip using video-microscopy (Tanaka and Yamauchi, 1999).

As the theoretical prerequisite for EWF, the development of a circular plastic deformation zone in the ligament before crack growth is required. Thus, sequential close-up temperature rise distribution images were examined for specimens of various ligament lengths (Tanaka and Yamauchi, 2000) and further the plastic deformation zone was theoretically estimated (Tanaka and Yamauchi, 1997). As a result, the plastic deformation zone appears in three ways: 1. Type (i) appearing through the whole ligament in a vague manner and developing into a circular (or oval) zone even before or at the maximum load point; (see Fig. 16 and 17 for UKP-sack paper in cross direction loading: CD); 2. Type (ii) appearing from the notch tip and amalgamating into a circular (or oval) zone after the maximum load point (see Fig. 18 and 19 for UKP-sack paper in CD loading); and 3. Type (iii) appearing from the notch tip

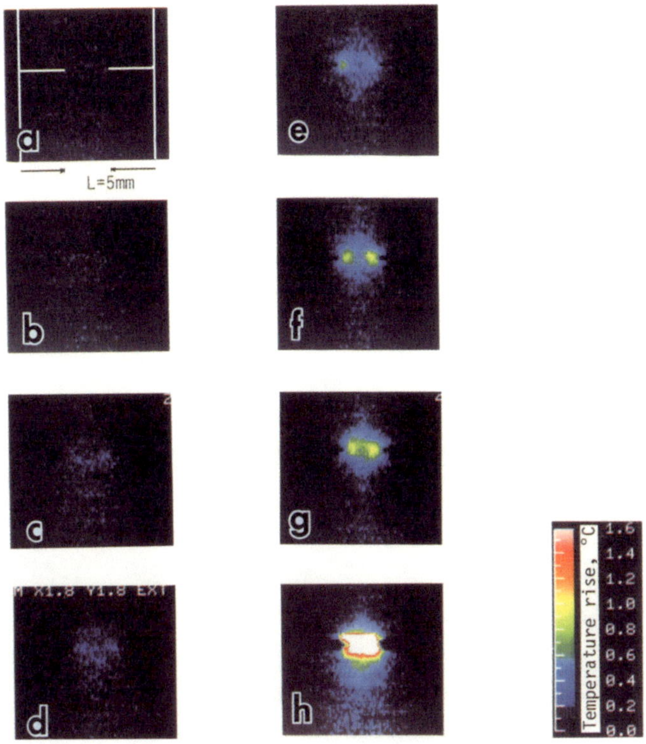

Fig. 16. A series of close-up temperature rise distribution images of UKP-sack paper (CD/*L*: 5mm); (a)—(h) correspond to the positions on the load-displacement curve in Fig.17 (Tanaka and Yamauchi, 2000).

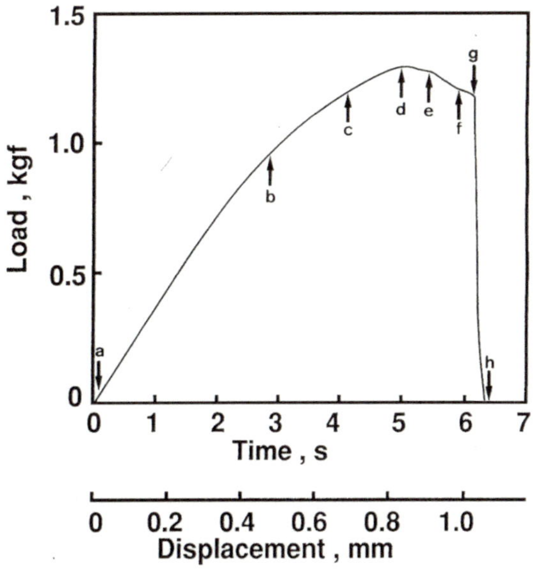

Fig. 17. Load-displacement curve for the UKP-sack paper (CD/*L*: 5mm) (Tanaka and Yamauchi, 2000).

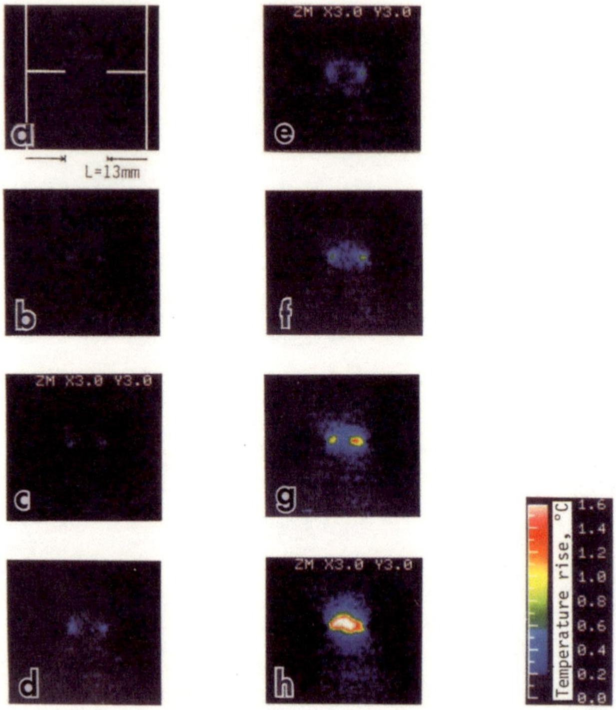

Fig. 18. A series of close-up temperature rise distribution images of UKP-sack paper (CD/**L**: 13mm); (a)—(h) correspond to the positions on the load-displacement curve in Fig. 19 (Tanaka and Yamauchi, 2000).

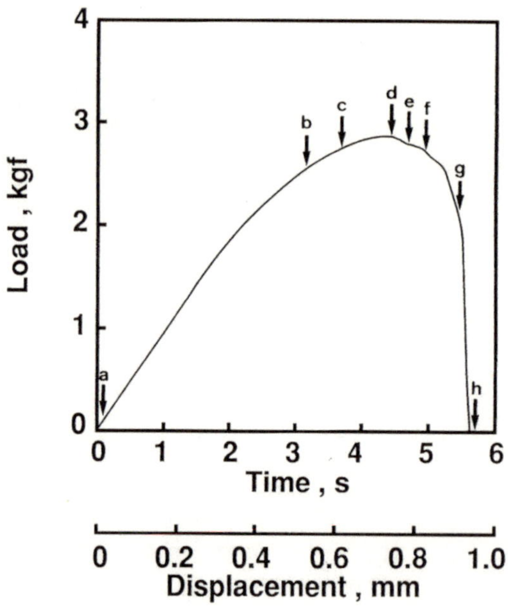

Fig. 19. Load-displacement curve for the UKP-sack paper (CD/**L**: 13mm) (Tanaka and Yamauchi, 2000).

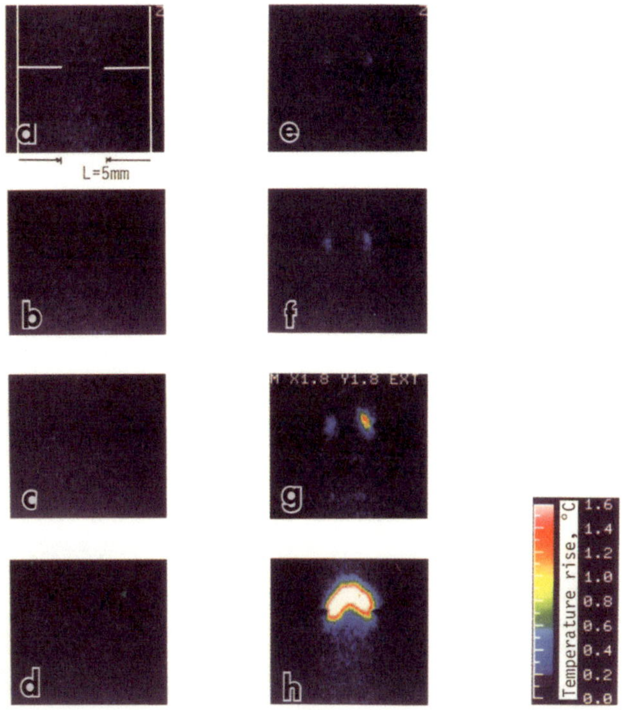

Fig. 20. A series of close-up temperature rise distribution images of UKP-sack paper (MD/***L***: 5mm); (a)–(h) correspond to the positions on the load-displacement curve in Fig. 21 (Tanaka and Yamauchi, 2000).

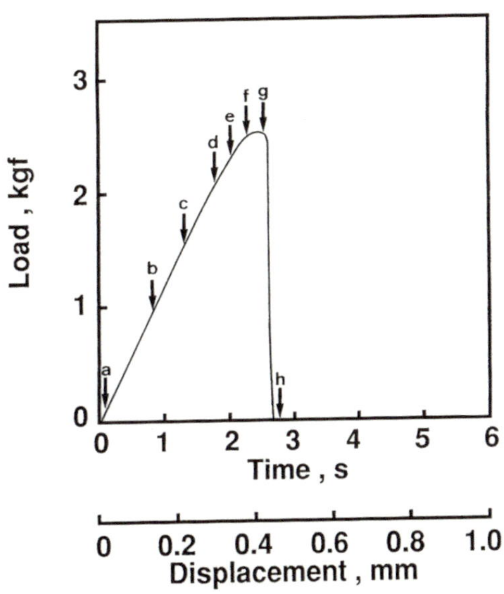

Fig. 21. Load-displacement curve for the UKP-sack paper (MD/***L***: 5mm) (Tanaka and Yamauchi, 2000).

and not amalgamating into a circular (or oval) zone until sheet failure (see Fig. 20 and 21 for UKP-sack paper in machine direction loading: MD). Table 1 shows the type of plastic deformation zone appearance for representative commercial papers of various specimen widths. Paper specimens with a small ligament length (***L***) are likely to belong to type (i), while those with a large ***L*** to type (ii) and (iii). Among these three types, type (i) fulfills the original assumption of the EWF method best in terms of complete ligament yielding before crack initiation. Thus, the specific EWF determined by using the extrapolation of the linear relation of type (i) to ***L***=0 should be correct, although the estimated work (EWF) is a little smaller than that from the linear relation of type (ii) and (iii) without exception, as shown in Fig. 22. Furthermore, the corrected experimental plots based on theoretical calculation for large ligament specimens had almost the same linear relation as small ligament specimens, as shown in Fig. 23 (The scale of ordinate spread to three times. Tanaka et al., 1998).

		L=1mm	L=2mm	L=3mm	L=4mm	L=5mm	L=9mm	L=13mm	L=17mm	L=21mm
UKP -sack	MD	(i)	(i)	(ii)	(iii)	(iii)	(iii)	(iii)	(iii)	(iii)
	CD	(i)	(i)	(i)	(i)	(i)	(ii)	(ii)	(ii)	(ii)
Machine grazed	MD	(i)	(i)	(i)	(ii)	(iii)	(iii)	(iii)	(iii)	(iii)
	CD	(i)	(i)	(i)	(i)	(ii)	(ii)	(ii)	(ii)	(ii)
News print	MD	(i)	(i)	(ii)	(iii)	(iii)	(iii)	(iii)	(iii)	(iii)
	CD	(i)	(i)	(i)	(i)	(ii)	(iii)	(iii)	(iii)	(iii)
Filter	MD	(i)	(i)	(i)	(i)	(ii)	(iii)	(iii)	(iii)	(iii)
	CD	(i)	(i)	(i)	(i)	(i)	(ii)	(iii)	(iii)	(iii)

Table 1. Type of plastic deformation zone appearance for representative commercial paper specimens of various specimen widths (***L***=1/3 width)

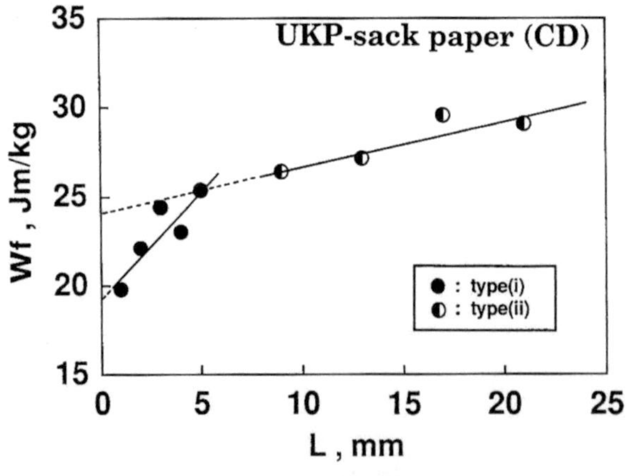

Fig. 22. Experimental determination of the specific essential work of fracture by extrapolation of the linear relation between work of fracture (w_f) and ligament length (***L***) for UKP-sack paper (CD) (Tanaka and Yamauchi, 2000).

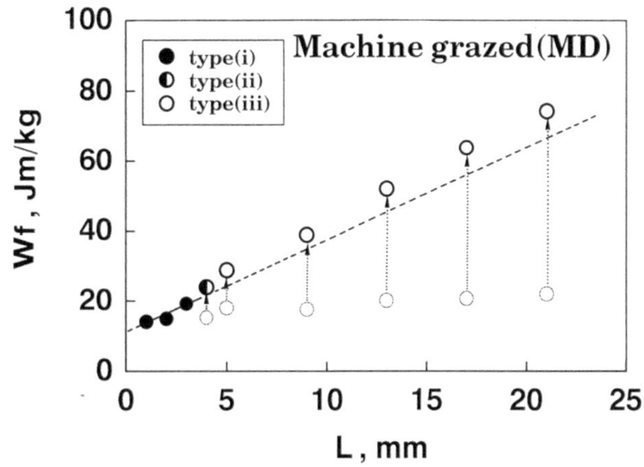

Fig. 23. Corrected experimental plots between work of fracture and ligament lengths for machine grazed paper (MD). Broken arrows show corrections (The scale of ordinate spread to three times Tanaka et al., 1998).

For in-plane fracture toughness testing, J-integral and crack tip opening displacement other than EWF have also been introduced for paper materials. The thermal images during these testing gave much knowledge on the examination of these methods (Hirano and Yamauchi, 2000). Further, the sequential thermal images gave some suggestions for studying the effects of notch application and its geometry (Yamauchi, 2004).

8. Application of thermography to tearing test of paper

As described above, fracture toughness is estimated as the work required to fracture a notched specimen, and the fracturing of planar materials including paper has two modes, in-plane and out-of-plane. In the former mode, stresses are applied along the plane of the paper sheet. On the other hand, stress is applied perpendicular to the plane of the paper sheet in the latter mode, which is identical to out-of-plane tearing, known as Elmendorf tearing. Compared with in-plane and out-of-plane tearing, the cumulative energy of the micro failures that occurred during out-of-plane tearing was markedly larger than that during in-plane tearing, as shown in thermal distribution images on tearing (Yamauchi, 2005). The development of stress concentration at the notch tip was observed as close-up sequential thermal distribution images and 3-D images of temperature rise on tearing, as shown in Fig. 24 (Tanaka and Yamauchi, 2005). The size of the area where heat was generated during tearing decreased, i.e. the degree of stress concentration increased with an increase of the beating degree for a paper sheet made from longer fibers. The ratio of heat generation to tearing work was ~10%, which is smaller than the corresponding ratio of 40 to 80 % for tensile testing (referring to the end of first paragraph of **3**. Temperature (heat) change, see Fig. 6). Furthermore, heat generation was classified as that attributable to damage and plastic deformation. In order to evaluate heat generation caused by damage (fiber pull out and breakage) around the crack tip, a threshold Tb showing the end of plastic deformation (yielding) and onset of damage was applied to the above 3D-image of temperature rise on tearing. Although some damage occurred slightly before paper breakage at maximum loading, most of the damage occurred at or after the maximum load

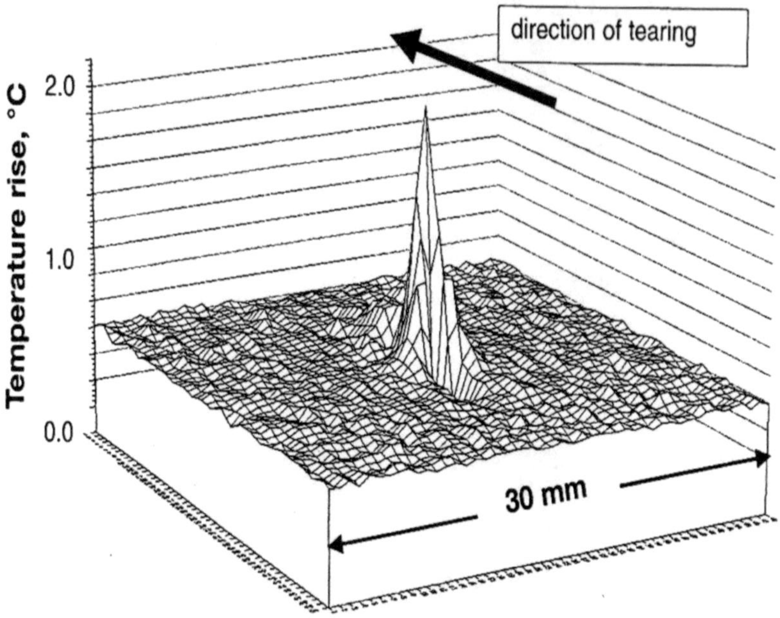

Fig. 24. 3D-image of temperature rise on tearing. obtained by image subtraction of handsheet from beaten pulp (Tanaka and Yamauchi, 2005).

point (Tanaka and Yamauchi ,1999). The schematic image of such thresholds is shown in Fig. 25. The tearing energy of paper from long—fibered pulp correlates well with heat from the damage.

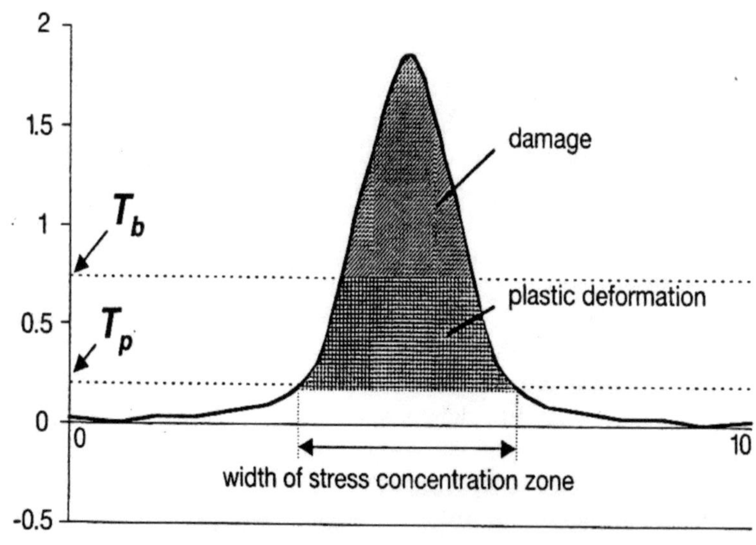

Fig. 25. Schematic image of threshold temperature. The curve shows the cross section of the 3-D image. Tp: threshold for plastic deformation and Tb: threshold for damage (fiber pull out or breakage) (Tanaka and Yamauchi, 2005).

9. Subsequent studies on paper materials using thermography

Thermography is applicable to identify hidden air-filled defects within the thickness of laminated paperboard (Sato and Hutchings, 2010). In a thermally non-stationary state, the region over a defect can be visually distinguished from neighboring defect-free areas by its thermal contrast. For quantitative determination, the pulsed thermography method was used and temperature history curves were examined in order to characterize the unknown defects. It was found that a defect can be quantitatively evaluated in a reasonable time at a depth of less than 2mm beneath the surface. Compared with many other materials, paper has higher specific heat and lower thermal conductivity, and thus the thermal contrast persists for a longer time.

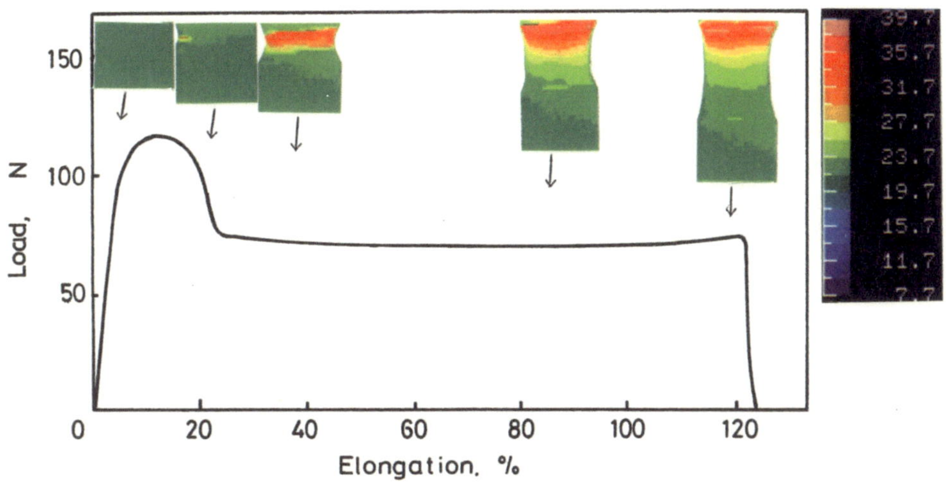

Fig. 26. Relationship of load-elongation, and some selected temperature distribution images of drawn PP film (Yamauchi, 2006).

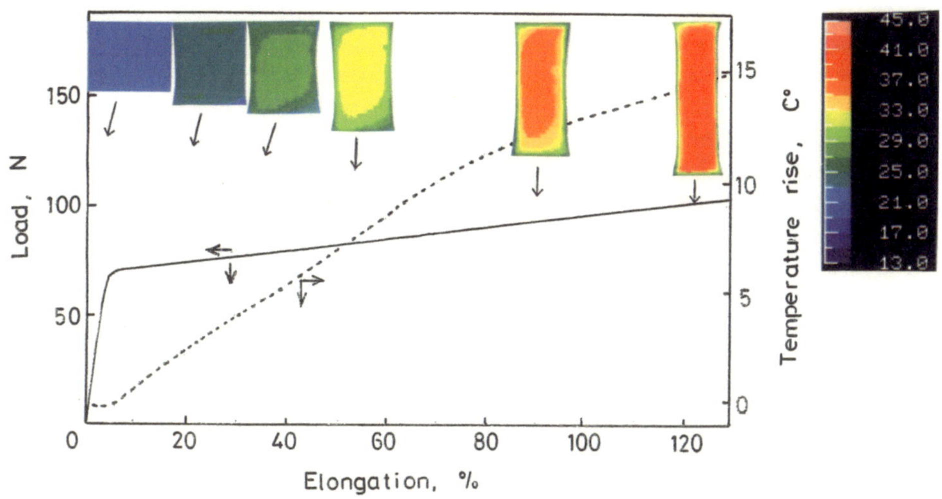

Fig. 27. Relationships of load and temperature rise—elongation, and some selected temperature distribution images of drawn PET film.

Additionally, heat generation that accompanied tensile drawing for other planar material, polymer film, was successfully observed using thermography (Yamauchi, 2006). The sequential temperature rise distribution images showed that the high temperature region, where rearrangements of molecular orientation and microcrystallines occurred, was located mainly at the upper end of the necked part for films that exhibited necking as shown in Fig. 26, or appeared uniformly throughout films for the polymers that did not exhibit necking behavior as shown in Fig.27.

10. Summary

Study on deformation and fracture of paper is very important as a fundamental of its strength properties. Thermal behavior with its deformation and fracture as shown in chronologically sequential surface temperature images can be successfully investigated by using IR thermography, since paper is a two-dimensional planar material and its thermal conductivity is low and further its specific heat is high. As the thermodynamics of paper deformation, heat absorption caused by Kelvin's thermoelastic effect at the elastic deformation region and heat generation in the plastic deformation region were observed, and further suggested that elastic deformation based on energy elasticity occurred from the beginning of elongation and plastic elongation partly arose from deformation based on entropy elasticity. Paper is a fairly homogenous material as shown in uniform planar mass distribution and its thermal change under loading occurred in fairly uniform manner throughout the specimen up to the moment of its breakage. On the other hand, poorly formed paper, which has heterogeneous mass distribution, showed an uneven distribution in the higher temperature region at the early stage of plastic deformation, but at a later stage of plastic deformation the temperature image became homogenous. The extent of uniformity in temperature rise distribution during straining, rather than that in mass distribution, was more directly related to the degree of stress concentration within the specimen and thus to the breaking load of paper. On essential work of fracture (EWF) toughness testing as a one of the in-plane fracture toughness testing, the stress concentration around the tip of notches and the appearance of a circular plastic deformation zone centered on the ligament (between double notches, one third of the width) from halfway through the plastic deformation region could be visualized using thermography as sequential temperature rise distribution images, and further movement of the crack was shown as the shift of the highest temperature point during testing. Paper specimens with a small ligament fulfilled the original assumption of the EWF method best in terms of complete ligament yielding before crack initiation. Thus, the specific EWF determined by using the extrapolation of the linear relation between ligament and work of fracture for the specimens with a small ligament should be correct, although the estimated EFW is a little smaller than that for the specimen with bigger ligament. Thermal distribution images on paper tearing showed that the cumulative energy of micro failures that occurred during out-of-plane tearing was markedly larger than that during in-plane tearing. The size of the area where heat was generated during tearing decreased, i.e. the degree of stress concentration increased with an increase of pulp beating for a paper sheet made from longer fibers. The tearing energy of paper from long-fibered pulp correlates well with heat from the fiber pull out and breakage around the crack tip. Thermography is a unique and versatile technology, and thus applicable to study on water absorption and moisture profile of paper, except for study on its deformation and fracture as described above.

11. Acknowledgement

The author expresses his huge thanks to co-worker, Dr Atsushi Tanaka (presently VTT Finland), and to the Laboratory of Wood Processing in the Graduate School of Agriculture, Kyoto University for permission to use the thermogrphy system, and also thanks to Arakawa chemical, Rengo and Showa products Co. Ltds for their financial support.

12. References

Chao, Y.J. and Sutton, M.A. (1988). Measurement of strains in a paper tensile specimen using computer vision and digital image correlation Part 2:Tensile specimen test. *Tappi J.*,Vol. 71, No.4, pp.153-156

Cotterell, B. and Reddel; J.K. (1977). The essential work of plane stress ductile fracture. *Int. J. Fracture*, Vol. 13, pp. 267-277

Dumbleton, D.P., Kringstad, K.P. and Soremark, C. (1973). Temperature profiles in paper during straining. *Svensk Papperstid*, Vol.76, No.14. PP.521-528

Ebeling, K.I, Swanson, J.W. and Van den Akker, J.A. (1974). Microcalorimeter for measuring heat of straining or destraining of sheet-like materials. *Rev. Sci. Instrum.*, Vol.45, No.3. pp. 419-426

Ebeling, K.I. (1976).Distribution of energy consumption during the straining of paper, In *The fundamental properties of paper related to its uses*, F. Bolam(ed), pp.304-343

Hirano, H. and Yamauchi; T. (2000). J-integral Jc and crack tip opening displacement CTODc as a material property evaluating the fracture toughness of paper. *J. Tappi Jpn.*,Vol.54, No.5, pp 75-80

Koizumi, T. (1983) .On thermo-elasticity measurement. *NDT Jpn.*, Vol.32, No.7, pp.61-565

Muller, F.H. (1969). Thermodynamics of deformation: Calorimetric investigations of deformation processes, In *Rheology, Theory and Application* Vol.5 , F.R. Eirich (Ed), pp.417-489, Academic Press

Niskanen, K. (1993). Strength and Fracture of paper, In *Products of Papermaking*, C.F. Baker (Ed) pp.641-725

Reifsnider, K.L. and Williams, R.S. (1974). Determination of fatigue-related heat emission in composite materials. *Experimental Mechanics*, Vol.14, pp.479-485

Sato, J. and Hutchings, I.M. (2010).Non-destructive testing of laminated paper products by active infra-red thermography. *Appita J.*, Vol.63, No.5, pp. 399-406

Seth, R.S. and Page, D.H. (1981). The stress-strain curve of paper, In *The Role of Fundamental Research in Paper Makin,g* F. Bolam (Ed), pp. 421-452

Tajima, T., Ueno, S., Yabu, N. and Sukigara, S. (2011) Fabrication and characterization of Poly-γ-glutamic acid nanofiber with the difunctional epoxy cross-linker. *Seni' Gakkaishi* , Vol.67, No.8, pp. 169-175

Tanaka, A., Othuka, Y. and Yamauchi, T. (1997). Observation of in-plane fracture toughness testing of paper by use of thermography. *Tappi J.*, Vol.80, No.5, pp. 222-226

Tanaka, A. and Yamauchi, T. (1997). Size estimation of plastic deformation zone at the crack tip of paper under fracture toughness testing. *J. Pack Sci .& Tech. Jpn.*, Vol.6, No.5, pp. 268-276

Tanaka, A., Matsumoto, T. and Yamauchi, T. (1998). Calculative correction of the essential work of fracture method for paper. *Seni' Gakkaishi* , Vol.54, No.12, pp. 134-140

Tanaka, A. and Yamauchi, T. (1999). Measurement of crack tip opening displacement of paper using CCD camera. *J. Pack Sci. &Tech. Jpn.*, Vol.8, No.1, PP. 11-18

Tanaka, A. and Yamauchi, T. (2000). Deformation and fracture of paper during the in-plane fracture toughness testing —Estimation of the essential work of fracture method—. *J. Materials Sci.*, Vol.35, No.7, PP. 1827-1833

Tanaka, A., Kettunen, H. and Yamauchi, T. (2003). Comparison of thermal maps with damage revealed by silicone impregnation. *Paperi ja Puu,* Vol.85, No.5, pp.287-290

Tanaka, A. and Yamauchi, T. (2005). A thermographic observation of out-of- plane tearing process of paper. *Appita J.*, Vol.58, No.3, PP. 186-189. 217

Vickery, D. E., Luce, J.E., and Atkins, J.W. (1978). Infrared thermography An aid to solving paper machine moisture profile problems. *Tappi* Vol.61, No.12, pp.17-22

Yamauchi, T., Okumura, S. and Noguchi, M. (1990). Acoustic emission as an Aid for investigating the deformation and fracture of paper. *J. Pulp Paper Sci.,* Vol.16, No.2, pp.44-47

Yamauchi, T. and Murakami, K. (1992). Observation of deforming and fracturing processes of paper by using infrared thermography. *J. Tappi Jpn.*, Vol.46, No.4, pp. 178-183

Yamauchi, T., Okumura, S. and Noguchi, M. (1993). Application of thermography to the deforming process of paper materials. *J. Materials Sci.*, Vol.28, No.17, PP. 4549-4552

Yamauchi, T. and Murakami, K. (1993). Observation of the deforming and fracturing processes of paper using thermography, In *Products of Papermaking,* C.F. Baker, (Ed.), pp. 825-847

Yamauchi, T. and Tanaka, A. (1994). Experimental detection of entropy elasticity occurred during the plastic deformation of paper. *J. Applied Polymer Sci.*, Vol. 53, No.8. pp. 1125-1127

Yamauchi, T. and Murakami, K. (1994). Observation of deforming process of poorly formed paper sheet by use of thermography. *Seni Gakkkaishi* , Vol.50, No.9, PP.424-425

Yamauchi, T and Hirano, H. (2000). Examination of the onset of stable crack growth under fracture toughness testing of paper. *J. Wood Sci.*, Vol.46, pp. 79-85

Yamauchi, T. (2004). Effect of notches on micro failures during tensile straining of paper. *J. Tappi Jpn.,* Vol.58, No.11, pp. 1599-1606

Yamauchi, T. (2005). Differences between in-plane and out-of-plane tensile fracturing of notched paper. *J. Pack Sci. & Tech. Jpn.,* Vol.14, No.5, PP. 329-340

Yamauchi, T. (2006). Observation of polymer film drawing by use of thermography. An introductory investigation on the thermodynamics. *J. Applied Polymer Sci.,*. Vol.100, pp. 2895-2900

2

Application of Thermography in Materials Science and Engineering

Alin Constantin Murariu, Aurel - Valentin Bîrdeanu, Radu Cojocaru, Voicu Ionel Safta, Dorin Dehelean, Lia Boţilă and Cristian Ciucă
National R&D Institute of Welding and Materials Testing – ISIM Timişoara
Romania

1. Introduction

The chapter presents the main applications of IR thermography in material science and engineering with focus on modern methods for examination of materials and applications in new processing technologies development and to the improvement of existing processing technologies by implementing process control. Beside a synthesis of the applications presented in the existing literature, applications of IR thermography developed by the chapter's authors are presented. The following subchapters will present various applications of IR thermography in materials science and engineering.

Infrared thermography is a one of the modern imaging methods which allows the temperature assessment of an object by contactless testing, with a wide range of possible applications both in materials science and in engineering. Most applications are related to non-destructive testing and monitoring of equipments, components or different technological processes.

In material science, destructive test methods provide information regarding the initial structure and material's strength characteristics or an estimation of these at different operating durations, in order to assess the service life of the components in safety critical conditions. For the modern new product's development strategies, the selection of the engineering material process has to take into account the material's in-time-stability characteristics. Due to this trend, new examination techniques, which are based on material fracture concept and theories, have been developed for studying and complete characterization of the new materials types.

2. Application of thermography in materials science

The relationship between temperature and material deformation was recognized in 1853 by Kelvin and then developed by Biot, Rocca, and Bever in the 1950s (Yang, B. et al., 2003). In 1956, Belgen developed IR radiometric techniques for detecting temperature changes and in the 1960s, Dillon and Kratochvil developed the thermoplastic theory that directly relates the temperature with the material internal stress-strain state, which, in turn, controls the mechanical and fatigue behaviour (Yang, B. et al., 2003). Active thermography allows the detection and the characterization of exfoliation between layers (Rajic, 2004) in different stages of testing and the passive one allows the localization of crack initiation. In case of composite materials, especially for "sandwich" structure type, active and passive infrared thermography (Balageas, et Al., 2006) became a current technique to monitor the mechanical testing.

Elaboration of the new materials or the properties improvement of the existing ones represents one of the major priority directions in the research field of thermoplastics and composite materials. Before launching of the new product or the material on the market, they must be tested in order to assess their in-service usage ability for the specific exploitation conditions. Recent research (Avdelidis et al., 2010; Bremond, 2010; Choi et al., 2008; Galietti & Palumbo 2010; Huebner et al., 2010; Kruczek, 2008; Murariu et al., 2010; Pieczyska et al. 2008; Sahli et al. 2010; Thiemann, 2010) has shown the potential of thermography in the monitoring of the mechanical damage but, at the same time, more detailed investigations and analyses are needed in order to develop a more practical thermography method for characterizing the materials' proprieties. In the following section some applications of thermography in materials science are presented.

2.1 Application of active thermography for quality assessment of PEHD pipes

The use of thermoplastic materials in the pipes networks was developed due to their advantages while compared to the use of metallic materials: greater corrosive resistance, low weight, good mechanical strength, high resistance to the aggressive fluids, low hydraulic losses, flexibility etc. A special utilization is given to the high density polyethylene (HDPE) for assemble of pipes networks for gases or water distribution.

Using the active thermography, a method for quality assessment of HDPE pipes was developed and, in order to reveal the new method's sensitivity, an experimental program was designed using pipe specimens with longitudinal simulated imperfections. The imperfections were performed using Laser Simulated Imperfection - LSI method (Murariu & Bîrdeanu, 2007), and activation was done with water at 80°C. For simulating the imperfections in the base material a Nd:YAG laser Trumpf HL 124P LCU (Trumpf GmbH, Ditzingen, Germany) with a fibreglass connected cutting head was used (figure 1).

Fig. 1. LSI experimental setup

For simulating the imperfections, the laser process parameters were specific to pulsed laser beam cutting, i.e. short width rectangular pulses, relative high pulse peak power and high pulse repetition rate (55Hz) in respect to the travelling speed (3.73mm/s). As processing gas, Ar99% at 6bar pressure was used. Experiments were performed to establish the process parameter adjustment space according to the desired penetration depths for the specimens. The data showed an appropriate linearity for penetration depending on the pulse peak power, so a programmed experiment was performed to establish the equation that correlates the penetration depth to the peak power pulse.

The data was graphically fitted (figure 2) and the corresponding equation describing the penetration depth variation as a function to the pulse peak power for the targeted domain was established. Using the mathematical equation it was possible to calculate the pulse power peak necessary to simulate the required imperfections in the material. Furthermore, the calculated parameters were verified by doing a set of experiments and measuring the penetration depths and it was determined that the measured data corresponded to the calculated one in respect to the experimental errors.

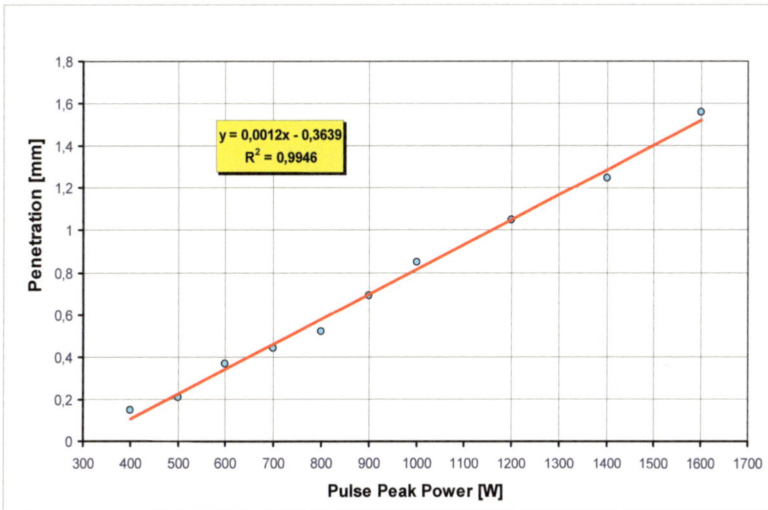

Fig. 2. LSI calibration - fitted experimental data

In order to estimate the sensitivity of the method, an experimental program was performed using PE80, DN 32, SDR11, GAS pipes and the longitudinal notch with depths between 0.15 mm and 1.5 mm have been created using LSI method. Thermographic examination results are shown in Figure 3. The thermographic images were analyzed and the imperfections were revealed. The minimum notch depth which could be detected was h=0.15mm for a pipe with wall thickness e=3.0mm (imperfection has a relative size of A=5% of the component wall thickness) but one could consider that the use of an automated computer analysis of the thermographic images, can improve the limit of the detection.

After the calibration and the evaluation of the method's sensitivity, experimental measurements were performed on pipes with real flaws obtain by hydrostatic inner pressure at constant temperature after different testing durations. The results are shown in Figure 4.

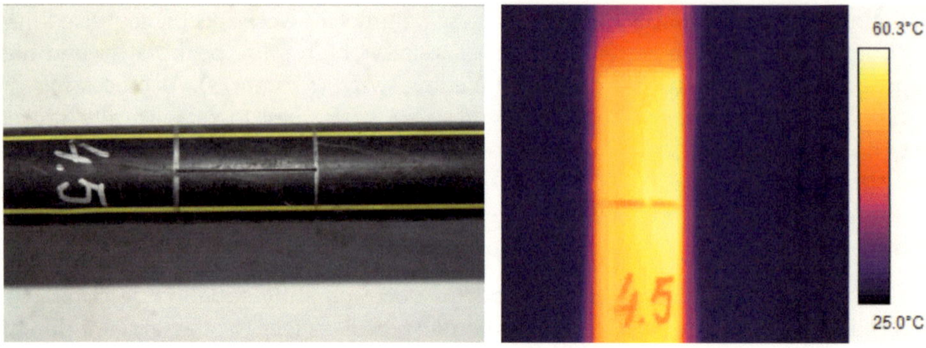

a)

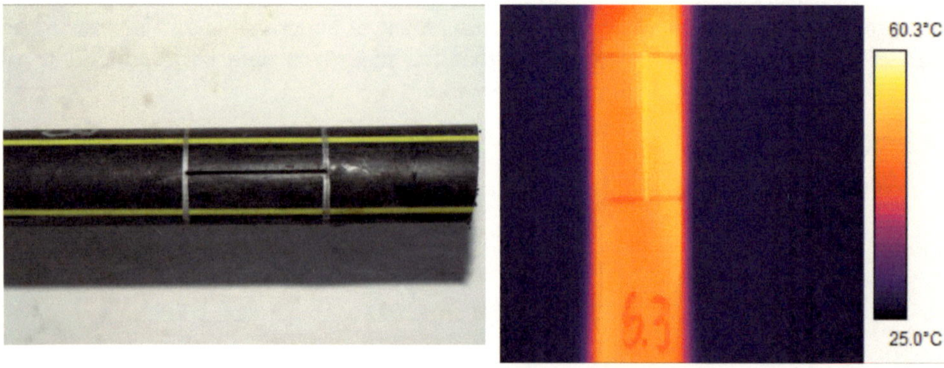

b)

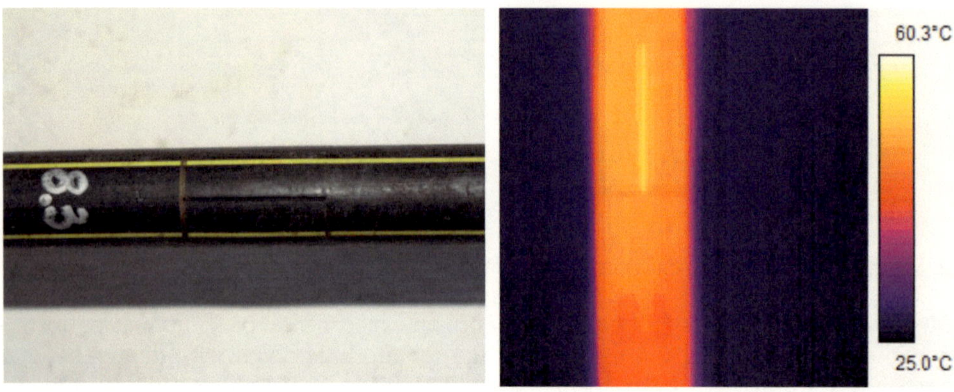

c)

a) relative notch depth A=5% of the pipe wall thickness b) relative notch depth A=10% of the pipe wall thickness c) relative notch depth A=40% of the pipe wall thickness

Fig. 3. PE 80 pipes with LSI simulated imperfections examined by thermography

The experiment did show that by using thermography method and transient temperature regime obtained by thermal activation with warm water, filiform defects of thermoplastic pipes can be detected. This rather fast and accurate method can also detect defects which are sometimes more difficult to identify using other NDT methods, e.g. ultrasound, X-Ray.

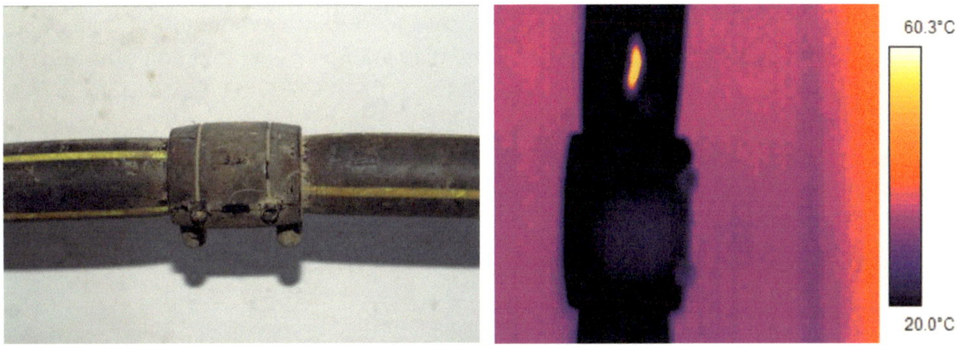

a)

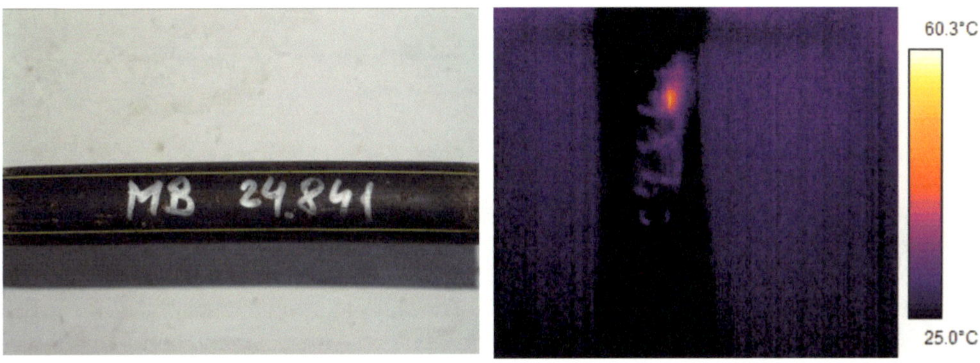

b)

Fig. 4. PE 80 pipes with flaws obtained on hydrostatic inner pressure test at a constant temperature (80°C and 10 bar); a) sleeve welded joints with flaws in the MB, b) PE 80 pipe cracked after 24841 hours of testing

2.2 Application of TT - IRT hybrid technique on thermoplastic materials

To assess the deformation and fracture behaviour of polymers, hybrid test techniques that combine the classical mechanical tests with the non-destructive examinations (i.e. acoustic emission, thermography, laser extensometry) can be applied. By applying these hybrid techniques, quantitative properties - morphology correlations can be made. Most of the materials heat up when they are loaded beyond elastic limit. Starting from this observation, for studying the polyethylene's behaviour in the presence of simulated plane imperfections, thermography method was used to display the heat emission of the specimen. A chapter presents the application of Tensile Test monitored by Infrared Thermography hybrid technique TT-IRT (Murariu, 2007) in order to put in evidence the noxiousness of geometric imperfections that can appear on pipes networks. Thermography method allows analyzing the polyethylene fracture behaviour due to simulated imperfections. The temperature changes in thermographic images of specimens during the tensile test allow studying of local deformation process of thermoplastic polymer.

The following is an example of the TT-IRT hybrid technique use for fracture behaviour study of PE 80 type thermoplastic material. The method may be also applied to other thermoplastics or advanced materials for stress raiser sensibility evaluation of materials or products in the elaboration/fabrication phase and also after an in-exploitation duration. The experimental program used samples extracted from PE 80 SRD 11 GAS ϕ160×15.5 mm polyethylene pipes. Strip samples with width b=20.0 mm and thickness a=15.5 mm were cut from sample pipes with length L=200 mm. Simulated imperfections were realized by mechanical processes (milling) and consisted in transversal cuttings with angle of the flanks of 45° and with top radius of 0.25 mm (figure 5). Five sets of ten samples with notch depth h=1.0, 1.5, 2.0, 2.5 and 3.0mm were prepared.

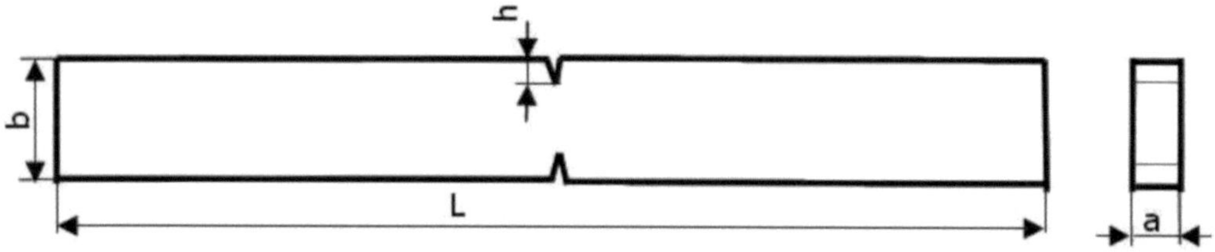

Fig. 5. Blueprint of the samples with bilateral V notch

The experimental setup included a universal mechanical testing machine, an air conditioning room and an infrared thermography camera. The clamping dies speeds of the tensile loading machine were, v=1, 5, 10, 25, 50, 75, 100, 125, 150 and 650mm/min. The test was performed at room temperature (23±2°C). The thermographic camera did record the caloric energy emission during the tensile loading process of the sample and temperatures of the interest zones (stress raisers zone created by the presence of the plane imperfections) were monitored. The method did allow the analyses of the thermal image of the tested samples, which distribution is modified during the testing. The surface temperature values in the imperfection simulated zone were recorded and correlated with the applied stress levels. Thus the method allows studying of the defects' severity and their evolution together with the reaction of the adjacent material due to loading process. The tests proved the tensile strength, the identification of the fracture type and the fracture position, under controlled tests condition: temperature, loading rate. It was determined that the recorded temperatures during the tensile tests are correlated with the local stress condition induced by the loading of the specimen which has geometrical discontinuities. Also it was highlighted that the sample failure is determined by the appearance of "hot" spots in the breaking cross-section. This spots are specific to each type of imperfection and do determine the sample's failure dynamic. Figure 6 shows the possibility of using infrared thermography to failure dynamics analysis of samples with simulated defects.

Next, to prove the way in which the temperature modifications during the testing do influence the failure's character, the aspect of the tested samples' surface failure was analyzed in cases of samples for which the notched area represents 10 and 30% of the transversal cross section of the sample. Figure 7 presents the surface aspect of a sample fracture by axial loading with clamping dies speed of 650mm/min. The sample presents bilateral V notches with h=1mm depth, occupying 10% from the transversal cross section area. Figure 8a, presents the fracture surface of a sample with bilateral V notches with h=3mm depth, affecting 30% of the transversal cross-section area. The tests were similar to the previous ones. One can observe that, depending on the fracture's character, the sample

Frontal view | Lateral view | Frontal view | Lateral view

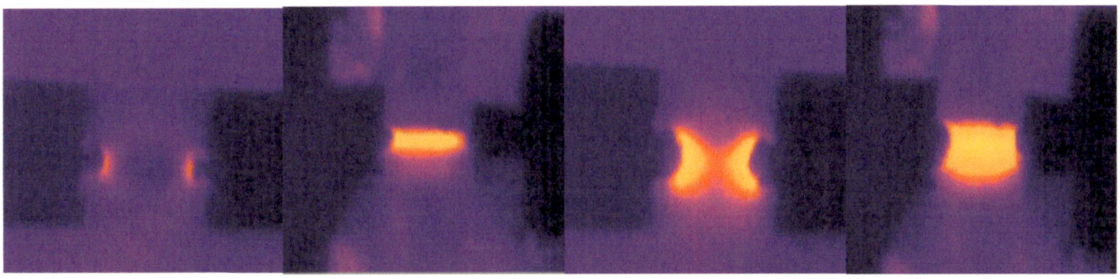

a - The sample heating in the notch and its deformation

b - The heat propagation to the centre of the sample until the thermal fields join

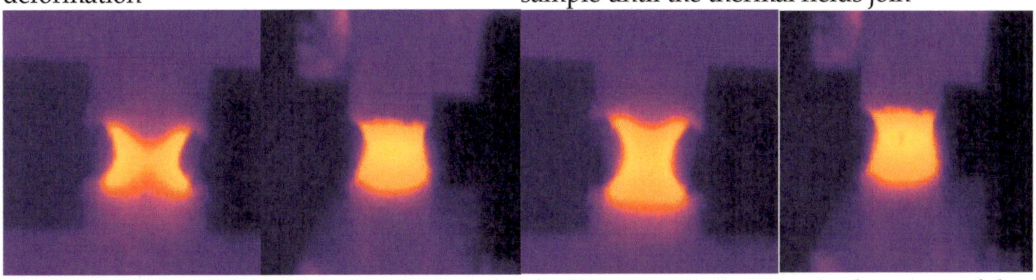

c - The temperature equalization in minimal cross-section of the sample

d - Temperature increase in the centre of the sample in minimal cross section and the leak formation

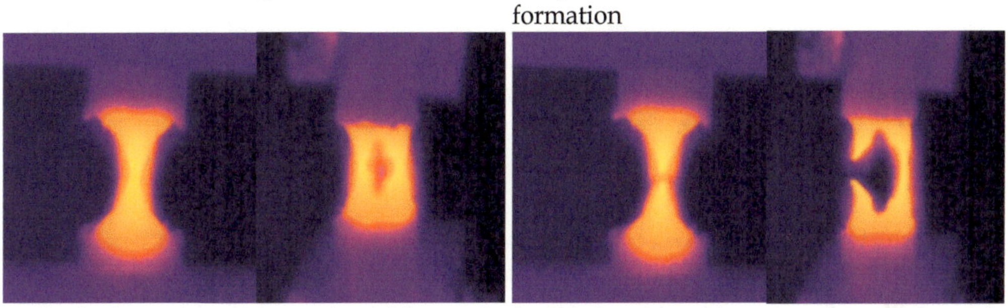

e - Leak elongation, propagation of the crack, initiated from the leak

f - Crack propagation up to the exterior and sample failure

Fig. 6. The fracture dynamic of a sample with V notch

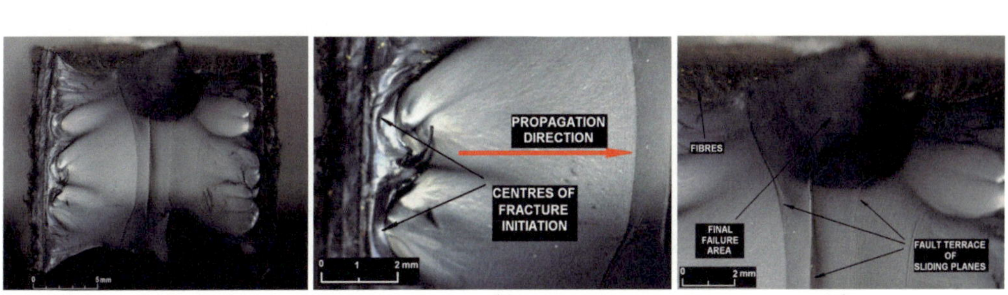

a) b) c)

Fig. 7. The breaking surface of a sample with bilateral V notches with h=1mm depth and v=650 mm/min.; a - failure surface, b - aspect of the fracture initiation zone, c – aspect of the final fracture zone.

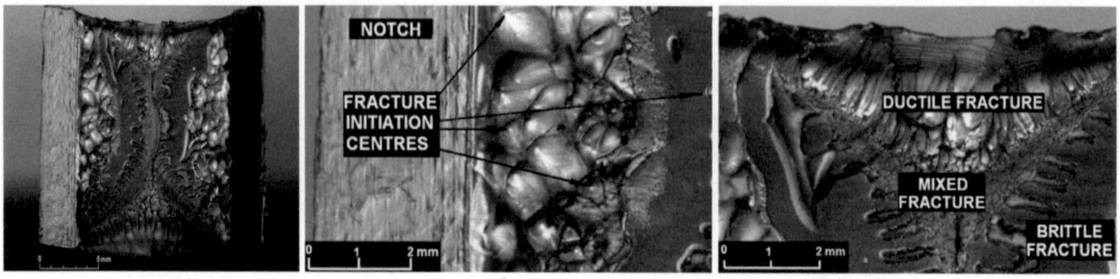

a. b. c.

Fig. 8. The breaking surface of a sample with bilateral V notch with h=3 mm depth and v=650mm/min.; a - The breaking surface, b - The breaking initiation centres, c -The areas with ductile, fragile or mixed breaking

presents bi-axial symmetry. In the transversal fracture surface of the sample multiple fracture initiation centres (figure 8b) are observed with a spherical shape when they are single, or elongated if they are grouped. In the fracture surface both ductile fracture areas, with significant deformations and slipping areas near the breaking initiation centre, and fragile fracture areas without significant deformations (figure 8c) are present.

The modification of the fracture's character is explained by the difference in the sample temperatures in the transversal cross section, during the tensile test. Thus, the sample's temperature is higher in the V notch which is a stress raiser and in the centre of the lateral surfaces, a fact that was confirmed by thermography. The apparition of the fragile failure is explained by the fact that the test was made at high rate (v=650mm/min.) in the conditions of a 30% decrease of the fracture's section. Because the time was insufficient to achieve a uniform temperature in the transversal cross section, the sample fracture took place by different failure mechanisms; the fracture's character was influenced by the local stress distribution and temperature.

While the presented above examples do reveal the practical use of the TT-IRT hybrid technique for studying the fractures behaviour for PE 80 type thermoplastic material, the method may be also applied to other thermoplastics or advanced materials for stress raiser sensibility evaluation of materials/products in the elaboration/fabrication phase and after an in-exploitation duration. Also the results may be useful for studying a fracture behaviour for typical planar imperfection from the butt welded joints of thermoplastic pipes, under different loading conditions, in order to assess the critical imperfection which leads to the joints failure.

3. Application of thermography in engineering

Infrared thermography is a modern imaging method which allows the assessment of the temperature of an object by contactless testing, with multiple possible applications in the field of non-destructive testing and monitoring of equipment, component and technological processes.

3.1 Lack of adhesion defect identification in thermal spraying layer

The section presents an application of active thermography for the process quality evaluation of thermal spraying coated surfaces. The thermal sprayed layers are commonly used in industrial applications aiming to protect certain severely loaded components. Depending on the specific application, they may be used as thermal barriers for base material's protection, for increasing the wear resistance or corrosion resistance in aggressive environments. A layer deposited by thermal spraying is usually characterized by: adhesion, structure, density and porosity. The adherence is one of the basic characteristics of the deposited layer which insures the strength, the durability and the protection level of the coated components. The evaluation of the thermal sprayed layers using the active thermography method is based on the physical phenomenon according to which any natural object is emitting a thermal radiation consisting of the radiation emitted at rotational and vibrational quantum level transitions, and of the reflected radiation, due to other thermal sources. The visualization technique of the infrared images obtained based on the characteristics of the thermal emission of objects is generically known as the thermo vision or thermography technique.

By thermography one does put in evidence the thermal field on the surface of the tested component. The presence of an imperfection located on the component's surface or in its vicinity acts as a thermal barrier that influences the thermal field's propagation at the surface. As a result, the inhomogeneity of thermal properties of the tested component, or the presence of some imperfections, is revealed as temperature variations of the component's surface. When heating and testing are performed on the same side of the component, the flaw indication appears as a hotter spot (hot flaws). When heating is performed on one side and testing is done on the other side, the flaw indication appears as a colder spot (cold flaws). Thus, the temperatures' distribution on a tested surface can be used to identify and localize the flaws hidden inside the coating material (lack of adherence between the coating layer and the base metal), excessive thickness of the coating and air infiltrations between the components of a sandwich-type structure and for the case with imperfections hidden in the various components' joints.

The following is an example of active thermography used for quality assessment of thermal spray coated surfaces. The proposed activation system (figure 9) is composed of 2 generators and two flashbulb based lamps with the necessary accessories, which ensure the activation energy and its uniform distribution on the tested object. To evaluate the quality of the layers deposited by thermal spraying, using the active thermography method, artificial defects were machined on testing specimens according to an experimental program using

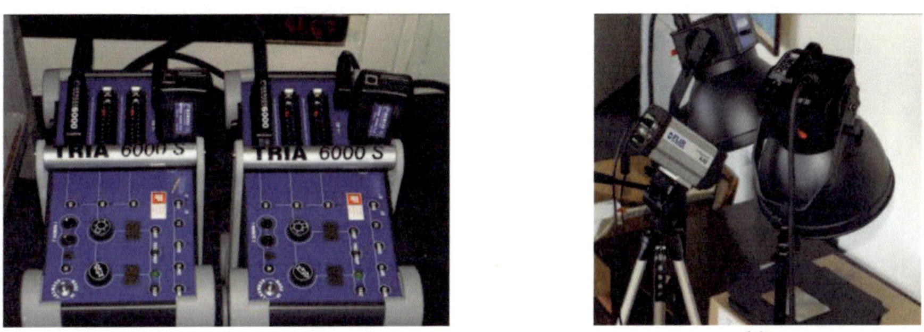

a) b)

a) Generators; b) Flash lamps and thermographic camera

Fig. 9. Activation system - experimental setup

electric-arc thermal spraying process. The base material consisted of S235J2 steel plates, 4mm and 10mm in thickness.

The test pieces with artificial imperfection were realized according to the sketches presented in figures 10. The holes and notches were filled with a Poxiline-type polymeric material, the surfaces being afterwards polished, sand blasted and cleansed with technical alcohol. Using the electric-arc thermal spraying procedure, a metallic layer was deposited on the surfaces with artificial flaws, using the following materials: Metco/ZnAW zinc wire, Ø1.62 mm and Metco/Tin Bronze wire, Ø1.62 mm.

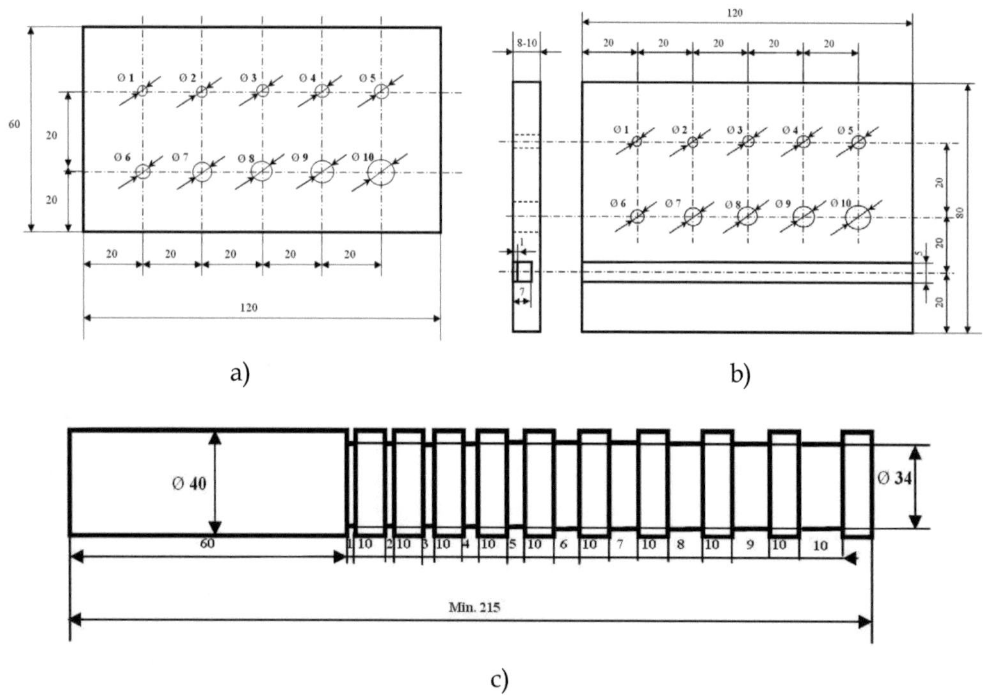

a) Type 1 test piece cu with Ø1 to Ø10 mm hole-type simulated defects
b) Type 2 test piece with Ø1 to Ø10 mm hole-type simulated defects and a notch (channel) with variable depth between 1 and 7 mm.
c) Type 3 test piece with 1 to 10 mm wide, simulated notches

Fig. 10. Test piece for thermographic test of the layers deposited by thermal spraying

Figure 11 presents, comparatively, the images of the test pieces with simulated flaws and their thermal images.

One could observe that the examination's sensitivity is influenced by the temperature of the examined part and by the coating material type. A lower temperature of the tested part leads to a more accentuated thermal contrast and a better flaw's detection. It was also established that a maximal thermal contrast is obtained after a certain period after the thermal activation (triggering the flashbulbs), and both the maximal temperature registered in the flaw area and the time after which the contrast is maximal depend on the flaw dimension.

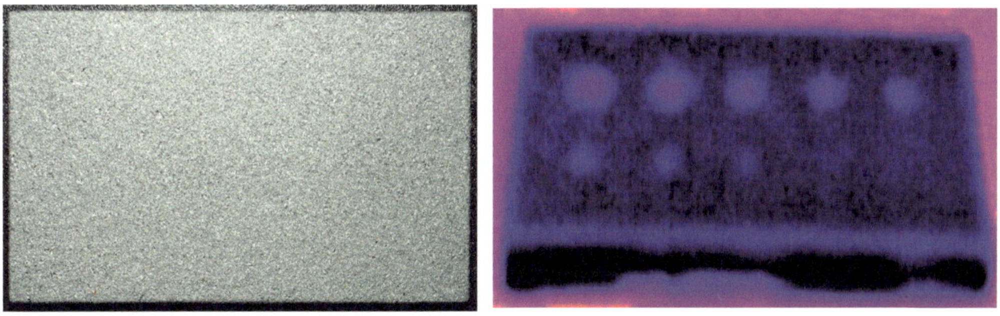

a)

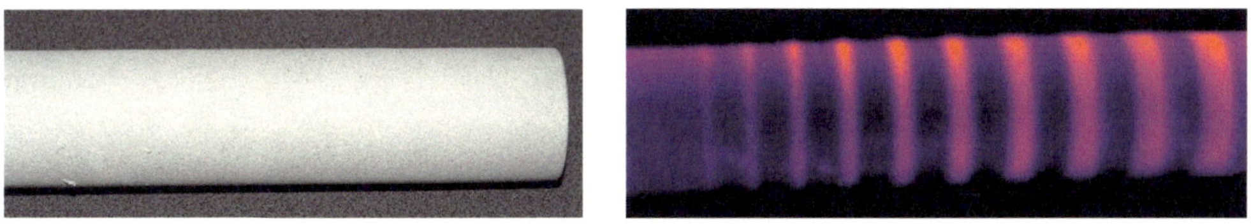

b)

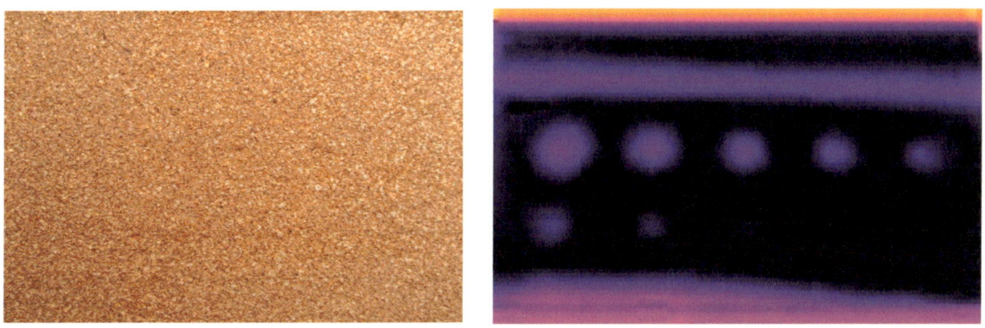

c)

a) Test specimen - zinc coating,
b) Test specimen - zinc coating,
c) Test specimen - bronze coating

Fig. 11. Active thermography testing of metallic coatings;

The small flaws, close to the sensitivity limit of the active thermography method, could be identified by maximal contrast after a relatively short activation time following thermal activation and the large flaws can be best identified after a longer time interval following the activation period of the piece. The response time could be correlated with equivalent dimension of the imperfection.

Figure 12 presents the sensitivity curve for imperfections detection related to a thermal spraying coated surface obtained using electric-arc thermal spraying procedure, S235J2 steel for substrate and bronze wire (Metco/Tin Bronze cu Ø 1.62mm). Figure 13 presents the correlation between maximal surface temperatures recorded during the thermographic examinations versus the imperfections sizes, for the same type of coated surface.

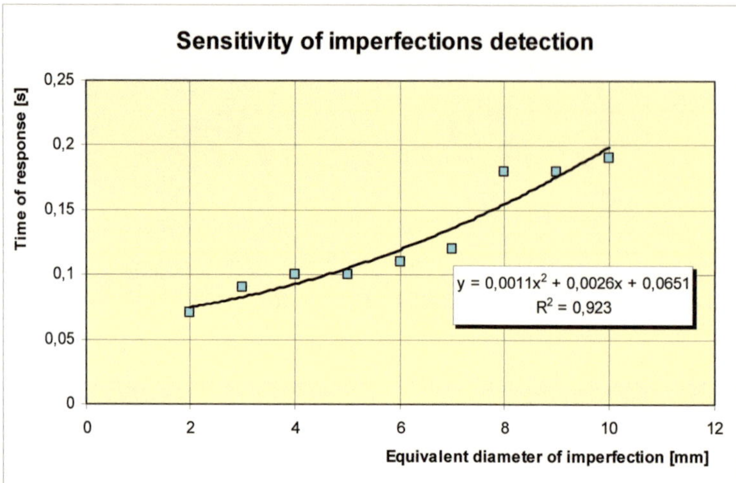

Fig. 12. Sensitivity of imperfections detection curve

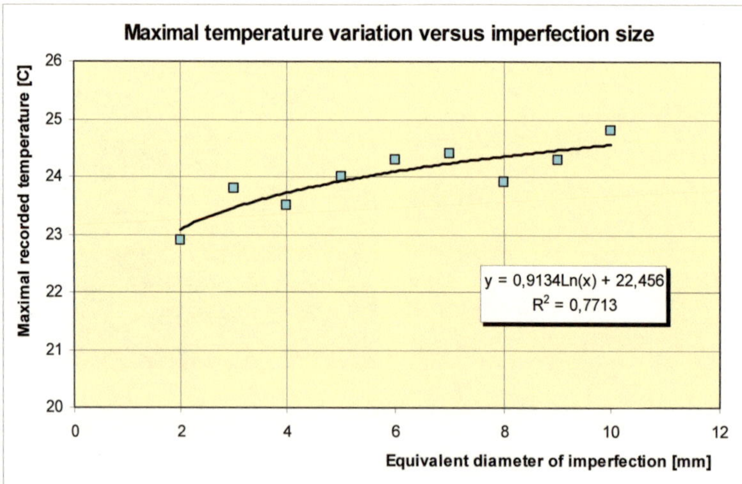

Fig. 13. Maximal temperature variation versus imperfection size

Experiments show that by using the active thermography method one can detect imperfection in coating layers deposited by thermal spraying. The notch-type flaws with the minimal width of 1 mm, respectively lack of adherence and porosity-type imperfection with equivalent diameters greater than 2 mm can be detected. Presented method is quantitative, and can be computerized by implementing of a complex informational system for intelligent data processing "artificial vision" type, based on the use of Hopfield and Kohonen neural networks (Bishop, 2006).

3.2 Studying a new laser-arc hybrid process dynamics

The use of IR thermography in welding process is mainly focused on developing process sensors with and without control (Huang et al., 2007; Mathieu et al., 2006), but also for process comparison (Mattei et al., 2009), as a tool for validating the process models (Ilie et al., 2007) or, with a more complex approach, into integrated systems for evaluating the materials' weldability and controlling the welding process (Bîrdeanu A.-V. et. al, 2011b).

LASER-ARC hybrid welding process is not a new process (Steen & Eboo, 1979), and it is characterized, in the last decade, by an accelerated and continuous development towards industrial implementation (Staufer, 2005) and extending its applicability (Asai et al., 2009) with research focused on understanding and modelling the complex phenomena that characterizes the hybrid welding process (Cho et al., 2011; Zhu et al., 2004) and developing prediction models and tools to control it (Bidi et al., 2011).

Following this trend in laser-arch hybrid welding process development, with focus onto thin sheets joining applications a new laser-arc hybrid welding process was proposed (Bîrdeanu et al., 2009) which combines and couples the use of a pulsed laser source and a pulsed TIG process. The new laser - arc hybrid welding process was characterized, statistically modelled in order to evaluate the important process parameters and welding technologies were developed in order to compare its applicability as a replacement for other classical arc welding technologies (Bîrdeanu et al., 2011a) in respect to energy efficiency and process productivity.

The first steps in the development of the new laser-arc hybrid process required to investigate the hybrid process dynamics by doing bead-on-plate experimental work and using two video acquisition systems (one co-axial with the laser beam and one perpendicularly on the welding direction) in combination with thermal imaging of the processed area (figure 14) which proved that thermal imaging can be a tool fit for necessary investigations.

Fig. 14. Experimental system with 3 video acquisition systems: CCD camera, 25fps video camera, IR thermographical camera

While the video acquisition system allowed to reveal the way the two processes interact (figure 15), the IR thermography revealed important information regarding the intensity of the process interaction, the stability of the process by means of relative temperature variation of the processed area and why specific phenomena takes place (figures 16 and 17).

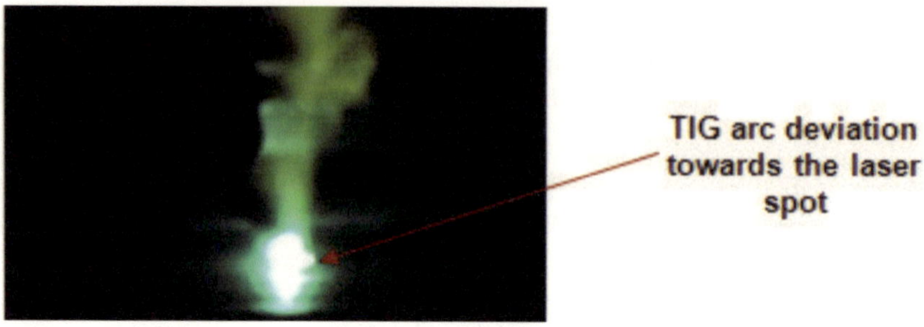

Fig. 15. LB (Laser Beam) generated plasma and TIG arc deviation toward the laser spot

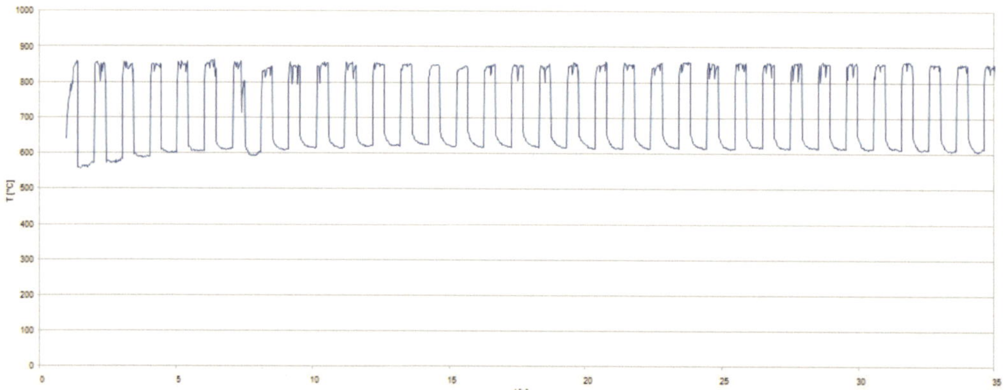

Fig. 16. Temperature variation of the monitored area (process variant: pulsed TIG-laser, Iav=28A, FrecvW=1Hz): Tmax=865.2°C; Tmin=591.807; Amplitude=273.4°C

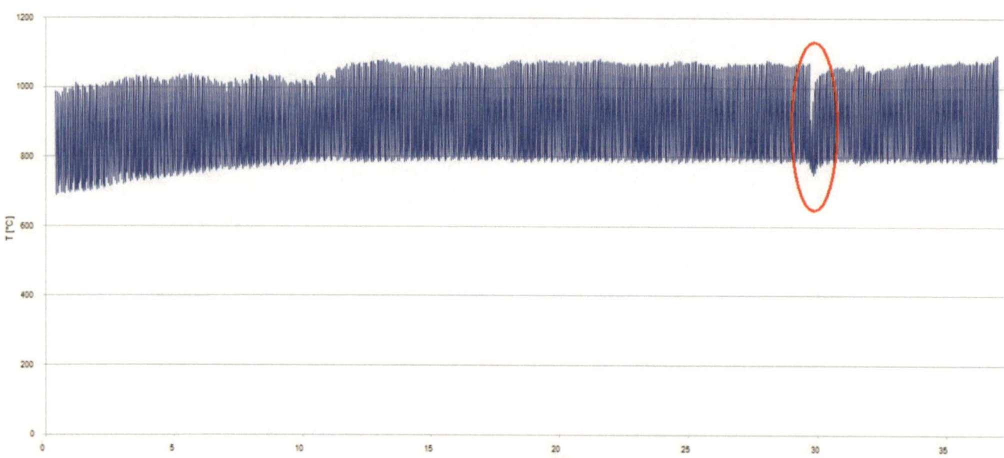

Fig. 17. Temperature variation of the monitored area for pulsed laser-TIG variant without laser protection gas at average TIG frequency (15Hz) - red circle indicates an instability

Though the temperature measurement was not calibrated, i.e. by determining a value for the emission coefficient of a point in the monitored area, the information provided by the thermal imaging acquisition system could be used to evaluate and characterize the process due to the fact that the thermal imaging recordings were done by using the same experimental setup configuration and the relative values of the temperatures were compared by using the same processed area.

The three specific coupling situations (Bîrdeanu et al., 2009) for the pulsed laser-TIG hybrid welding process - during the peak current, during the base current and during the base-peak-base current transition - could be also be identified, beside using the normal video acquisition system, using the thermal imaging with peaks of increases in the temperature of the monitored area, especially for the low TIG current frequencies (figure 18).

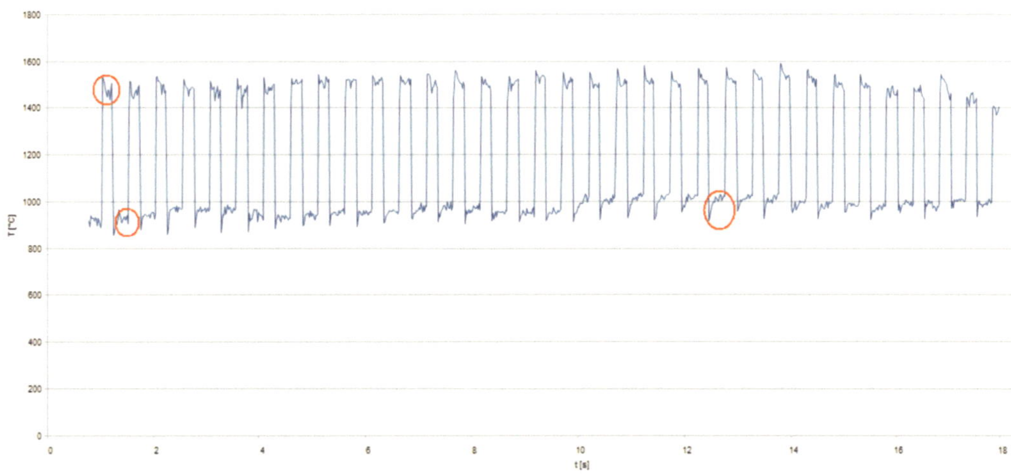

Fig. 18. Spikes in temperature variation (red circles) of the monitored area indicating the laser - arc interaction (process variant: pulsed TIG-LASER, TIG freq. 1Hz, ratTIG 40%)

The intensity of the process interactions (the hybrid coupling intensity) could be associated with the relative values of the temperature in the monitored area, i.e. higher average temperature of the monitored area were associated with an increase of the hybrid coupling and lower average temperatures were associated with lower hybrid coupling intensity. At the same time, sudden temperature changes of the monitored area could be associated with instabilities of the process (Fig. 17) while the overall temperature variation of the monitored area could be associated to the general degree of process stability.

Other important information regarding the process stability and the laser beam - electrical arc interaction was determined by evaluating the maximum temperature variation of the monitored area during the process and correlating that data with the video recordings of the two video acquisition systems. By processing the acquired data and evaluating the maximum temperature variation of the process, it was determined that the LASER-TIG variant (laser beam as leading process) was more stable and this could be observed both by using the visual video recording by observing the TIG arc stability due to its deviation towards the laser beam spot (Figure 19), by visual inspection of the weld (Figure 20), which could be correlated to a slight increase of the variation of the maximum temperature difference but at an increased absolute value of the temperature (Figure 21).

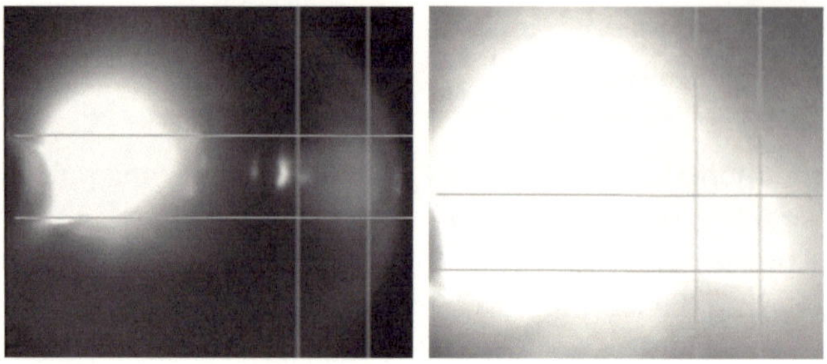

Fig. 19. Stronger influence on the pulsed TIG arc with pulsed laser beam as leading process both for base current and peak current regions (Iav=20A, TIG rat=40%, FTIG=1Hz) (Bîrdeanu et. al, 2009)

Fig. 20. Welded bead visual aspect for LASER-TIG variant

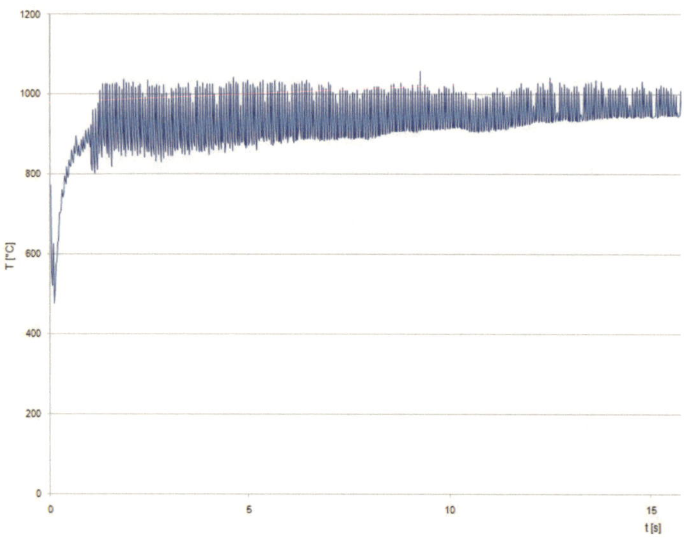

Fig. 21. Excerpt of temperature variation for LASER-TIG variant (Tmax: 1058.251°C, Tmin: 861.738°C, amplitude of temperature variation: 196.513°C)

Beside the presented data, which was correlated to the other information acquired during the experimental work, it was also possible to correlate the temperature variation with other types of instabilities, e.g. spattering, which were afterwards verified by visual inspection of the weld and macroscopic analysis of transversal and longitudinal sections of the bead on plate in order to establish the direction for optimization of the process and develop welding technologies.

3.3 Using the IR thermography for monitoring and controlling of the Friction Stir Welding (FSW) process

3.3.1 General considerations

The worldwide achievements for FSW process monitoring are rather well known and important results were achieved in particular by using complex systems that ensure monitoring of the forces developed during the process and acting on the welding tools.

In this subchapter results of FSW research team from National R&D Institute for Welding and Material Testing - ISIM Timisoara are presented, regarding the possibilities of using infrared thermographic technique for monitoring the friction stir welding process (FSW).

Scientific results presented were obtained through complex FSW experimental programs, which included:

- FSW experiments on a wide range of materials (alloys of Al, Mg, Cu, Ti, steels, dissimilar materials) and optimization of the welding technologies;
- assisting the welding process using infrared thermography technique;
- interpretation of the temperature diagrams in correlation with the complex characterization of the welded joints (PT - Penetrant Tests, RT - Radiation Test, macro and microscopic analysis, static tensile and bending tests, hardness, surface roughness of welded joint, etc.).

The main researches objective was to demonstrate that the infrared thermographic technique can be used for monitoring the FSW process and to establish the limits for its application. The following research methods were used:

- simulation method of various defects type, i.e. holes, slots and implants, having different sizes
- real time tracing method in of the welding process - welded joints with different materials were made using optimized process parameters in previous research programs, the diagrams of temperature evolution were analyzed.
- samples were taken from welded joints, which were non-destructive and destructive analyzed and controlled, etc.
- comparing the results of the analysis diagrams of the evolution temperatures measured during the welding process, with the results obtained during the non-destructive and destructive control and evaluation of welded joints, for a wide range of types and thicknesses of metallic materials.

3.3.2 Experimental conditions

The experimental program was conducted on a specialized FSW machine (figure 22), with the following main technical characteristics:

- adjustable welding speed range: 60-480 mm/min.
- adjustable speed of the welding tool in range: 300-1450 rot/min.
- usable stroke (welding distance): 1000 mm.

Welding parts were butt positioned and rigidly fixed on a steel backing plate. The topping up of the machine with a monitoring and control system of the FSW process using infrared

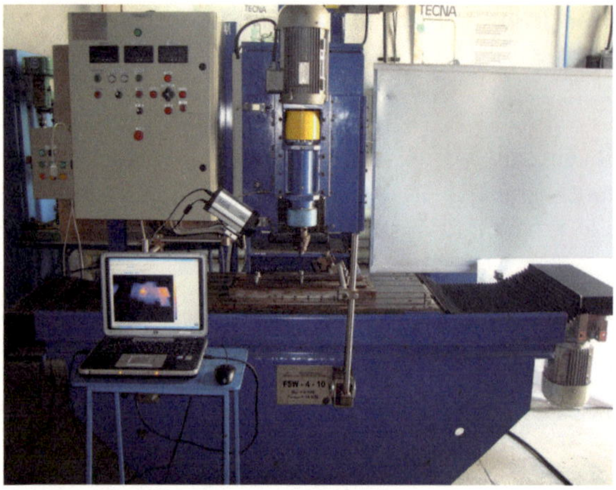

Fig. 22. FSW machine from ISIM Timişoara, Romania

thermography can provide information on process stability, welding parameters stability, the appearance of some imperfections and/or defects, and also quality analysis of welds through the thermal image, as well as the adjustment and the optimization of the welding parameters through feedback connections.

The check-up of the operating principle in terms of identifying imperfections during the welding process, revealed that they can be evidenced through thermographic method because they represent a thermal barrier which are preventing the heat propagation inside of the object examined in accordance with its thermal characteristics, by having a different thermal conductivity of imperfections compared with the homogeneous material.

The temperature recording was realized on-line, using a Thermo - Vision A 40 M camera, at an acquisition rate of 25 images/s. The camera was placed on the welding equipment, (figure 22) in order to trace the intersection zone between the tool shoulder and the weld surface, on the back semi-circle zone (π/2), figure 23. The temperature recordings and processing were done by using the Therma Cam Researcher Pro specialised software.

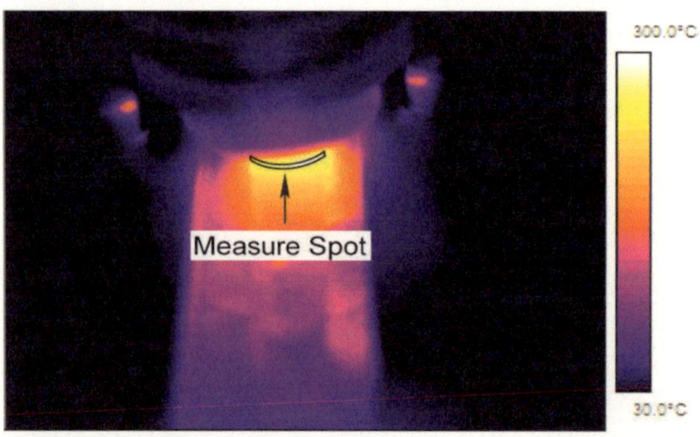

Fig. 23. Measure spot seen on the infrared image

Based on the recorded values from the temperature measurements, done with the infrared thermographic camera, the temperature evolution diagram during FSW welding process could be obtained. The measurements were made on the joint line at a distance of 1 mm behind the welding tool shoulder (figure 24).

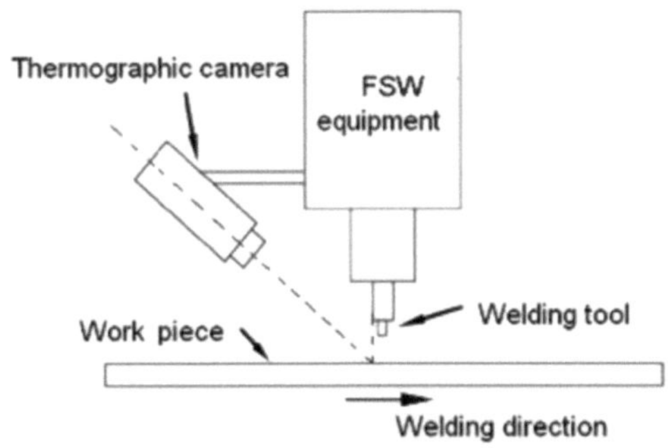

Fig. 24. Scheme of positioning for thermographic camera regarding to FSW machine

To determine how the temperature varies in material or in the materials (for the case of dissimilar materials) during welding, the measurements were made perpendicular to joint line at different distances from the weld start and the afferent variation graphs have been drawn.

3.3.3 Defects and simulation method

In order to verify the possibilities to apply the infrared thermography for on-line monitoring of the welded joints defects, an experimental program was developed, by using two types of artificially simulated defects, on EN AW 1050 sheet metal plates, having the dimensions 330x10 mm:

- inclusions , made perpendicularly on the plates and covered by FSW process, using a welding tool with the pin length bigger than the depth of the defects,
- defects made in the weld gap, on the frontal surface, in longitudinal direction of one of the plates which or butt welded by FSW process.

The experiments (Dehelean et al., 2008a) demonstrated the viability of infrared thermography in detection of the defects during the welding process. The experiments were based on using different forms and dimensions for the artificial defects made in the welded sheets. The sketch with the positions and the dimensions, for the case of artificially simulated defects having elliptical slits with variable width 2-6mm and constant depth h=4mm, are presented in figure 25.

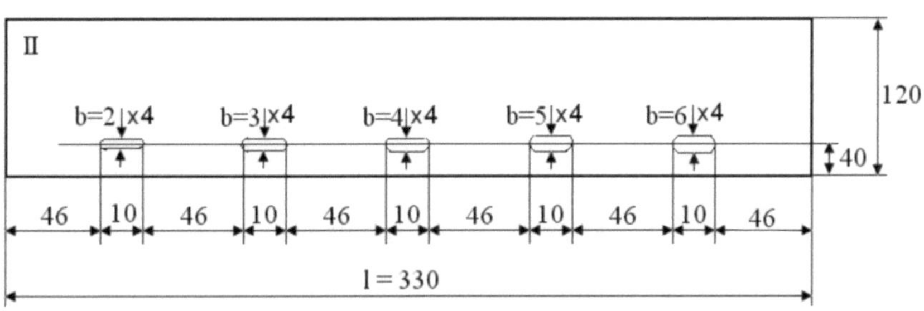

Fig. 25. Sketch of samples with simulated defects

In respect to the temperature evolution, recorded by the thermographic camera, for the welds done over the open slots, the recording T = f(l), presented in the oscillogram from figure 26 was obtained. Significant for the experiments are the "jumps" that appear on the temperature graphic, in front of the slits, due to local overheating.

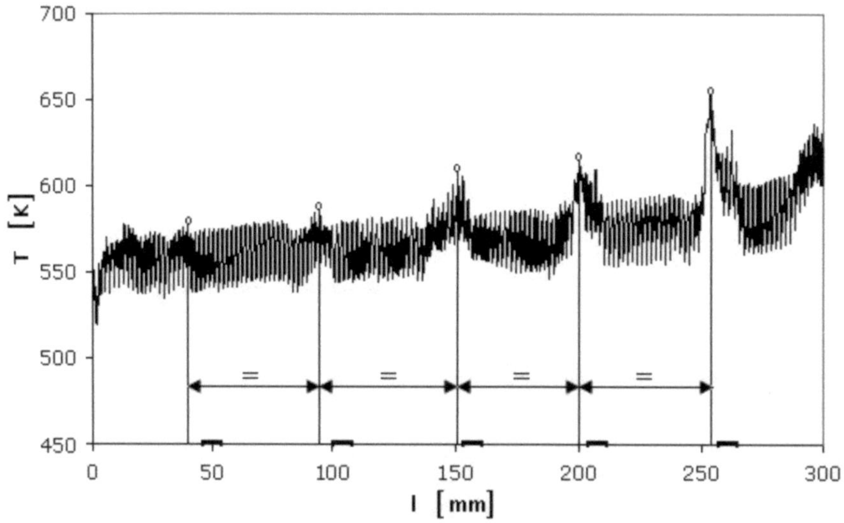

Fig. 26. MIT recording of the process

The performed researches reveal the real possibilities, both qualitative and quantitative, to detect the defects of FSW joints, using the infrared thermography method. Also, the good reproducibility of the results regarding localization of defects and dependence of the temperature variation with the dislocated volume by the defects, have been demonstrated.

The appearance of the peaks of temperature on the length of the welded joint and their systematic localizations in the defects zone, and in the working area of the welding tool respectively, was determined by the suddenly modification of the temperature gradient, caused by the thermal conductivity variation.

Also, it was concluded that, no matter the shape of the artificial defect, the minimum necessary volume to obtain a good evaluation of the thermographic recording can be determined (Safta, 2010).

3.3.4 Tracing method in real time of welding process for concrete applications

Based on the results obtained through the developed method by using simulated defects, one can mention good results in respect to the monitoring of the FSW process used for some concrete applications. By analyzing the temperature evolution diagram (figure 27), e.g. in case of welding of dissimilar alloys EN AW1200-EN AW 6082, uniformity is observed along the full length of the welded joint, after the welding process has been stabilized (after approximate 50-70 mm from the beginning of the welding process), (Cojocaru et al., 2011b) and, through X-ray analyse it was verified that the welded joint was free of defects. For such a welding application, after the stabilization process, the average temperature recorded in the joint line area was approximately 450°C.

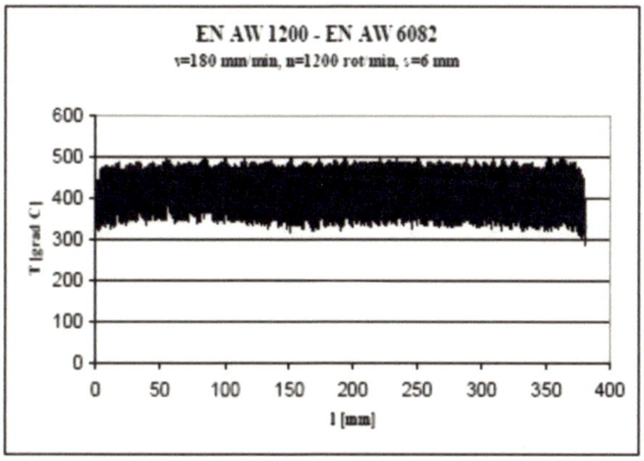

Fig. 27. Temperature evolution during FSW process

Comparing the temperature diagram evolution on a direction perpendicular to the joint line in the three moments of the realized measurements (figure 28) one can lay down several conclusions that may have an important role in the evaluation of welded joints and regarding the research results generally:

- before 50 mm of welding, the process is not yet stabilized and the optimum plasticizing temperature of materials (75-80% of melting temperature) was not reached.
- after the welding process stabilisation, the highest temperatures were recorded in the joint line area ≈ 465°C.
- in interference zone of the welding tool shoulder and welding materials, to ±11mm from joint line, the following temperatures were recorded: 250-300°C (EN AW 6082), respectively ~205-230°C (EN AW 1200), (Cojocaru et al., 2011b).

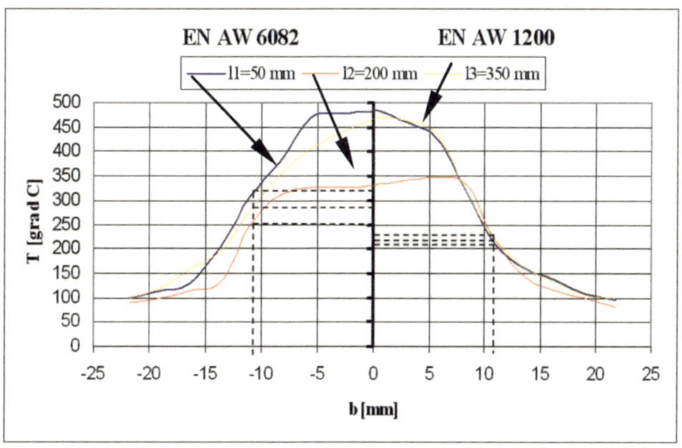

Fig. 28. Comparative evolution of temperatures

Another example is presented in figure 29 which shows the formation of a weld defect which occurred at friction stir welding of AZ31B magnesium alloy. The existence of the defect (revealed by X ray control) does determine a perturbation on the temperature diagram.

It was also observed an 80ºC decrease of the temperature along the zone in which the defect occurred (lack of penetration), followed by an increase towards the stable values of the temperature after the defect position was overcome (Dehelean et al., 2008b). The cause of the defect was also identified being a malfunction/fault of FSW machine, which was characterized by ~ 40% reduction of rotational speed of welding tool While the welding speed was kept at the prescribed values. The temperature developed during the process is directly proportional with the rotational speed, it decreased with about 80ºC, reaching values below the optimum temperature for plasticizing of the welded materials. As a result of the machine malfunction, there was an insufficient mixing of the materials, a consolidated nugget was not achieved and a tunnel type defect did appear (figure 29).

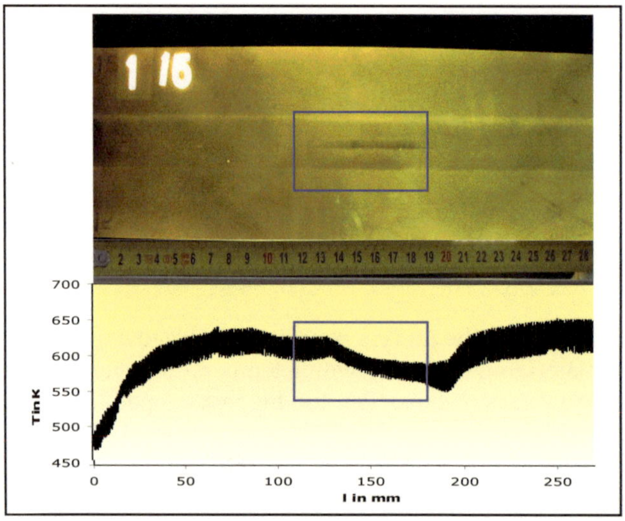

Fig. 29. Defect detection with radiographic control and correlation with temperature diagram

After remedying the machine's fault and returning to the initial set point of rotational speed, the welding process was carried out under optimal conditions, which could be revealed by the diagram of evolution of the temperature. This example also demonstrates the possibility to use infrared thermography for real time monitoring of the FSW process.

The same approach was used also for welding of steel, to obtain information about the temperature developed during the FSW process and for monitoring the process, by using infrared thermography. It was determined that for S235JR+N, S420MC and AISI 304L steels, after the 80-100mm from the beginning of the welding process, the temperature was constantly evolving around 980-1000ºC (stable welding process), figure 30a. For friction stir welding of S235 steel, macroscopic aspect presented in figure 31a demonstrates the lack of defects and formation of nugget well consolidated in centre of the weld.

Another particular case where IR thermography allowed the identification of a faulty condition occurred during welding experiments of AISI 304L stainless steel. Analyzing the evolution of temperature diagram from figure 30 b one can notice that approximately 150mm from welding start there was a disturbance (area A). The subsequent verifications of welded sample revealed that in that area a 20% pin damage occurred (a volume of approx.

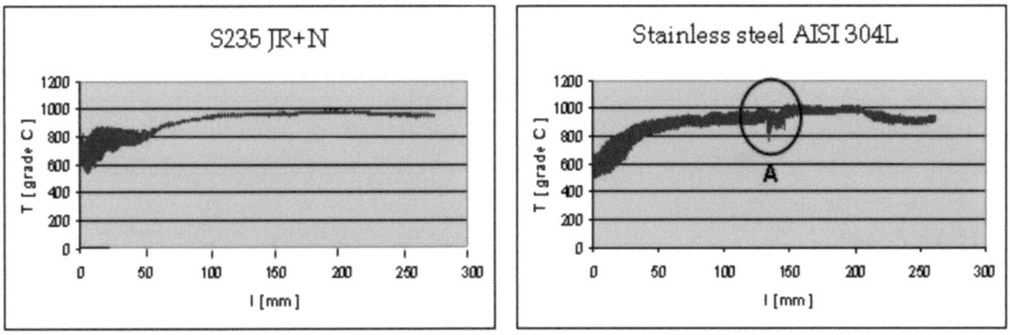

a) b)

Fig. 30. Evolution of temperature monitoring by thermographic camera

a) S 235 JR+N b) AISI 304L

Fig. 31. Macroscopic aspect of welded joints

20% of pin broke, which remained „implanted" in the welded material, figure 31b, the area marked B), (Cojocaru et al. 2011a).

This incident supports also the fact that the infrared thermography technique can be used for online monitoring of FSW welding process.

4. Conclusion

Worldwide, the infrared thermography usage is in continuous expansion and development both in material science and in engineering. Thermography method allows improving the characterization of new types of materials (composites, polymers), used more and more frequently in industrial products. As an example the polyethylene fracture behaviour in the presence of simulated imperfections was analyzed. It was found that recorded temperatures during the tensile tests are correlated with the local stress condition, induced by loading of the specimen with imperfections.

Using of thermography for process monitoring represents also actual trend in engineering. Beside its use as a monitoring or control tool, IR thermography proved to be also a good method to investigate and better understand the dynamics of a new laser-arc hybrid process and the acquired data could be correlated with the information and observations gathered using the usual video recording system and the specific analysis of the welding process results. It was also demonstrated that the system used for on-line monitoring of the FSW process has a good reproducibility for a large range of defects with different sizes and positions.

5. Acknowledgment

This work was partially supported by the Romanian National Authority for Scientific Research, project PN-09-160203, project PN 09-160101, PN-09-160 104 (2009-2010) and partially supported by the National Research Plan PNCDI 2: Frame 4: "Partnerships in priority areas", project 72174, acronym TIMEF, financed by the UEFISCDI (2008-2011).

6. References

Asai, S.; Ogawa, T.; Ishizaki, Y.; Minemura, T.; Minami, H. & Miyazaki, S. (2009). *Application of Plasma-MIG Hybrid Welding to Dissimilar Joint between Copper and Steel,* International Institute of Welding Doc. No. XII-1972-09

Avdelidis, N.P.; Ibarra-Castanedo, C.; Theodorakeas, P.; Bendada, A.; Saarimaki, E.; Kauppinen, T.; Koui, M. & Maldague, X. (2010). NDT characterisation of carbon-fibre and glass-fibre composites using non-invasive imaging techniques, *10th International Conference on Quantitative Infrared Thermography*, July 27-30, 2010, Qubec, Canada

Balageas, D.L.; Levesque,P.; Brunet,P.; Cluzel, C.; Déom A. & Blanchard, L. (2006). Thermography as a routine diagnostic for mechanical testing of composites, In: *Quantitative Infrared Thermography –The 8th QIRT,* Padova, Italy, June 27-30, 2006

Bidi, L.; Mattei, S.; Cicala, E.; Andrzejewski, H.; Le Masson, P. & Schroeder, J. (2011). *The use of exploratory experimental designs combined with thermal numerical modelling to obtain a predictive tool for hybrid laser/MIG welding and coating processes,* Optics & Laser Technology 43, pp. 537-545, ISSN 0030-3992

Bishop, C.M. (2006). *Pattern recognition and machine learning,* Springer, 2006, ISBN 978-0-387-31073-2, http://www.springer.com/978-0-387-31073-2

Bîrdeanu, A.-V.; Ciucă, C. & Iacob, M. (2011a). Pulsed LASER-TIG hybrid welding of coated unalloyed steel thin sheets, *Proceedings of The 5th International Conference Innovative technologies for joining advanced materials,* ISSN 1844-4938, Timişoara, Romania, June 16-17, 2011

Bîrdeanu, A.-V.; Ilie, M; Stoica, V. & Verbitchi, V. (2011b). Integrated experimental system for evaluating polymers laser weldability, real-time monitoring and control, *Proceedings of Modern Technologies, Quality and Innovation-ModTech International Conference,* ModTech Publishing House Chisinau, Republic of Moldova, 25 – 27 May, ISSN 2069-6736

Bîrdeanu, V.; Dehelean, D. & Savu, S. (2009). *Laser - TIG hybrid micro-welding process developments,* BID - ISIM Welding & Material Testing, No. 4 (December 2009), pp. 37-42, ISSN 1453-0392

Boțilă, L.; Murariu, A. C.; Cazacu, A. & Ciucă, C. (2011). Applications of infrared thermography in nondestructive examination of materials and welding, *Revista Sudura,* no. 1/2011, pp. 6-13, ISSN 1453-0384

Bremond, P. (2010). Full Field Experimental stress/strain analysis by Thermographic stress analyser for fatigue crack detection during HCF Testing. Examples in Automotive and Aircraft industry, *10th International Conference on Quantitative Infrared Thermography,* July 27-30, 2010, Qubec, Canada

Cho, Y.T.; Cho, W.I. & Na, S.J. (2011). *Numerical analysis of hybrid plasma generated by Nd:YAG laser and gas tungsten arc,* Optics & Laser Technology 43, pp. 711-720, ISSN 0030-3992

Choi, M.Y.; Park, J.H.; Kim, W.T.; Lee, S.S.; Kim, K.S.; Kang, K.S. (2008). Predicting the Dynamic Stress Concentration Factor Using the Stress Measuring Method Based on the Infrared Thermography, *9th International Conference on Quantitative Infrared Thermography*, July 2-5, 2008, Krakow, Poland

Cojocaru, R.; Boţilă L.; Ciucă, C. (2011a) Possibilities to apply the friction stir welding (FSW) to steels, *Revista Sudura*,No.1 (March 2011), pp.34-40, ISSN 1453-0384

Cojocaru, R.; Boţilă L.; Ciucă, C. (2011b) Friction stir welding of EN EW 1200-EN AW6082 aluminum alloys, *Proceedings of the 15th International Conference Modern Technologies, Quality and Innovation MODTECH New Face of T..M.C.R.*, Vol. I, pp.277-280, ISSN 2069-6736, Vadul lui Vodă, Chişinău, Republic of Moldova, 25-27 May, 2011

Dehelean, D.; Safta, V.; Cojocaru, R.; Hälker, T.; Ciucă, C. (2008a) Monitoring the quality of friction stir welded joints by infrared thermography, *International Conference Safety and Reliability of Welded Components in Energy and Processing Industry*, Graz, Austria, 2008

Dehelean, D.; Cojocaru, R.; Boţilă, L.; Radu, B. (2008b) Friction stir welding of aluminum-magnesium dissimilar joints, *International Conference Welding*, Subotica, Serbia, 2008

Galietti, U. & Palumbo, D., (2010). Application of thermal methods for characterization of steel welded joints, *10th International Conference on Quantitative Infrared Thermography*, July 27-30, 2010, Qubec, Canada

Grellmann, W.; Bierögel, C.; Langer, B. (2003). Modern mechanical methods of technical polymer diagnostics *In: Proceeding of Amsler Symposiums: World of Dynamic Testing*, pp. 117-126, Gottmardingen, Germany, 19-22 May, 2003

Grellmann, W. & Seidler, S. (2001). Deformation and Fracture Behaviour of Polymers, *Series: Engineering Materials*, ISBN: 978-3-540-41247-2, Springer-Verlag, New York

Huang, R.-S.; Liu L.-M. & Song G. (2007). *Infrared temperature measurement and interference analysis of magnesium alloys in hybrid laser-TIG welding process*, Materials Science and Engineering: A, Volume 447, Issues 1-2, 25 February 2007, pp. 239-243, ISSN 0921-5093

Huebner, S.; Stackelberg, B.V. & Fuchs, T (2010) Multimodal Defect Quantification, *10th International Conference on Quantitative Infrared Thermography*, July 27-30, 2010, Qubec, Canada

Ilie, M.; Kneip, J.-C.; Mattei, S.; Nichici, A.; Roze, C. & Girasole T. (2007). *Through-transmission laser welding of polymers - temperature field modeling and infrared investigation*, Infrared Physics & Technology, Volume 51, Issue 1, July 2007, pp. 73-79, ISSN 1350-4495

Kruczek, T. (2008). Particular applications of infrared thermography temperature measurements for diagnostics of overhead heat pipelines, *9th International Conference on Quantitative Infrared Thermography*, July 2-5, 2008, Krakow, Poland

Mathieu, A.; Mattei, S.; Deschamps, A.; Martin, B. & Grevey D. (2006). *Temperature control in laser brazing of a steel/aluminium assembly using thermographic measurements*, NDT & E International, Volume 39, Issue 4, June, pp. 272-276, ISSN 0963-8695

Mattei, S.; Grevey, D.; Mathieu, A. & Kirschner, L. (2009). *Using infrared thermography in order to compare laser and hybrid (laser+MIG) welding processes*, Optics & Laser Technology, Volume 41, Issue 6, September 2009, pp. 665-670, ISSN 0030-3992

Murariu, A. C. & Crâsteţi, S. (2011). Nondestructive testing options by active thermography of thermal spraying coated surfaces, *Revista Sudura*, no. 1/2011, pp. 14-18, ISSN 1453-0384

Murariu, A. C.; Safta, V. I. & Mateiu, H. S. (2010). Long-term behaviour of polyethylene PE 80 pressurized pipes, in presence of longitudinal simulated imperfections, *Revista Materiale Plastice*, Vol. 47, No.3, pp. 263-266, ISSN: 0025-5289

Murariu, A.C.; Safta, V.I. & Fleşer, T. (2008). Investigations regarding the thermoplastic resistance evaluation with simulated imperfections, *Strength of Materials Laboratory at 85 years International Conference*, University POLITEHNICA of Timisoara Faculty of Mechanical Engineering Department of Strength of Materials, 2008, In: Key Engineering Materials Vol. 399, pp. 131-138, Trans Tech Publications, Switzerland, 2009, ISSN 1013-9826

Murariu, A.C. & Bîrdeanu, V. (2007). Experimental Method (LSI) for Planar Simulated Imperfections for Remaining Life Assessment of the Thermoplastic Pipe Networks, In: *The International Conference on Structural Analysis of Advanced Materials - ICSAM 2007*, Patras, Grece, 2-6 sept., 2007

Murariu, A.C. (2007). TT-IRT Hybrid Testing Method Applied in the Study of PE 80 Polyethylene Behaviour in the Presence of Simulated Imperfection, *The 5th International conference Structural integrity of Welded Structures*, 20-21 Nov., Timişoara, Romania, 2007

Pieczyska, E.A.; Nowacki, W.K.; Tobushi, H. & Hayashi, S. (2008). Thermomechanical properties of shape memory polymer SMP subjected to tension and simple shear process , *9th International Conference on Quantitative Infrared Thermography*, July 2-5, 2008, Krakow, Poland

Safta, V. (2010) Application of infrared thermopgraphy in the non-destructive examination of friction stir welds, *Welding &Material Testing BID ISIM*, No1/2010, pp.29-40, ISSN 1453-0392

Sahli, S; Fissette, S. & Maldague, X. (2010). Infrared Image Processing for Online Quality Control in Laser Welding, *10th International Conference on Quantitative Infrared Thermography*, July 27-30, 2010, Qubec, Canada

Staufer, H. (2005) Laser hybrid welding and laser brazing: state of the art in technology and practice by the examples of the AudiA8 and VW Phaeton, in: *Proceedings of the third International WLT Conference on Lasers in Manufacturing*

Steen, W.M. & Eboo, M. (1979). *Arc Augmented Laser Welding*. Met. Constr. 11, H. 7, pp. 332-333, 335

Rajic, N. (2004). Modelling of thermal line scanning for the rapid inspection of delamination in composites and cracking in metals, *DSTO Publications Online*, Available from http://hdl.handle.net/1947/4097

Thiemann, C.; Zaeh, M.F.; Srajbr, C. and Boehm, S. (2010). Automated defect detection in large-scale bonded parts by active thermography, *10th International Conference on Quantitative Infrared Thermography*, July 27-30, 2010, Qubec, Canada

Yang, B.; Liaw, P.K.; Wang, H.; Huang, J.Y.; Kuo, R.C. & Huang, J.G. (2003). Thermography: A New Nondestructive Evaluation Method in Fatigue Damage, In: *TMS Online JOM-e: A Web-Only Supplement to JOM*, Available from
http://www.tms.org/pubs/journals/jom/0301/yang/yang-0301.html

Wang, H.; L. Jiang,L; He, Y. H.; Chen L. J. & P. K. Liaw, P. K. (2002). Infrared imaging during low-cycle fatigue of HR-120 alloy, Metallurgical and Materials Transactions A, Vol.33 pp. 1287–1292 DOI: 10.1007/s11661-002-0231-1

3

Thermographic Inspection at the Interface of Dry Sliding Surfaces

G. Cuccurullo, V. Spingi, V. D'Agostino, R. Di Giuda and A. Senatore
Department of Industrial Engineering, University of Salerno
Italy

1. Introduction

During the last decades, in order to clarify the coupled thermal and frictional aspects in dry sliding contacts, researchers have been involved in theoretical models as well as in experimental testing. In order to outline both the analytical and the experimental approach, a procedure for estimating the maximum temperature increase in dry sliding contacts is introduced. The procedure is based on an analytical solution for the two-dimensional temperature field in a slab subjected to a suitably shaped moving heat source. Experiments were carried out on a specifically designed pin-on-disk device with the aim of taking into account the unavoidable uneven friction distribution under the coupled surfaces. The subsequent data reduction led to quite satisfactory agreement with analytical predictions and provided a suitable shape for the heating source distribution thus allowing for proper maximum temperature rise estimate. Since measuring the interface temperature of a friction pair is a difficult task, temperature data were collected by means of infrared thermography; this technique seems to be the most effective and valuable due to its ability of performing continuous temperature map recording with relatively high resolutions when compared to other traditional methods. It is a matter of facts that high energy rates are dissipated by friction during short periods, thus transient and localized thermal phenomena with high thermal gradients are to be expected. The latter conditions still suggest the use of IR thermography.

2. Tribological remarks

It is well known that the dynamic temperature distribution arising at the interface of dry sliding contacts has a strong influence in friction phenomena, thus the interface temperature characterization, from an engineering point of view, has always been an imperative topic in machine design.

Today, the heat production assessment is classified among the essential problems in the tribological behaviour of a broad mechanical devices area yet. Among them, a relevant role is played by dry contact friction phenomena, that is the ones featured by the absence of coherent liquid or gas lubricant film between the two coupled solid body surfaces. In fact, dry contacts, while performing in many components (brakes, clutches) an active role, in many kinematics and under specific operating conditions play also a passive role; this often results in harmful processes, such as undesirable temperature rise that can induce thermo-mechanical stress. It follows, therefore, the need to reduce their effects by means of more sophisticated projects requiring, for instance, the use of lubricants, of more suitable materials, design and operating conditions. However, it's well known that the heat transfer has also notable impact for fully developed hydrodynamic lubrication since it strongly modifies the oil viscosity and in severe cases may lead to oil film rupture. Nevertheless, in the case of hydrodynamic lubrication, the modelling approach is completely different, since the heat transfer mechanism is mainly convective at the oil film interface.

3. Sliding contact interfaces

Since pioneering studies, it is well known that the friction at sliding contact interfaces where two surfaces come together generates heat and most of this heat is conducted away through local rubbing asperities. It then understands how temperature field at the interface of dry sliding contacts has a strong influence on friction phenomena. Thus, much work has been done in the past in order to investigate this aspect. The theoretical approach encompasses finite element analysis, (Kennedy, 1981; Salti & Laraqi, 1999) or analytical solutions, even transient ones based on a Fast Fourier Transform method, (Gao et al., 2000; Stanley & Kato, 1997). Blok (Blok, 1937) first proposed the concept of flash temperature and derived simplified formulas for the maximum temperature rise on moving surfaces. Jaeger (Jaeger, 1942) formalized the mathematical models for the flash temperature on a semi-infinite medium for moving uniform rectangular heat sources. Many other flash temperature models have already appeared in the literature. These studies extended Jaeger's theory to various heat source shapes and to multiple asperity contact based on steady-state conditions (Archard, 1959; Francis, 1971; Laraqi et al., 2009; Kuhlmann-Wisdorf, 1987).

With reference to the experimental approach, as expected, it is strongly desirable to measure interface temperature during actual friction braking tests so that precise operating conditions were known for design purposes. However, measuring the interface temperature of a friction pair is a difficult task. Several methods have been reported (Dinc et al.,1993), but infrared measurements seem to be the most effective (Anon, 1995, Cuccurullo et al., 2010).

Some conclusions are well established. It is a matter of facts that high energy rates are dissipated by friction during short periods, thus transient and localized thermal phenomena with high thermal gradients such as hot bands and hot spots are to be expected, (Anderson & Knapp, 1990; Panier et al., 2004); it is also clear that when surfaces slide over one another, the static contacts can change in time due to tangential load effects on junction growth, thermal expansion, wear, chemical oxidation or a variety of other physical phenomena (Vick et al., 1998), but also due to the actual complexity of the real area of contact between sliding surfaces (Vick & Furey, 2001). Many thermal problems associated with brake friction pairs, including performance variation (fade, speed sensitivity) and rotor damage (heat spotting and thermal cracking) can be analysed in terms of localized frictional heat generation (Day et al., 1991).

In addition, it has been known for some time that the friction intensity is not distributed evenly across the surface of friction pair and that the local friction intensity of an imaginary friction lining segment changes in the course of the friction process (Barber, 1967; Rhaim et al., 2005; Severin & Dörsch, 2001). Among the parameters yielding to uneven friction intensity, many authors ascribe great importance to the pressure distribution. It's well known that the pad wear is greater on the leading side and, according to the Reye's theory which states the proportionality between friction work and wear, the pressure distribution is expected to have its centre on the leading side. The calculations through free body diagrams of the dynamic centre of pressure (CoP) position have been shown (Fieldhouse et al., 2006) for a brake pad during a normal braking operation, both in radial and axial directions. In fact, there is an interaction between frictional effects at the pad abutment between pad backplate and caliper finger. It has also been shown that the position of reaction force of the pad, which identifies the center of pressure to provide equilibrium, depends on friction level at pad/disc interface and the one at pad/abutment side of the caliper. The combination of these parameters normally yields to a leading centre of pressure and this effect is more marked at low brake pressure levels. The same authors have realized the complex task of measuring the interface pressure distribution during braking by means of a pressure sensitive film within the pad. Further experimental results have been achieved on a modified 12-pistons opposed calliper equipment (Fieldhouse et al., 2008). Tests were carried out at different levels of speed by applying uniform actuation pressure on the calliper side. The results about the pressure maps at pad/disk interface have shown that the position of the centre of the pressure moves considerably during a braking event, both radially and axially along the pad. Furthermore, it has been shown that under light braking with an uniform actuation pressure, the centre of pressure will always tend to be leading. It has been also demonstrated that, for increasing pressure and speed, the CoP tends to move towards the central region of the pad; moreover, the CoP position is more influenced by the level of the pressure than by the speed one. Recent studies have shown that one of the possible reasons for the variation in contact pressure distribution during braking is due brake pad surface topography. Some authors, in particular, examined six different pad configurations (Rahim et al.); even in this work, in order to measure contact pressure distributions, a suitable type of sensor film for a defined range of a local contact pressure have been chosen. Then a linear gauge has been used to measure brake pad topography. The test results have been proved that each pad has a different surface topography despite being produced by the same manufacturer; this indicates that when the pairs of pads are fitted in the brake system, it may produce different contact pressure distribution and consequently may generate different braking torque.

Another essential aspect responsible for the uneven friction distribution is to be related to the metal particles that originate from the drum or disk and diffuse into the bound friction layer, (Severin & Musoil, 1995). Of course, the dynamical positioning and arrangement of the coupled surfaces can lead to unpredictable friction distribution during the occurrence of the sliding contact.

Since all the addressed issues can be related to uneven friction distribution, one of the major objective of the following work is to introduce a parameter, the engagement between the coupled surfaces, encompassing the addressed effects and being able to reproduce them in an equivalent fashion in the proposed theoretical framework. More general, the proposed analytical solution aims to explore the influence of the key problem parameters (geometry, materials, boundary and operating conditions) on the dry contact thermal response. Then a procedure has been setup in order to obtain a reliable estimation of the maximum temperature attained under the contact area. With reference to selected experimental evidences, interesting indications have been obtained since the agreement between analytical and experimental data is quite satisfactory.

4. Basic equations and analytical solution

The theoretical model used to predict the temperature rise due to a moving heat source is shown schematically in Fig. 1. A finite thickness slab is subjected to a heat source featured by linearly varying distribution, whose variable slope is the engagement, μ. The reference system is attached to the source moving with constant velocity U in the y-direction and supposed to be due to the frictional effect at the interface. The slab is cooled by radiative-convective heat transfer (h) to an ambient at uniform temperature while it is adiabatic on the lower surface. The slab is subjected to thermal coupling conditions on the remaining edges (y = ±L). For such a problem, if the observer is travelling on with the source and if a suitable time has elapsed, the energy balance equation and the related boundary conditions turn out to be time-independent and fully developed temperature field is attained. Then the problem can be written in dimensionless form as:

$$\frac{\partial^2 \vartheta}{\partial \xi^2} + \frac{\partial^2 \vartheta}{\partial \eta^2} + 2\omega \frac{\partial \vartheta}{\partial \eta} = 0 \tag{1}$$

$$\vartheta(\xi, \eta \to -\eta_L) = \vartheta(\xi, \eta \to \eta_L) \tag{2}$$

$$\left.\frac{\partial \vartheta}{\partial \eta}\right|_{\xi,-\eta_L} = \left.\frac{\partial \vartheta}{\partial \eta}\right|_{\xi,\eta_L} \tag{3}$$

$$\left.\frac{\partial \vartheta}{\partial \xi}\right|_{0,\eta} = \mathrm{Bi}\,\vartheta(0,\eta) - f(\eta) \tag{4}$$

$$\left.\frac{\partial \vartheta}{\partial \xi}\right|_{1,\eta} = 0 \tag{5}$$

where: $\xi = x/s_x$; $\eta = (y - U\,t)/s_x$; $\theta = (T - T_a)/(s_x\, \dot{q}_0 /k)$, k being the slab conductivity; $\omega = s_x U/(2\alpha)$ is the Peclet number, α being the slab diffusivity. The source intensity is assumed to be described such as $\dot{q}\,(y) = \dot{q}_0\, f(y)$, where $\dot{q}_0$ is the heat flux intensity at the origin and f(y) represents its spatial distribution which, for the present purposes, is assumed to be linear: $f(y) = 1 + \mu y/s_y$, the parameter μ being responsible for the variable slope and thus for the pin to disk engagement. Finally, $\mathrm{Bi} = h\, s_x/k$ is the Biot number.

According to Blok's postulate, i.e. the energy balance at the contact interface, the total heat generated by friction is the sum of the heat flux entering the pin and the one entering disk. Thus, $\dot{q}\,(y)$ represents the local heat flux entering the disk. Wishing to extend the problem to encompass the pin thermal behaviour, the heat flux entering the pin is usually modelled assuming 1D heat transfer according to the thin rod model, which is suggested by the typical pin geometries. Thermal coupling conditions are usually considered imposing the continuity of the average temperature at the interface contact area, i.e. the perfect thermal

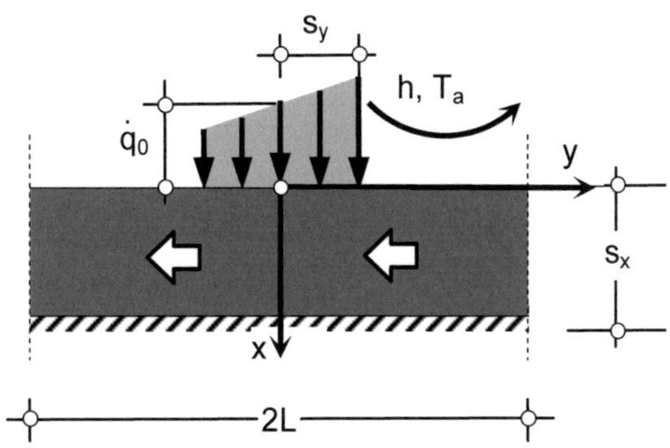

Fig. 1. Sketch of the problem

contact between the coupled surfaces. For usual geometries and materials, it can be stated that, the heat flux evacuated by the disk is much greater than that evacuated by the pin (Laraqi et al., 2009).

The two-dimensional temperature field has been solved analytically in closed form by Fourier series; in particular the structure for the temperature field has been sought to be:

$$\vartheta(\xi,\eta) = \sum_{k=-\infty}^{\infty} Q_k(\xi)\ e^{i\Omega k\eta} \tag{6}$$

where $\Omega = \pi\, s_x/L$. The unknown $Q_k(\xi)$ functions are derived by imposing the assumed structure to satisfy the above set of equations:

$$Q_k'' + (2\omega\ i\ \Omega\ k - \Omega^2\ k^2)Q_k = 0 \tag{7}$$

$$Q_k'(0) = \mathrm{Bi}\, Q_k(0) - F_k \tag{8}$$

$$Q_k'(1) = 0 \tag{9}$$

where :

$$F_k = \frac{s_x}{2L}\int_{-L/s_x}^{L/s_x} f(\eta)e^{-i\Omega k\eta}d\eta \tag{10}$$

is the k-th component of the transformed wall heat flux.

5. Experimental setup

In order to characterize the friction between solids, all the tests were performed on a pin-on-disk braking system specifically designed; it is a classical wear rigs and essentially consists in a cylindrical pin in eccentric contact with a rotating disk, Fig. 2.

During each test, the disc is rotated at constant speed by an electric motor controlled by an hydraulic regulator; two stationary pin specimens are symmetrically pressed against the

Fig. 2. Experimental setup

disk. The pins are made of a commercial braking material and are featured by cylindrical shape with 5.5 mm radius. Two 20 cm radius discs made of polycarbonate (5 mm thickness) and bakelite (4 mm thickness) were used. In order to closely recover the model assumptions, both disk surfaces where insulated with the exception of a path allowing pin motion. A weight device allows to control the axial pin load. Each test was configured by setting pin-load and disk angular speed. A large value of the pin eccentricity was considered in any test taking care to realize a sufficient distance from the outer radius such as to avoid edge effects. It can be shown that, under such operating conditions, the curvature effects are negligible and thus the model at hand can be properly used (Laraqi, 2009).

Provided a suitable emissivity calibration, the thermal pattern left on the disk by the track due to the relative pin motion is detected by an IR camera. The IR equipment (Thermacam Flir P65) exhibits a spatial resolution of about 30 dpi, due to 320x240 pixel matrix, the operating distance and optics in use. A specifically designed software allows to extract from the IR image the temperature profile to process: this latter is pointed out as the one where the radial slope is zero along the thermal pattern left on the disk.

A typical infrared image is reported in Fig. 3. Here it is possible to appreciate that the pin-holder partially covers the pin-track thermal patterns. As a consequence, it was possible to

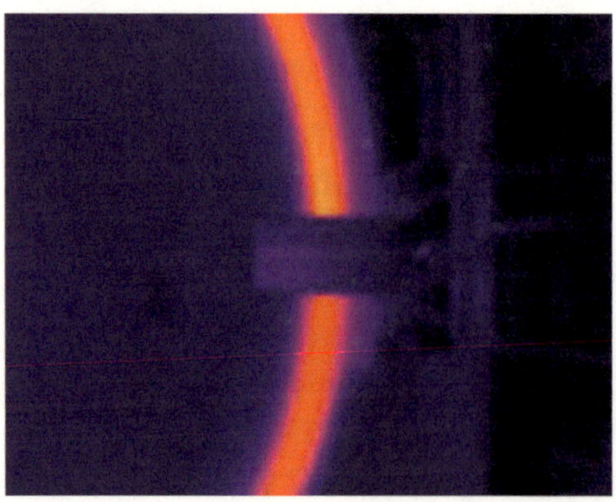

Fig. 3. A typical thermogram

detect temperatures along a track with the exclusion of an area 3 cm wide along the circumferential pin-track and placed around the pin itself. It can be argued that the presence of the pin-holder always interferes with optical measurements; however it will be shown later that a suitable data reduction can be done to reconstruct the hidden portion of the temperature profile at hand.

6. Data reduction and simulation

When performing measurements, it is assumed that temperature rise on the loaded surface is described by the temperature distribution given by the above model, after thermal equilibrium is attained. Due to the complexity of the response model, the Levenberg Marquadt technique χ^2 based fitting method has been selected. The technique enables to process non linear models with an arbitrary number of parameters. Thus, the optimal choice for matching experimental and theoretical data is accomplished by minimizing the χ^2-merit function:

$$\chi(\underline{a}) = \sum_{i=1}^{N} \left[T_i - T(y_i, \underline{a})\right]^2 \tag{11}$$

where the N experimental data points, (T_i, y_i), are to be detected by means of IR thermography along the circular pattern left on the disk surface by the pin; the function T(y_i, $\underline{a}$) is the functional relationship given by the model for the disk surface temperature, $\underline{a}$ = (h, μ, $\dot{q}_0$) being the unknown-parameters vector.

The accuracy of the fitting in estimating the unknown thermal parameters was tested on Montecarlo simulated thermograms to obtain confidence interval width of the fitted parameters for fixed operative conditions.

Simulated thermograms were obtained by perturbing reference analytical surface temperature profiles with a noise level due to a randomly generated maximum uncertainty of ±1°C. As an example in Fig. 4, a fixed reference temperature profile (continuous line) has been perturbed to obtain a discrete number of simulated experimental temperatures, i.e. the dots in Fig. 4; processing the perturbed profile with the χ^2 fitting, a set of the three unknown parameters was estimated, thus allowing to reconstruct analytically the reference profile, i.e. the dashed curve in the figure.

In order to realize the reference profile, some parameters were fixed having in mind the experimental setup, namely: T_a = 23.5 °C, s_x = 2.5 mm, s_y = 5.5 mm, h = 32.5 W/(m^{2}K), L = 56 cm, U = 0.28 m/s while the engagement was set equal to 0.7. Spatial resolution was assumed to be fixed at the best allowed by the camera and optics in use, i.e. 30 dpi. Then, the simulation took place by considering twenty different configurations for performing data reduction: each configuration was based on processing pin-track bands all starting from -(s_y + 3 s_x), i.e. the first detectable point free from the pin-holder shielding, and increasing widths by s_x-steps. Suitable dimensionless temperature profiles were processed, in such way only two unknowns are to be estimated, i.e. the Biot number and the pin-disk engagement. Results are shown in Figs. 4 and 5 where the mean values of the Biot number and of the engagement parameter are reported for the twenty increasing bandwidth extensions, each one identified by a progressive band index. For each band index, figures also report the

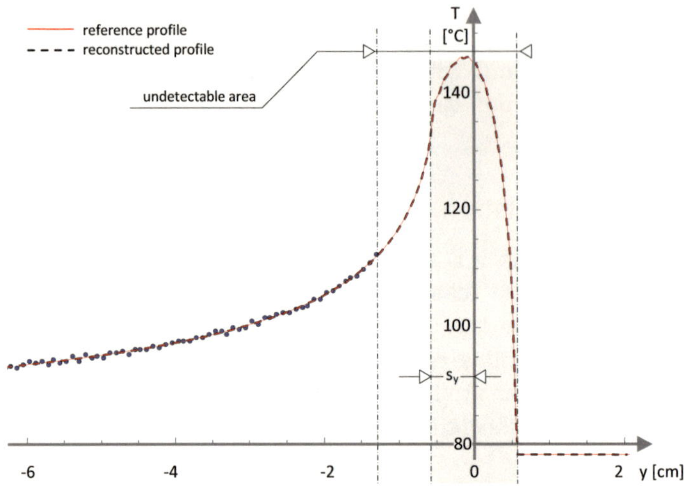

Fig. 4. Reference, simulated and resulting temperature profiles

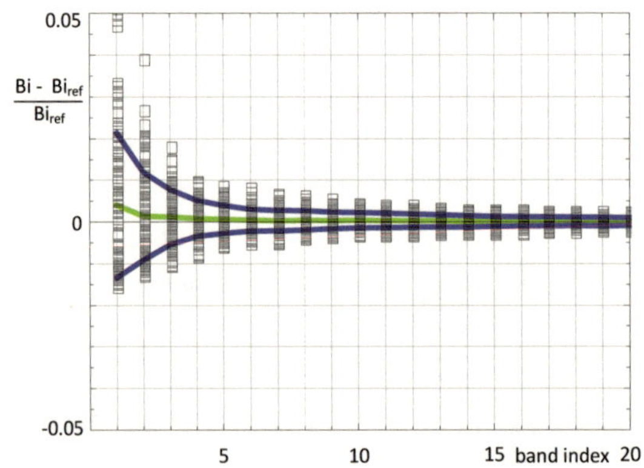

Fig. 5. Simulation results for Bi

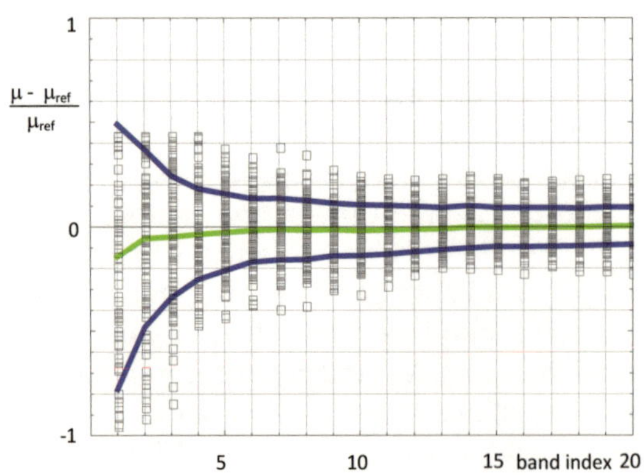

Fig. 6. Simulation results for μ

estimated percentage parameters deviation from the corresponding reference values and the 63% confidence limit. Since the resulting standard deviation practically attains fixed values, it can be concluded that both for the former and the latter parameter there is no improvement if bandwidths larger than 8 s_x are considered. The sensitivity on the Biot number seems to be higher than the one on the pin-on-disk engagement, at least around the reference preset values. The latter effect can be explained considering that Biot number controls the slab energy level.

7. Selected experimental tests

Experimental tests presented in the following were carried on by considering two angular disk-speeds corresponding to peripheral values of U_1 = 0.28 m/s and U_2 = 0.57 m/s. The latter values were calculated considering the radial collocation of the pin cylindrical axis, that is 17.7 cm. The pin-load considered was 30 kg/cm^2 for the former test and 20 kg/cm^2 for the latter. The circumferential thermal pattern extracted according to the procedure reported in paragraph 5, after data reduction procedure gave for the polycarbonate disk: $\underline{a}_1 = \{h_1 = 32.5\ W/(m^2\ K),\ \mu_1 = 0.7,\ \dot{q}_{0,1} = 1.97\ 10^5\ W/m^2\}$ for the 30 kg/cm^2 load; $a_2 = \{h_2 = 32.5\ W/(m^2\ K),\ \mu_2 = 0.7,\ \dot{q}_{0,2} = 3.11\ 10^5\ W/m^2\}$ for the 20 kg/cm^2 load. Such values were used to build the analytical profiles in Figs. 7 and 8: the agreement with experimental points seems to be quite satisfying. It has to be underlined that the estimated maximum values attained by temperature under the contact area, hidden to the IR camera view, are 145.6 and 203.1°C, respectively. Both values are far beyond the nearest detectable temperatures, namely 111.4 and 157.1°C. It appears that the region under the contact area is a critical one due to the occurrence of high thermal gradients which could lead to get wrong temperature estimates. Finally, a first check about $\dot{q}_0$ -values shows that they satisfyingly agree the expected dry friction characteristic.

The same trends outlined before were recovered for bakelite disk undergoing different average operating pin-loads, namely 0.93 and 1.4 MPa, for the two selected speeds fixed before. Figures from 8 to 12 report the data reduction output related to the four different combinations of the addressed parameters. It is to be underlined that the resulting values of the engagement parameter attain values very close to the unity for all the tests, thus showing that heat dissipation profile is shaped so to exhibit higher values toward the leading front.

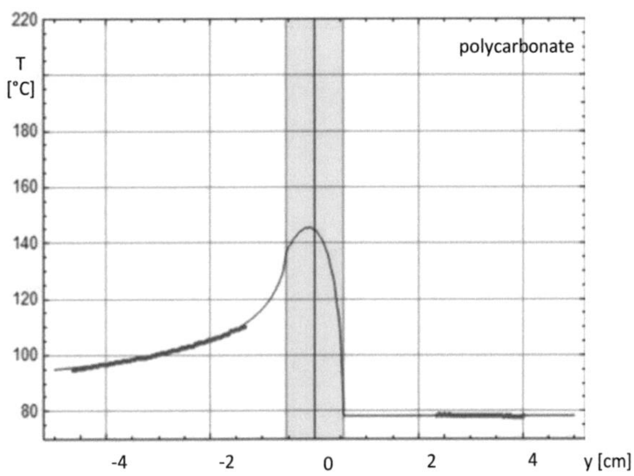

Fig. 7. Results for 0.28 m/s, 30 kg/cm^2

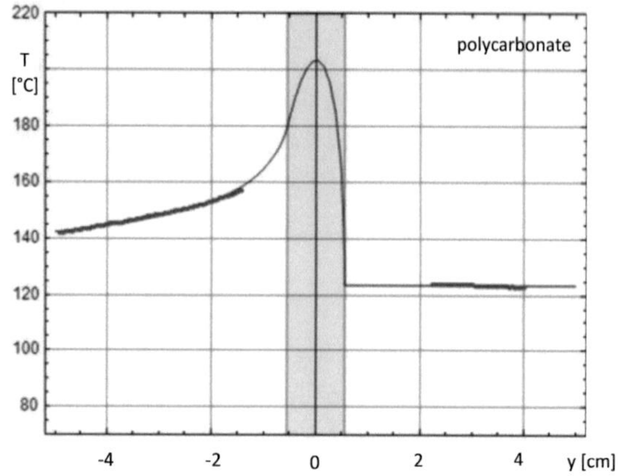

Fig. 8. Results for 0.57 m/s, 20 kg/cm2

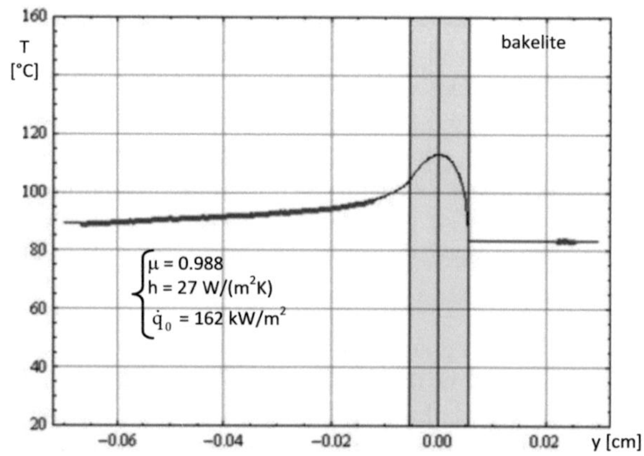

Fig. 9. Results for 0.57 m/s, 0.93 MPa

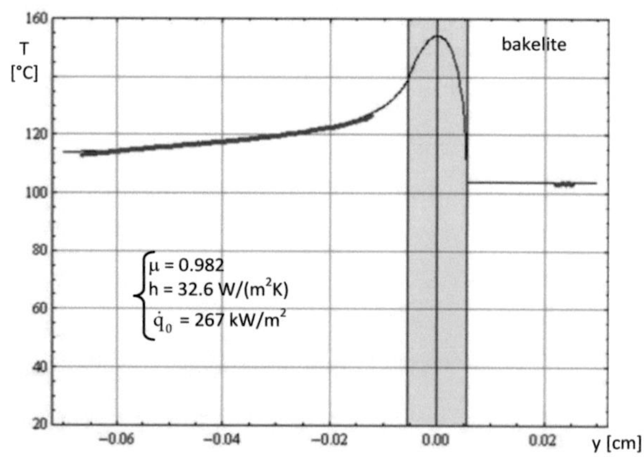

Fig. 10. Results for 0.57 m/s, 1.4 MPa

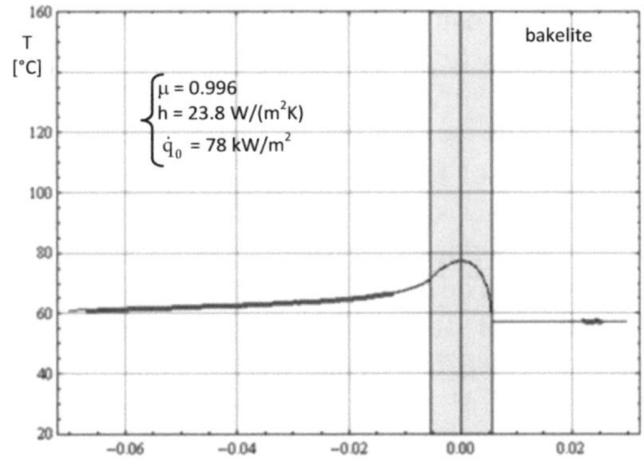

Fig. 11. Results for 0.28 m/s, 0.93 MPa

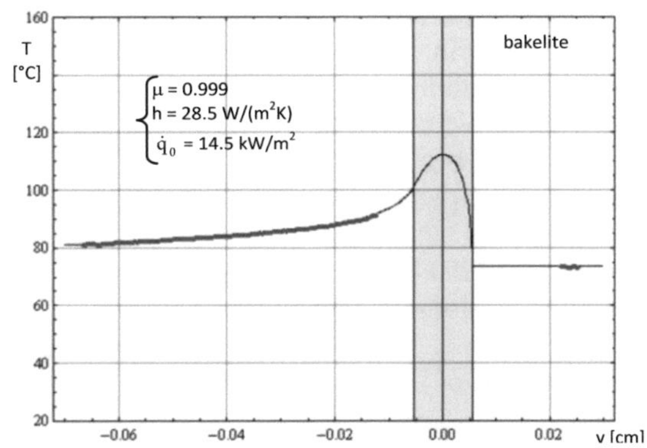

Fig. 12. Results for 0.28 m/s, 1.4 MPa

8. Conclusion

Having in mind to feature the maximum temperature rise in dry sliding contacts, both the classical experimental and analytical approaches have been run. The dimensionless analytical model allows to take in to account the effect of the relative speed, thermal boundary conditions, disk thickness and material making use of only few parameters. It is simple to encode in any commercial software since the rectilinear motion description involves the use of trigonometric functions.

From experimental point of view, IR thermography revealed itself to be a valuable tool while an attempt to take into to account the uneven friction distribution has been done by introducing the engagement parameter. The latter is able to realize a suitable heating distribution shape. A quite satisfying agreement between analytical and experimental predictions was realized, thus a reliable estimate has been obtained for the maximum temperature under the contact area, where direct measuring is always critical.

9. References

Anderson, A.E., & Knapp, R.A. (1990). Hot spotting in automotive friction systems, *Wear*, 135, 319-337.

Anon (1995), Material development using infrared thermography, Metallurgia 62, 409–410.
Archard, J.F (1959). The temperature of rubbing surfaces, *Wear*, 2, 438-455.
Barber, J.R (1967). The influence of thermal expansion on the friction and wear process, *Wear*, 10: 155–159.
Blok, H. (1937). Theoretical study of temperature rise at surfaces of actual contact under oiliness lubricating conditions, *Proc. of the Institute of Mechanical Engineers General Discussion of Lubrication*, London: Institute of Mechanical Engineers.
Cuccurullo, G., D'Agostino, V., Di Giuda, R. & Senatore, A. (2010). An Analitycal Solution and an Experimental Approach for Thermal Field at the Interface of Dry Sliding Surfaces, *Meccanica*, 46, 589-595, ISSN 0025-6455.
Day, A.J., Tirovic M. & Newcomb T.P. (1991). Thermal effects and pressure distributions in brakes, *Proceedings of the Institution of Mechanical Engineers*,205.
Dinc, O.S.C., Ettles, M., Calabrese, S.J. & Scarton, H.A. (1993). The measurement of surface temperature in dry or lubricated sliding, *Journal of Tribology*, 115, 78–82, ISSN 0022-2305.
Fieldhouse, J.D, Ashraf, N. & Talbot, C. (2006). Measurement of the dynamic centre of pressure of a brake pad during a braking operation, *SAE Technical Papers*, 2006-01-3208.
Fieldhouse, J.D., Ashraf, N. & Talbot, C. (2008). The measurement and analysis of the disc/pad interface dynamic centre of pressure and its influence on brake noise, *SAE Technical Paper*, 2008-01-0826.
Francis, H.A. (1971). Interfacial temperature distribution whitin a sliding hertzian contact, *ASLE Trans.*, 14, 41-54.
Gao, J., Lee, C., Ai, X. & Nixon, H. (2000). An FFT-Based Transient Flash Temperature Model for General Three-Dimensional Rough Surface Contacts, *Journal of Tribology*, 122, 519-523, ISSN 0022-2305.
Jaeger, J.C. (1942). Moving sources of heat and the temperature at sliding contacts. *Journal Soc. NSW*, 76, 203–24.
Kennedy, F.E. (1981). Surface temperatures in sliding systems: a finite element analysis. *Journal of Lubr. Technol.*, 103, 90-96.
Kuhlmann-Wisdorf, D. (1987). Temperatures at interfacial contact spots: dependence on velocity and one role reversal of two materials in sliding contact, *Journal of Tribology*, 109,321-329.
Laraqi, N., Alilat, N. Garcia De Maria, J. M. & Bairi, A. (2009). Termperature and division of heat in a pin-on-dsik frictional device-exact analytical solution, *Wear* ,266, 765-770.
Panier, S., Dufrenoy, P. & Weichert, D. (2004), An experimental investigation of hot spots in railway disc brakes, *Wear*, 256, 764–773.
Rhaim, A., Bakar, A. & Ouyang, H. (2005). Brake pad surface topography part I: contact pressure distribution, *SAE Technical Paper*, 2005-01-3941.
Salti, B. & Laraqi N. (1999). Surface temperatures in sliding systems: a finite element analysis, *Int. Journal of Heat and Mass transfer*, 42, 2363-2374.
Severin, D. & Dörsch, S., Friction mechanism in industrial brakes, Wear, 249: 771–779, 2001.
Severin, D., Musiol, F. (1995). Der Reibprozess in trockenlaufenden mechanischen Bremsen und Kupplungen, Konstruktion, 47, 59–68.
Stanley, H.M. & Kato, T. (1997). An FFT-based method for rough surface contact, *T. ASME Journal of Tribol.*, 119, 481-485, ISSN 0022-2305.
Vick, B. & Furey, M.J. (2001). A basic theoretical study of the temperature rise in sliding contact with multiple contacts, *Tribology International*, 34, 823–829.
Vick, B., Furey, M.J. & Iskandar, K. (1998). Surface temperatures and tribological behavior of pure metallic elements, *Proc. of the Fifth International Tribology Conference*, Austrib'98, Brisbane (Australia).

4

IR Support of Thermophysical Property Investigation – Study of Medical and Advanced Technology Materials

Andrzej J. Panas
Military University of Technology,
Air Force Institute of Technology
Poland

1. Introduction

In spite of a huge advance in thermophysical property [TP] investigation technology and instrumentation there are still areas where we experience lack of precise information concerning heat transfer characteristics of certain materials or structures. Many biological or medical technology materials (see eg. Lin at al., 2010a or O'Brien, 1997) can be indicated as good examples. The same concerns some of the advanced technology materials and structures. The reasons for that are difficulties in obtaining the adequate specimens as well as complicated composition and structure of the studied materials which make the standard measurements more difficult. Since standard technologies of thermophysical property investigation are based on analytical models of regular geometry, they need regular specimens of a certain dimension (comp. Maglić et all, 1984, 1992). This is not always possible regarding the discussed materials.

For that reason, other technologies for TP investigation of non-typical materials are still being developed. Many of them take advantage of contactless temperature measurements inherent in infrared (IR) technology. Non-intrusive examination can be especially beneficial in the case of investigations of materials available only in form of thin sheets and foils, specimens of anisotropic materials, composite samples, etc. For specimens of irregular shape the most important is the possibility of inspecting a specific area with a certain spatial resolution. To perform such an inspection an IR camera is usually applied.

The infrared technique is so effective that it creates opportunity for unconventional (see eg. Perkowski, 2011) arrangement of TP investigation. However, in most cases IR cameras are applied for TP investigation just to replace traditional contact sensors (see eg. Bison at all., 2002; Miettinien at all., 2008). Using conventional methods can be justified because they are reliable and because their increased performance is usually expected when using thermal imaging. By studying high resolution data of steady-state temperature distribution instead of discrete temperature recordings one can widen the range of the on-going analysis. In the case of transient temperature measurements the spatial temperature distribution provides additional opportunity for data completion or correction. In all the cases it seems that experiments are facilitated but that is not always true.

Quantitative analysis of TP of the investigated structure should account for some negative effects inherent in IR technology like open area heat losses and systematic errors of temperature measurements because of local differences in surface properties. Violation of heat transfer phenomena symmetry should also be kept in mind concerning anisotropic, composite or irregular in shape specimens. Some of the systematic errors could be corrected by introducing additional data processing techniques ie. numerical modelling. But the most important is compensation of some IR technique disadvantages by a proper selection of the experimental TP measurement method. It seems that transient methods prevail in that domain over stationary ones because of an extended range of possibilities based on both spatial and temporal temperature distribution. Regarding transient techniques regular heating regime procedures assure less problematic temperature field excitation (comp. eg. Maglić, 1984). There are three modes of regular thermal operation (Volkhov & Kasperovich, 1984): step heating, linear heating or temperature oscillation. Usually these methods are applied for the thermal diffusivity [TD] investigation which is the key parameter in the transient thermal conductivity. It has been shown that the combination of the linear heating approach with the temperature oscillation technique (Panas & Nowakowski, 2009) provides an interesting possibility for additional study of TD dependence on the temperature. Both of them exhibit better metrological conditioning than other techniques. And last but not least, they can be easily implemented utilising relatively simple instrumentation. It should be noticed that this does not concern problems of measurements performed on micro or nano-sized particles (comp. eg. Wang & Tung, 2011). In such an instance microscope apparatus makes measurements high-priced. But it is not the reason for which we will focus our attention on macro-sized specimens only. Microscope technology measurements evoke the problem of validity of classical heat transfer theory application in such circumstances. It concerns both the issue of application of Fourier vs. non-Fourier law and, in wider perspective, even the problem of local thermal equilibrium definition.

2. Theory

Regular heating regime refers to a situation when the initial condition has negligible effect on the actual thermal state of the body (Kondratiev, 1954, as cited in Lykov, 1967). The theory of such a regime is the generalisation of well-known quasi-stationary conditions. Usually the first, second, and third kind of regimes are distinguished (Volokhov & Kasperovich, 1984; Platunov, 1992). The first two are also referred to as monotonic heating regimes. At monotonic heating the investigated specimen is heated at the constant ambient temperature (1st) or heating proceeds at the constant rate (2nd). The third kind of regular regime includes periodic heating also referred to as temperature wave technique (Phylippov, 1984). The temperature oscillation was introduced for the thermophysical property investigation of good conductors by Ångström (Ångström, 1861) but its application has also been extended to insulators (Belling & Unsworth, 1987).

There are available many mathematical models of regular heating regime problems that have been formulated in all three basic coordinate systems and solved for different dimensions (see eg. Carslaw & Jaeger, 2003; Maglić et al., 1992, 1984; Lykov, 1967). In the present study two models of the temperature oscillation are discussed: classical Ångström's solution of the longitudinal temperature oscillation in semi-infinite rod with side surface convection heat losses and modified Ångström's model (Belling & Unsworth, 1987) with additional, linear in time temperature change imposed onto basic boundary temperature oscillation (Panas & Nowakowski, 2009). In addition a generalised model of the 1st kind monotonic heating is recalled.

For the purpose of this study we restrict to one-dimensional Cartesian coordinate system O–x formulation in all cases. To benefit from linear character of mathematical problems we also assume that the basic thermophysical properties and characteristics are constant. In practice it means that although the density ρ, specific heat c_p, and the thermal conductivity λ can be temperature dependent the changes are negligible regarding the expected maximum temperature difference within the analyzed object. The same assumption concerns the heat transfer coefficients (surface conductances): h - describing the convection heat losses taken present in the governing equation and H - present in boundary condition [BC]. Thus, the generalized governing equation for all the analyzed problems is as follows

$$\frac{\partial \theta}{\partial \tau} = a\frac{\partial^2 \theta}{\partial x^2} - \nu\theta ; \quad \theta(x,\tau) = T(x,\tau) - T_0 ; \; a = \frac{\lambda}{\rho\, c_p} \, , \quad \nu = \frac{hp}{\rho c_p S} \, , \tag{1}$$

where

$$\theta(x,\tau) = T(x,\tau) - T_0 \tag{2}$$

denotes the temperature referring to a certain T_0 value, τ is the time, a is the thermal diffusivity and ν is a constant describing the intensity of internal convection heat loses.

2.1 One-dimensional temperature oscillation in semi-infinite rod

The solution of Eq. (1) is sought within the interval $[x, \infty)$ for the rod treated as a thermally thin body of the perimeter p and cross-section area S that defines ν constant as

$$\nu = \frac{hp}{\rho c_p S} \, . \tag{3}$$

Regarding possibility of Fourier series representation of any periodic temperature variation of the angular frequency ω, the analysis in this instance can be restricted to such BCs:

$$\theta(0,\tau) = A\sin(\omega\tau + \varepsilon) \, , \qquad \lim_{x\to\infty} \theta(x,\tau) = 0 \, . \tag{4}$$

The solution, describing a developed thermodynamic process, is (Carslaw & Jaeger, 2003):

$$\theta(x,\tau) = A\, e^{-qx} \sin(\omega\tau - q'x + \varepsilon) = A\, e^{-qx} \sin(2\pi f\tau - q'x + \varepsilon) \, , \tag{5}$$

where f is the frequency, and

$$q = \sqrt{\frac{\sqrt{\nu^2 + \omega^2} + \nu}{2a}} \, , \quad q' = \sqrt{\frac{\sqrt{\nu^2 + \omega^2} - \nu}{2a}} \, . \tag{6}$$

There are several possibilities to arrange an experiment utilising the above solution for the thermal diffusivity identification (Bison at al., 2002; Phylippov, 1984). One of the arrangement benefits from examination of the temperature oscillation attenuation ψ and the phase shift φ when passing from x_1 to x_2, $x_2 > x_1 \geq 0$ (Fig. 1):

$$\psi = \frac{\max\limits_{\tau}\left[\theta(x_2,\tau)\right]}{\max\limits_{\tau}\left[\theta(x_1,\tau)\right]} = \frac{\exp(-qx_2)}{\exp(-qx_1)} = \exp(-ql), \quad \varphi = q'(x_2 - x_1) = q'l, \; l = x_2 - x_1. \tag{7}$$

Hence the thermal diffusivity a and the convection heat transfer constant ν are given by:

$$a = \frac{\pi f}{q\,q'}, \qquad \nu = \frac{\pi f}{q\,q'}\left(q^2 - q'^2\right). \tag{8}$$

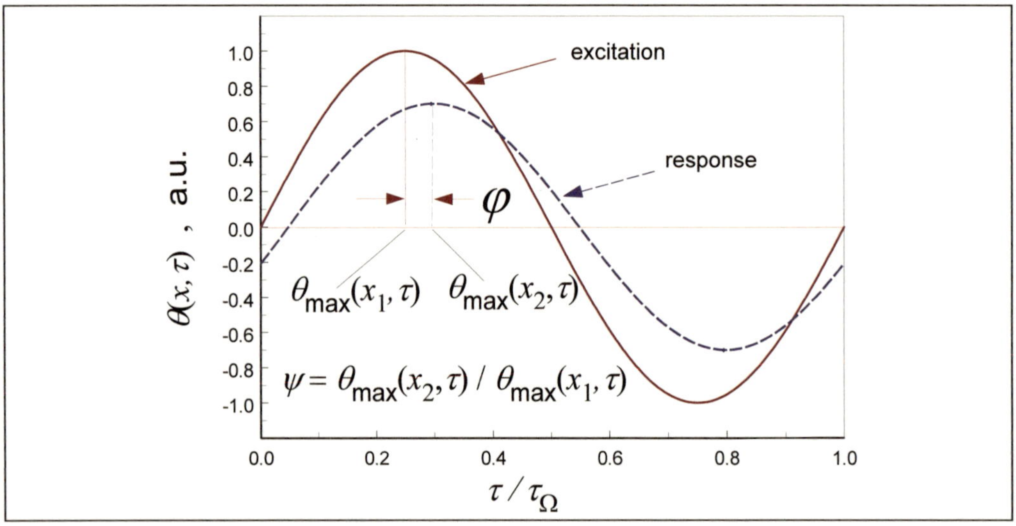

Fig. 1. Illustration of the sinusoidal temperature oscillation attenuation and lagging

Applying Eq. (7) the first relation from the two above can be rearranged into the form:

$$a = \frac{\pi f}{q\,q'} = \frac{\pi f}{\left(\frac{1}{l}\ln\frac{1}{\psi}\right)\left(\frac{\varphi}{l}\right)} = \sqrt{\frac{\pi f l^2}{\ln^2\frac{1}{\psi}}}\sqrt{\frac{\pi f l^2}{\varphi^2}} = \sqrt{a_\psi\, a_\varphi}\,, \tag{9}$$

where:

$$a_\psi = \frac{\pi f l^2}{\ln^2\frac{1}{\psi}} = \frac{\pi l^2}{\tau_\Omega \ln^2\frac{1}{\psi}}, \qquad a_\varphi = \frac{\pi f l^2}{\varphi^2} = \frac{\pi\, l^2}{\tau_\Omega\, \varphi^2}, \qquad \begin{cases} a_\varphi > a_\psi & \text{for } \nu > 0 \\ a_\varphi = a_\psi & \text{for } \nu = 0 \end{cases}. \tag{10}$$

Parameters a_ψ and a_φ represent the amplitude and phase apparent thermal diffusivity values. In the case when the side surface of the analysed rod is perfectly insulated, the thermal diffusivity can be obtained independently applying the amplitude attenuation and the phase shift data.

2.2 One-dimensional temperature oscillation within a slab with a linear boundary temperature drift

Developing this model we exclude internal convective heat losses (ν=0) and assume that such modified governing Eq. (1) is valid within the interval $0 \le x \le L$ where L is the slab thickness. Regarding the linearity of the discussed heat transfer problems, the generalised boundary conditions defined by the following formulae

$$\frac{\partial\theta(0,\tau)}{\partial x} = 0, \quad \theta(L,\tau) = A\sin(2\pi f\tau + \varepsilon) + b\,\tau, \tag{11}$$

where b is the temperature change rate, can be decomposed into two separate BC sets: the first taking into account periodic component and the second considering the right-hand boundary linear temperature change. The behaviour of the system under these conditions is illustrated in Fig. 2. The appropriate solutions for the slab with zero initial temperature can be found in monograph by Carslaw Jaeger (Carslaw & Jaeger, 2003; pp. 104-105). The final solution disregarding the initial irregular heating regime can be expressed as (Panas & Nowakowski, 2009)

$$\theta(x,\tau) = A\ \psi\sin(2\pi f\tau - \varphi + \varepsilon) + b\tau + \frac{b\left(x^2 - L^2\right)}{2a}. \tag{12}$$

The parameters ψ and φ have the same interpretation as in the previous case - they represent the oscillation amplitude attenuation and the oscillation phase shift respectively. However, the relations between ψ, φ and the thermal diffusivity a are now far more complicated (Carslaw & Jaeger, 2003):

$$\psi(x) = \sqrt{\frac{\cosh 2kx + \cos 2kx}{\cosh 2kL + \cos 2kL}}, \qquad \varphi(x) = \arg\left[\frac{\cosh kx(1+i)}{\cosh kL(1+i)}\right], \tag{13}$$

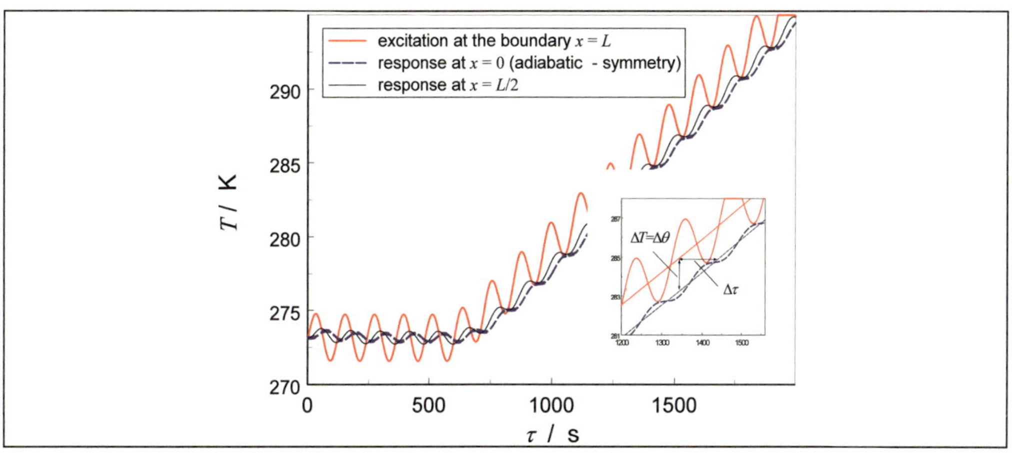

Fig. 2. Illustration of the sinusoidal temperature oscillation with a linear scan imposed from τ = 600 s (numerical modelling results of the appropriate heat transfer within the slab; Panas & Nowakowski, 2009)

where

$$k = \sqrt{\frac{\pi f}{a}} = \sqrt{\frac{\pi}{a\,\tau_\Omega}}\,. \tag{14}$$

It also should be noted that in present formulation the "thermal wave" propagates in opposite direction (the amplitude attenuates while x is decreasing from L to 0). The thermal diffusivity can be calculated on the basis of known ψ and φ by solving the transcendental Eqs. (13). The independently obtained a_ψ and a_φ when compared, can be used for cross-validation of the developed experimental procedure.

For $x = 0$, which is a typical situation during IR investigation, one can rearrange Eqs. (13) to

$$\frac{\sqrt{2}}{\psi} = \sqrt{\cosh\left(2\sqrt{\frac{\pi f}{a}}\cdot L\right) + \cos\left(2\sqrt{\frac{\pi f}{a}}\cdot L\right)}\,,\quad \tan\varphi = \tan\left(\sqrt{\frac{\pi f}{a}}\cdot L\right)\tanh\left(\sqrt{\frac{\pi f}{a}}\cdot L\right). \tag{15}$$

It was shown that these formulae can be reduced to much simpler explicit approximations:

$$a \cong \frac{\pi f L^2}{\ln^2 \dfrac{2}{\psi}}\,,\qquad a \cong \frac{\pi f L^2}{\varphi^2}\,, \tag{16}$$

if the oscillation frequency is high enough to fulfil the following condition (Bodzenta, 2006):

$$kL = \sqrt{\frac{\pi f}{a}}\,L = \sqrt{\frac{\pi}{a\,\tau_\Omega}}\,L > K_{min}\,. \tag{17}$$

The value of K_{min} for 2% and 1% discrepancy was stated at $K_{min\,1\%} = 1.78$ and $K_{min\,2\%} = 2.25$ respectively (Panas & Nowakowski, 2009). The discrepancy concerns the relative difference between the approximate thermal diffusivity value a_ψ or a_φ derived from Eqs. (16) in relation to the appropriate value obtained from Eqs. (15).

Analysis of linear components of the excitation and the response provides additional opportunity for the thermal diffusivity evaluation from the time $\Delta\tau$ or the temperature lag ΔT (comp. Fig. 2). The procedure is equivalent to application of 2nd kind regular heating regime methodology (Volokhov & Kasperovich, 1984). Directly from Eq. (12) one gets

$$a = \frac{b\left(x^2 - L^2\right)}{2\Delta T} = \frac{x^2 - L^2}{2\Delta\tau} \;\wedge\; x = 0 \quad\Rightarrow\quad a = \frac{-bL^2}{2\Delta T} = \frac{-L^2}{2\Delta\tau}\,. \tag{18}$$

The major advantage from imposing a linear scan is that the temperature function of the thermal diffusivity can be obtained within one continuous experiment without necessity for successive changing and stabilizing the sample temperature. However, it should be underlined that the investigation needs to be planned and performed with a special care in view of the assumption concerning linearity of the problem. This can be done preserving the appropriate thermal/temperature resolution of the experiment (see eg. Panas, 2010).

2.3 Generalized model for the 1st kind regime of monotonic heating of a slab

Mathematical formulation of this heat conduction problem requires , as in the previous case, putting ν=0 into the governing Eq. (1). The problem is formulated for the interval $0 \le x \le L$, applying homogeneous initial temperature distribution $\theta(x,0)=0,\ \ 0 \le x \le L$. For the BC the following expressions apply:

$$\frac{\partial\theta(0,\tau)}{\partial x}=0,\quad \frac{\partial\theta(L,\tau)}{\partial x}=\frac{H}{\lambda}\left\{\theta_D\left[1-\exp\left(-\frac{\tau}{\tau_{ch}}\right)\right]-\theta(L,\tau)\right\}. \tag{19}$$

where τ_{ch} is the (characteristic) time constant of exponential surface treatment. For $\tau_{ch} \to\infty$ we get an asymptotic approximation of classical step heating function (comp. Volokhov & Kaperovich, 1984). The solution is given by (Lykov, 1967):

$$\theta(x,\tau)=1-\frac{\cos\frac{x}{\sqrt{a\tau_{ch}}}}{\cos\frac{L}{\sqrt{a\tau_{ch}}}-\frac{\lambda}{H\sqrt{a\tau_{ch}}}\sin\frac{L}{\sqrt{a\tau_{ch}}}}\exp\left(-\frac{\tau}{\tau_{ch}}\right)-\sum_{n=1}^{\infty}\frac{A_n\cos\frac{\mu_n x}{L}}{1-\frac{a\tau_{ch}\mu_n^2}{L^2}}\exp\left(-\mu_n^2\frac{a\tau}{L^2}\right), \tag{20}$$

where μ_n are roots of

$$\tan\mu=-\frac{\mu}{\mathrm{Bi}-1},\quad \mathrm{Bi}=\frac{HL}{\lambda}, \tag{21}$$

and constants A_n are given by

$$A_n=(-1)^{n+1}\frac{2\mathrm{Bi}\sqrt{\mu_n^2+(\mathrm{Bi}-1)^2}}{\mu_n^2+\mathrm{Bi}^2-\mathrm{Bi}}. \tag{22}$$

The above presented model has been included into consideration mostly for comparison reasons and because of the needs for referring to some previous results of TP investigation applying IR imaging technique. Analysing the result given by Eq. (20) one can notice that, similarly as in the previous cases, a long term approximation can also be derived by neglecting all but the first one (n=1) series components. However, this solution does not describe a developed (stabilised) thermodynamic process of a finite "amplitude"[1]. It results in lowering values of sensitivity coefficients[2] with increasing time when the temperature of the body equilibrates.

2.4 Comments on the implementation of the TD measurement procedure applying IR imaging technique

There are two basic prerequisites for a successful implementation of a certain heat transfer model for TP/TD investigation. The first is precise reconstruction of all the model

[1] It can be the amplitude, the time lag or the temperature lag as shown in Fig. 2.

[2] For more detailed description of the sensitivity issue see Özişik & Orlando, 2000.

assumptions in the performed experiment. The second is that the heat flow should not be affected by the measurement procedure.

Implementation of the infrared imaging technique creates unique opportunity for non-intrusive analyses of the temperature distribution and its evolution in time. However, in many cases these studies are limited to qualitative studies. Investigations of the hard tooth tissues performed by Sakagumi & Kubo, 2002 or by Panas at all., 2007 can serve as an example of that. The major problems arise from not so good performance in quantitative measurements of the absolute temperature. The measurements are strongly dependent on the surface condition and influenced by surroundings. Usually precise estimation of the surface emissivity, which is the crucial parameter for the temperature assessment, could be problematic. For this reason, investigations are frequently planned to account for the temperature differences or for the temperature evolution only. Cooling fin (Miettinen at all., 2008; Rémy at all., 2005), cooling block (Perkowski, 2011) or flash experiments (Bison at all., 2002) can serve as typical examples of that. The temperature oscillation technique exhibits a unique performance in that domain. Being not dependent on the absolute temperature measurements, it reduces major IR technique implementation problems. The TD investigation procedure is based on the amplitude ratio and on the phase lag evaluation. This is why we notice a growing popularity of this methodology both in qualitative and in quantitative investigations (see eg. Bison et all., 2002; Muscio at all., 2004).

Performing measurements one still should take into consideration increased heat losses from the IR monitored surface. These losses can affect the applied theoretical model conformity. The investigated object needs also to be separated from the surrounding irradiation. If the above mentioned problems cannot be avoided while performing the experiment they need to be corrected applying other techniques and/or accounted for in the course of the uncertainty budget analysis.

3. Experimental

Developing an experimental system for the TD investigation at different heating modes one should keep in mind the limitations imposed by the applied mathematical models. However, even in this case the system can be adapted for multi-mode operation. Such versatility is not only economically reasonable but is judicious from metrological point of view. By comparing the data from different experiments the appropriate procedures can be tested and the experimental results can be verified. Of course, the experiment management and data processing procedures should be also accommodated. Realization of this concept for the 1st, 2nd and 3rd kind of regular heating regime TP investigation is described below.

3.1 Experimental setup

The actual experimental system configuration depends on the type of the investigated specimen. There are two basic modes of the system operation: for measuring in-plane (longitudinal) TD of bars, plates and sheets/foils (Fig. 3.a,b,c) and for transversal TD measurement of plates (Fig. 3.d). The in-plane measurements are further differentiated respecting the investigated material anisotropy to ensure effective heat transfer into the body of the sample (see Fig. 3.c). The specimen is usually pressed between two thermoelectric Peltier elements or pressed to the elements' unipolar side surface. If necessary, thermally

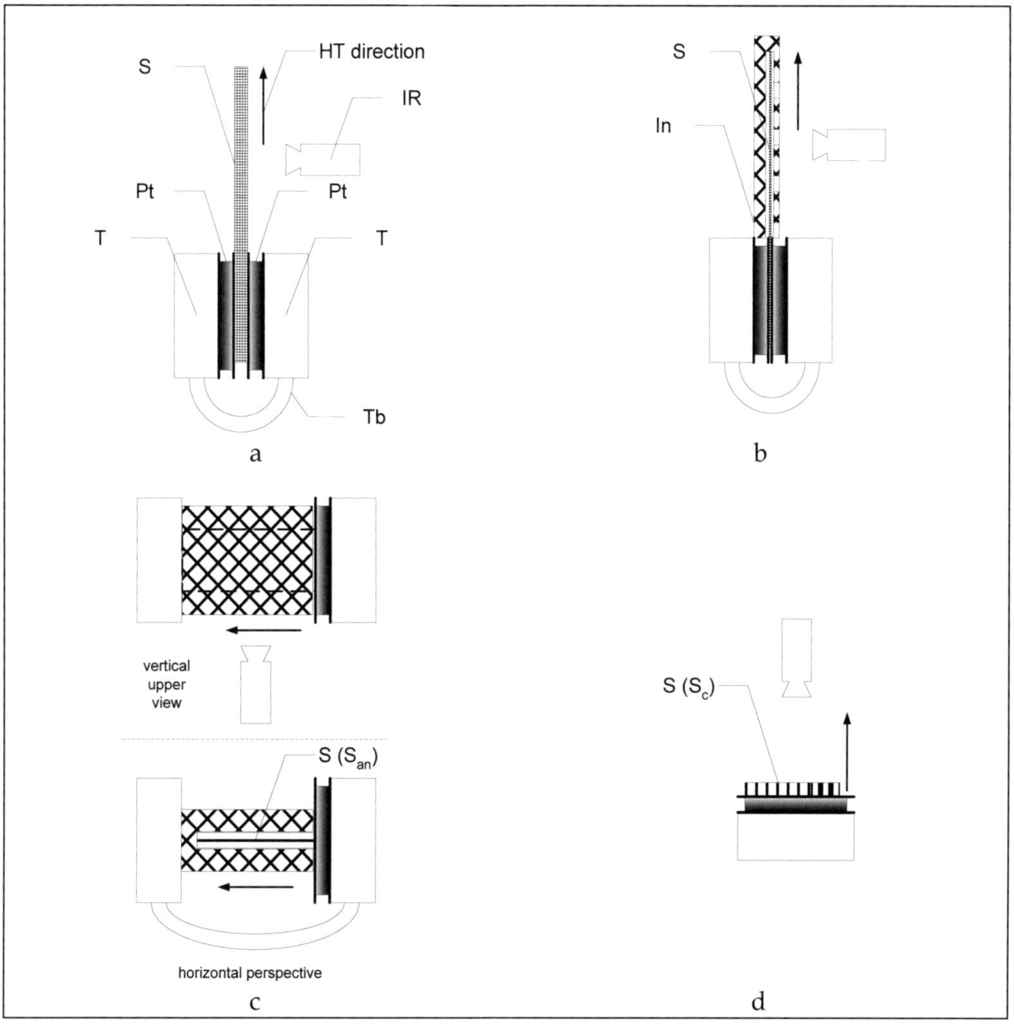

Fig. 3. Typical arrangements of longitudinal TD measurement applying an IR camera for: a - slabs, b - thin sheets and foils and c- bars of anisotropic materials, d - for transversal TD investigation of slabs/composite slabs (d); S - specimen, S_{an} – anisotropic specimen, S_c - composite non-homogeneous specimen, HT - heat transfer direction, Pt - Peltier thermoelectric device with the polarity indicated with shadowing, IR - infrared camera, T - thermostatic block, Tb - tubing, In - insulation

conductive paste is used to improve the thermal contact. Slab specimens of bad heat conductors studied for transversal TD are usually fastened with a thermal conductive resin or paste onto an additional conductive plate (usually made of copper). This is to homogenize the lateral temperature distribution. The periodic or exponential boundary excitation is performed by the appropriate powering of Peltier elements. As it is shown in Fig. 4 thermoelectric elements are powered by a DC power supply (Amrel PPS 1320). In typical measurements the power supply voltage is changed every 1 s to accommodate the demanded power changes according to the following formulae:

$$U(\tau)=U_0+U_A(1+\sin\omega\tau) \quad \vee \quad U(\tau)=\begin{cases}0 & \tau<0 \\ U_A(1-\delta\, e^{-\frac{\tau}{\gamma}}) & \tau\geq 0\end{cases}, \tag{23}$$

where U_A is the amplitude, U_0 - offset voltage (that can be changed programmatically), δ=0 or 1, and γ is the appropriate time constant. The temperature of every Peltier element opposite end is stabilized by thermostatic blocks fed from a thermostat (Lauda RL6CP). The Lauda device can operate in a programmable mode. This option is applied to ensure a linear change of the oscillation offset. The time-dependent temperature field in the sample is recorded with an infrared camera (Flir SC5600) and stored into the PC memory. The specimen attached to the heating (cooling) unit is separated from enviroment by an illumination protective shield that simultaneously fulfils the role of a thermal insulating guard. The instrumentation is supplemented by an 8 channel temperature recorder (NI SCXI 1000 system). The additional measurements of the temperature are made for control and verification purposes and are performed with thermocouples. The whole system is controlled from the PC through GPIB (SCXI), Fast Ethernet (IR camera) and RS232 (Lauda) buses. For this function special virtual instrumentation software has been implemented.

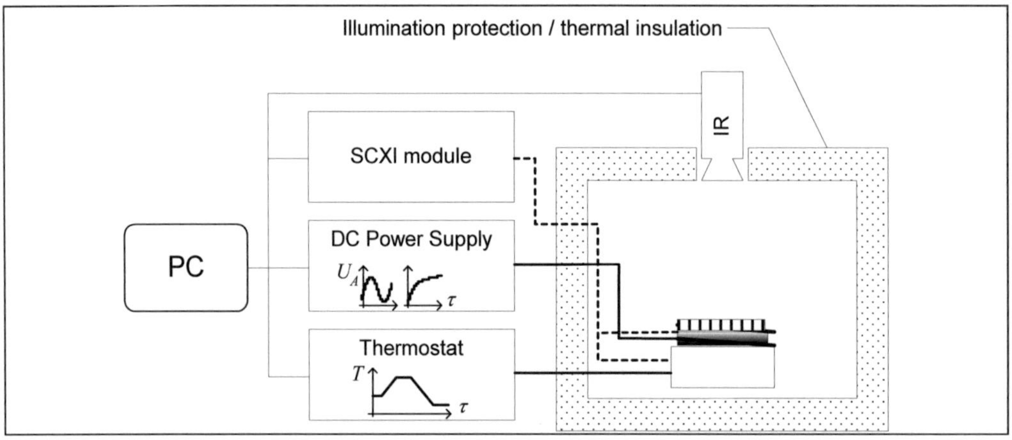

Fig. 4. Schematic diagram of the experimental setup for thermal diffusivity measurements by temperature oscillation in scanning mode

3.2 Experimental procedures and data processing

Before mounting the specimens into the measuring unit, they are usually painted black with a Graphite 33 (Kontakt Chemie) graphite spray. The estimated thickness of the graphite layer is about 15 µm. The black layer is used to improve the surface emittance, homogenize surface emissivity and to prevent the recording from the effects of translucency (Fig. 5 - translucence effects are clearly seen at not painted strips). The effect of the graphite layer presence has been expected to be negligible because individual effects on the temperature excitation and temperature response compensate each other (see Eqs (7)).

During measurements the temperature of the thermostatic blocks is stabilized or being changed programmatically. In the case of step or exponential heating experiments, the whole process of the specimen temperature disturbance and stabilization is being recorded.

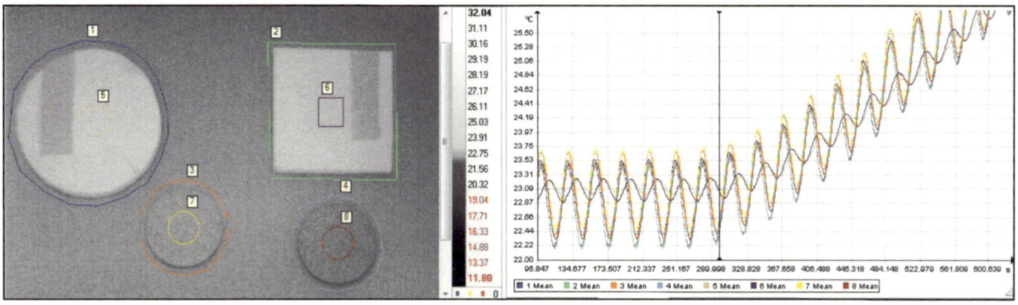

Fig. 5. Illustration of the IR data management: the IR image of polytetrafluoroethylene (PTFE; large disc), polymethylmethacrylate (PMMA; square) and two Al_2O_3-SiO_2 composite ceramic specimens during oscillation treatment (on the right) with a time history diagram of temperatures taken from selected lines (on the left)

While operating in the oscillation heating mode at least five to ten subsequent periods of the temperature change are recorded but usually the examination lasts much longer. The thermocouple temperature measurements are performed simultaneously with the IR camera recording. The sampling rates are adjusted to different study conditions.

At the end of measurements the time histories of the temperature/mean temperature taken from selected points, lines (see Fig. 5) or areas are processed with a specially developed software of a non-linear curve fitting. Prior to final calculation the aerial temperature data were inspected for any disruptions in uniform temperature distribution. The procedure of the step heating and exponential heating data approximation with a function:

$$f(\tau) = A_e \exp\left(-\frac{\tau}{B_e}\right) + C_e, \tag{24}$$

where A_e, B_e and C_e are the best fit coefficients, is similar to that described by Panas et al., 2005; 2007. The data from oscillation excitation experiments, are fitted with a function:

$$f(\tau) = A_s \sin(\omega\tau + B_s) + C_s + D_s\tau , \tag{25}$$

where A_s, B_s, C_s and D_s are the best fit coefficients. The constant D_s accounts for a linear component of the temperature changes. The approximate is usually obtained for every single "period". Next, the coefficients ψ and φ are calculated from the appropriate values of A_s and B_s of the two selected experimental time histories. Finally two complementary values of a_ψ and a_φ are obtained by solving the Eq. (13). At this stage of calculation the appropriate distances or dimensions are being used. When Ångström's model assumptions apply, ie. in the case of a short "thermal wave" (comp. Carslaw & Jaeger, 2003), the appropriate amplitude a_ψ and phase a_φ TD values are calculated directly from Eq. (10) and the thermal diffusivity a is obtained from Eq. (9).

4. Typical results and discussion

The presented investigations have been selected mostly to illustrate typical experimental procedures and to show performance of the developed instrumentation and methodology.

The considered materials and structures are (Fig. 6): electrolytic pure copper [Cu], Ni_3Al nano-structural alloy [Ni3Al], anisotropic pyrolitic graphite [PG], polymethylmetacrylate [PMMA], polytetrafluoroethelene [PTFE] and sliced tooth structures (comp. Żmuda at all., 2005 and Panas at all., 2007). The Cu, PMMA and PTFE have been included mainly for testing and reference purposes. In all the cases the attention was focused on IR imaging methodology and such physical effects as anisotropy and non-homogeneity of the investigated specimens.

As was mentioned before, prior to the experiments the investigated structures had been coated as shown in Fig. 6 a, b and d. This did not concern the graphite specimen which did not need such treatment. During the measurements the temperature IR recordings were taken at stated sampling rates. Afterwards, the IR images were inspected for the temperature distribution non-homogeneity. Next, and the temperature histories were taken from previously selected objects/sections: point/spots, lines, segment lines, circle lines, circles or rectangular areas. The location of control sections is shown in Fig. 7.

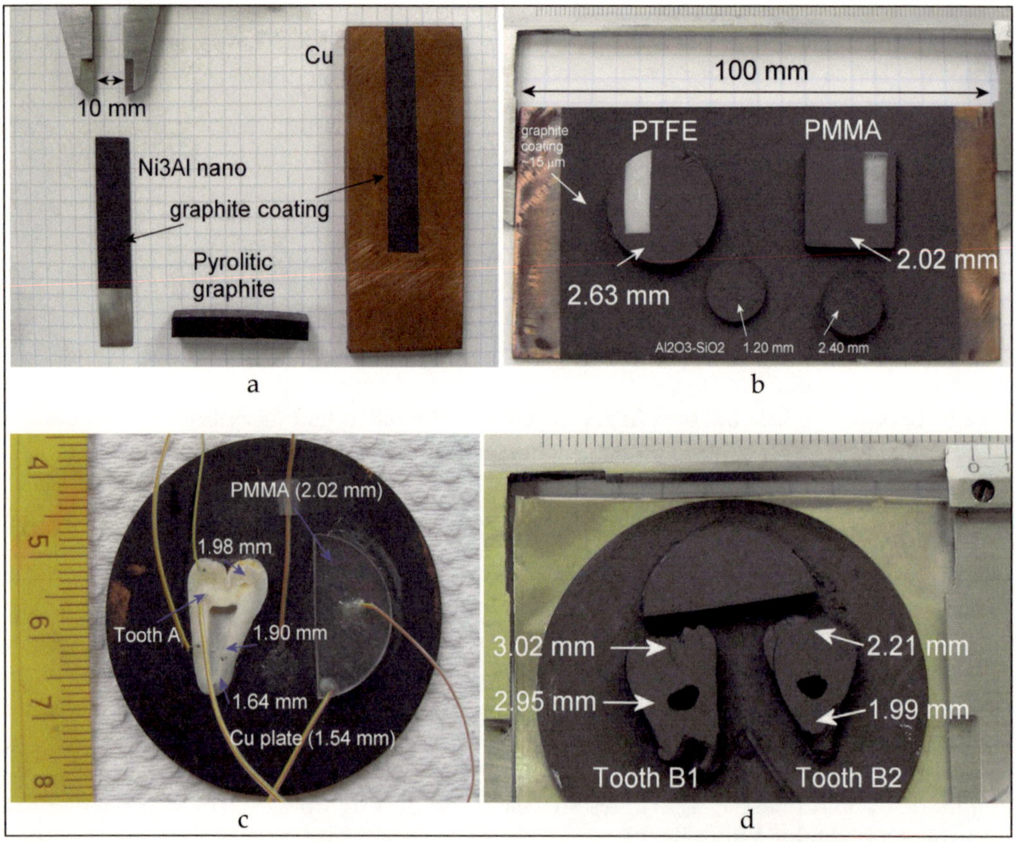

Fig. 6. Investigated specimens and structures: a - slabs (bars) of electrolytic pure copper (left), pyrolitic graphite (middle), and thin sheet of nano-structural Ni_3Al alloy studied for in-plane TD; b - plates of PMMA and PTFE; c - sliced tooth A with a reference PMMA half-disk just before applying a graphite layer, d - sliced tooth B structures coated with graphite. In the cases b, c, and d the specimens were investigated for their transversal TD.

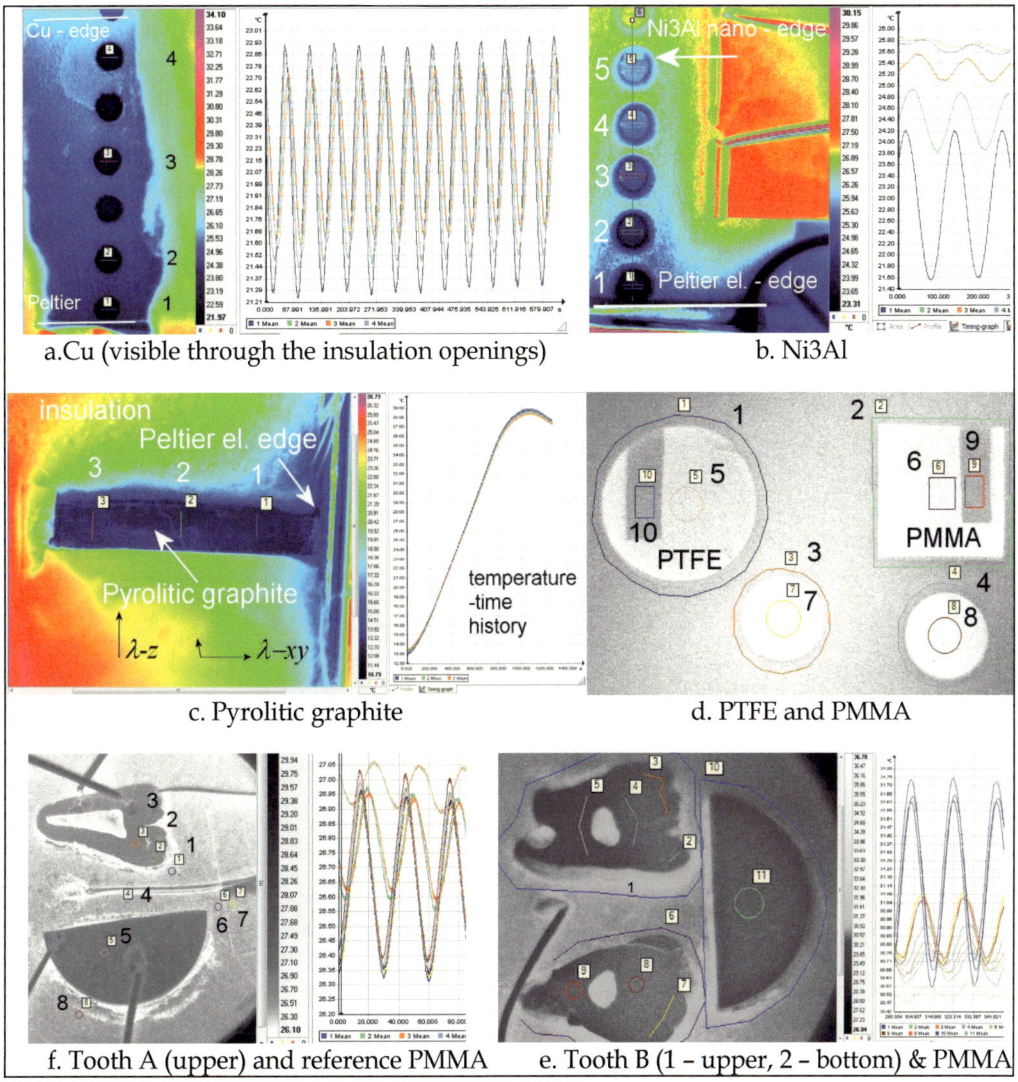

Fig. 7. Illustration of the IR images of the investigated specimens with indication of control sections for the temperature history acquisition

4.1 Copper slab under long "thermal wave" treatment

The major problem with application of Ångström's method is to comply with the semi-infinite rod model. In practice it means that the investigated object should be long enough to damp down the temperature oscillations before reaching the far end of the specimen (Muscio at all., 2004; Phylippov, 1984). It makes the investigations of materials and objects which are available only in small pieces more difficult. This is the case of many advanced technology structures.

The measurements performed on a 100.0 mm × 38.0 mm × 5.80 mm slab of electrolytic pure copper were planned mostly to face this problem. They were carried out at sampling rate equal to 2 Hz for the temperature oscillation over 22.2 °C with the period $\tau_\Omega = 60$ s. Application of long "thermal waves" not only violates the classical Ångström's assumptions but exceeds limits of conventional implementation of the modified method (comp. eg. Bodzenta at all., 2006). The parameter *kL*, defined by Eq. (15), amounted only to 0.42 that is much below the stated K_{min} limit for validity of simplified formulae given by Eq. (18). For that reason the data were processed utilising exact relations for a slab model given by Eq. (13).

Recordings of the specimen temperature were done through openings in the 5 mm thick polyurethane foam separating the Cu specimen from ambient air convection (Fig. 7.a). Twelve subsequent periods of temperature changes were recorded. The time curves for preparatory data processing - approximation according to Eq. (25) - came from lines i = 1, 2, 3 and 4 (Fig. 7.a). Next, the values

$$\psi_{ij} = \frac{A_{s,j}}{A_{s,i}}, \quad \varphi_{ij} = B_{s,j} - B_{s,i}, \quad T_{mean,\,ij} = \frac{C_i + C_j}{2} \quad , i > j, i = 1,2,3,\ j = 2,3,4\,, \tag{26}$$

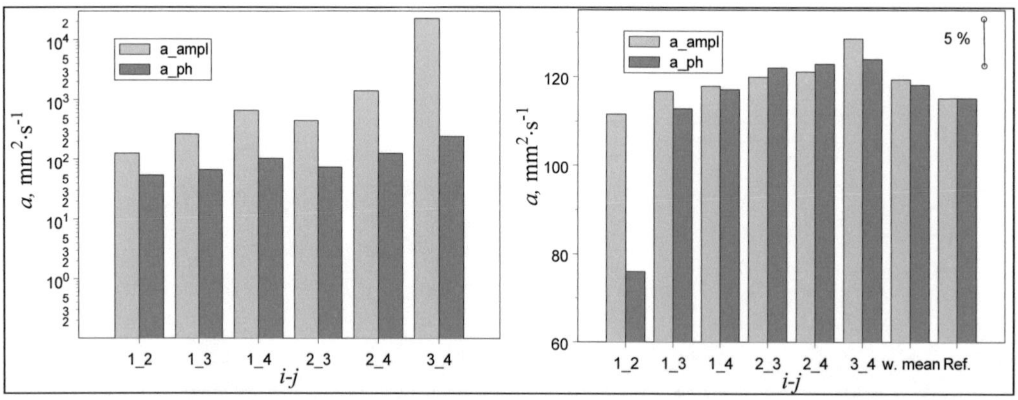

Fig. 8. Comparison of the apparent TD of Cu calculated from Ångström's formulae (9) – on the left – with values obtained by solving Eqs. (13) – on the right. The measurements were performed applying "long thermal wave" (at τ_Ω=60 s). The reference data is taken from Ražnievič, 1966.

were obtained from the approximation best fit parameters of corresponding single periods. This means that all possible mutual configurations of control lines were taken into consideration. Next, the appropriate: a_ψ - the amplitude and a_φ - phase values of apparent TD were derived by solving Eqs. (13) with

$$x = l = x_{ij} = \left| x_i - x_j \right|, \quad L = \left| x_{edge} - x_i \right|. \tag{27}$$

where x_i is a relevant control line coordinate and x_{edge} is the specimen free edge coordinate. It should be noticed that the same parameter l=x is applicable for classical Ångström's model data processing according to Eq. (9). Such processing was conducted in parallel calculations to provide illustrative comparable data. Finally, the obtained amplitude and phase TD values were averaged over all 12 periods of stable oscillations providing appropriately

$\bar{a}_{\psi,ij}$ and $\bar{a}_{\phi,ij}$ mean values. The time averaged data are shown in Fig. 8 for both models: the semi-infinite rod (on the left) and the slab (on the right). With no doubt the "classical" results are unacceptable which was expected regarding the applied long "thermal wave" excitation. The quality of these results worsens while approaching the specimen free edge. On the contrary, the modified procedure results are in a good agreement with the reference data taken from Ražnievič, 1966. The results of comparison are even better when the following weighted mean value

$$\bar{a}_{w\_mean} = \sum_{i=1}^{M-1} \sum_{j=i+1}^{M} \left(x_i - x_j\right)^2 \bar{a}_{ij} \Bigg/ \sum_{i=1}^{M-1} \sum_{j=i+1}^{M} \left(x_i - x_j\right)^2 , \tag{28}$$

is considered. In Eq. (28) the M stands for a total number of subsequent control sections - in the present case M=4. The weights, in a form of squared distances, account for dimensional dependence of the calculated TD. The obtained $\bar{a}_{\psi,w\_mean}$ and $\bar{a}_{\varphi,w\_mean}$ values are shown in Table 1. The discrepancy between global weighted mean values and the reference data is not greater than 3.7%.

Analysing the individual data one can even identify some specific effects like nonconformity with one-dimensional model assumptions. This is the most probable reason for underestimation of $\bar{a}_{\psi,12}$ and $\bar{a}_{\varphi,12}$ results.

Concluding, it should be pointed out that the experiment has proved a good metrological conditioning of the temperature oscillation technique combined with the IR imaging technology.

i-j	1-2	1-3	1-4	2-3	2-4	3-4	w. mean	Ref.
l, mm	9.78	29.21	49.65	19.43	39.87	20.44	n.a.	n.a.
a_ψ, $mm^2 \cdot s^{-1}$	111.6	116.7	117.9	119.9	121.1	128.6	119.4	115.2
a_φ, $mm^2 \cdot s^{-1}$	76.0	112.8	117.1	122.0	122.9	123.9	118.2	

Table 1. Results of electrolytic pure copper TD investigation with modified Ångström's procedure compared with the reference data according to Ražnievič, 1966

4.2 Nano-structural Ni3Al intermetallic alloy

The major challenge of the TD investigation of intermetallic Ni_3Al alloy is that the nanostructural material is available only in the form of thin metallic sheets. In such a situation the non-intrusive character of the IR temperature measurements is essentially beneficial (comp. Nakamura, 2009). However, it should be underlined that infrared measurements can be effectively performed only in a narrow range of temperature changes.

The procedure that was tested on Cu specimen has also been applied for a Ni_3Al alloy investigation. The specimen was delivered in the form of a sheet 0.63 mm thick, 11.00 mm wide and about 66 mm long (Fig. 6.a - on the left). The measurements were carried out at sampling rate equal to 1 Hz for the temperature oscillation over 23 °C with the period τ_Ω = 120 s (Fig. 7.b). The obtained results are shown in Fig. 9 and displayed in Table 2. They were supplemented with one more parameter that is a geometric mean of the amplitude and phase apparent TD values:

$$a_{g\,mean} = \sqrt{a_\psi\, a_\varphi}\ , \tag{29}$$

The discrepancy between a_ψ and a_φ can be attributed to heat losses from the specimen. Taking into account relatively high ratio of the specimen perimeter p to its cross-section area S, they are extremely influential. The heat losses resulted in a high ν parameter value (comp. Eq. 3), but they are not considered in a slab model. However, regarding the expected asymptotic convergence of the appropriate solutions when increasing the L parameter of a slab model one can assume that a_ψ is the underestimate and a_φ is the overestimate of the real TD value as in the case of semi-infinite rod (comp. Eqs. 6, 9 and 10). This was confirmed in the course of numerical simulation. Moreover, it was stated that the geometric mean value defined by Eq. (29) is a good approximation of the real TD value for certain parameters of combined conductive and convective heat losses. The detailed discussion is outside the scope of this review so we restrain to this communication only.

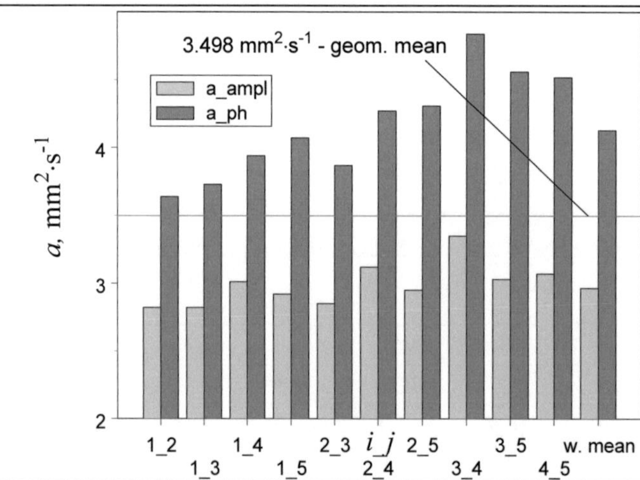

Fig. 9. Results of TD investigation of Ni_3Al nanostructural intermetallic alloy

i-j	1-2	1-3	1-4	1-5	2-3	2-4	2-5	3-4	3-5	4-5	w.mean	g.mean
l, mm	9.90	19.98	29.47	39.72	10.08	19.57	29.82	9.49	19.75	10.25	n.a.	n.a.
a_ψ, mm²·s⁻¹	2.82	2.182	3.01	2.92	2.85	3.12	2.95	3.35	3.03	3.07	2.964	3.498
a_φ, mm²·s⁻¹	3.64	3.73	3.94	4.07	3.87	4.27	4.31	4.84	4.56	4.52	4.128	

Table 2. Results of TD studies of Ni_3Al nano-structural intermetallic alloy specimen

4.3 Pyrolitic graphite

Pyrolitic graphite (PG) is one of a few commercially available highly anisotropic materials. Typically, its in-plane thermal conductivity λ_{xy} exceeds transversal (out-of plane) λ_z value for at least two ranges (comp. Heusch at all., 2002). The same concerns also thermal diffusivity. Because of that the PG specimen was investigated applying the system arrangement as shown in Fig. 3.c. The in-plane dimensions of the specimen were 44.0 mm and 9.6 mm for x and y direction respectively. Transversally the investigated bar was 6.5

mm thick. The measurements were performed at 2 Hz sampling rate for the temperature oscillation period $\tau_\Omega = 10$ s. The offset temperature was quasi-linearly changed within the interval from about 12°C to 38 °C (Fig. 7.c). The results of measurements are shown in Fig. 10. Contrary to three previously discussed cases, the obtained data have not been averaged. Instead of that, linear regressions were derived for both the amplitude and phase apparent TD. The appropriate coefficients are given in Table 3.

In spite of a relatively high scatter of individual data the discrepancy between a_ψ and a_φ characteristics is not so significant. It is because of a relatively small τ_Ω value which is the major cause of the scatter. But proving the effectiveness of modified linear scanning procedure was the most important outcome of the investigation (Panas & Nowakowski, 2009).

Regarding the quantitative aspect of the results it should be noticed that they exceed those reported by Maglić & Milošević, 2004 by about 2.5 times and are at least 8 times lower than presented by Heusch at all., 2002. This is not a surprise considering the differences of PG manufacturing technology. The tendency of the TD changes is exactly as expected: it decreases with the increasing temperature.

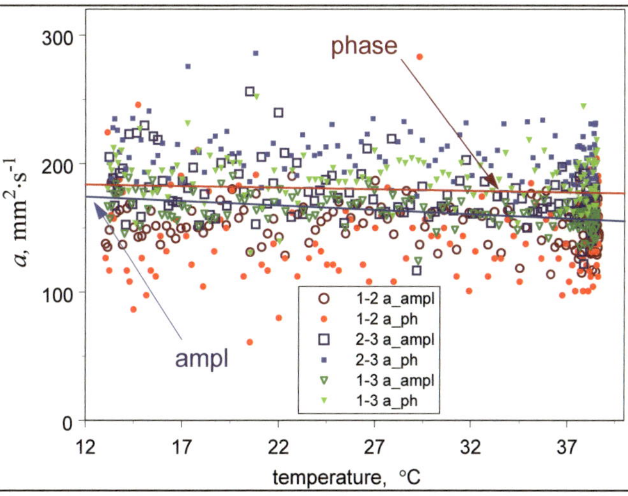

Fig. 10. Results of investigation of the TD temperature dependence of pyrolitic graphite

a_ψ, mm^2·s^{-1}	182.4640 - 0.6800363 ·t/°C	13 °C ≤ t ≤ 38 °C
a_φ, mm^2·s^{-1}	186.5935 - 0.2439950·t/°C	

Table 3. Linear regressions for pyrolitic graphite TD (t - temperature in °C)

4.4 PMMA and PTFE slabs

The aim of this part of research was to test and to verify procedures for investigation of transversal TD of slabs. The PMMA and PTFE were selected because literature data on their properties are easily available (Panas at all., 2003b; Tye & Salmon, 2005; Blumm at all., 2010). The experiments were conducted in the system configuration as shown in Fig. 3.d. The investigated specimens are shown in Fig. 6.b. All measurements were performed at 2 Hz sampling rate for the temperature oscillation period $\tau_\Omega = 30$ s. They started from steady oscillations over 22.5 °C. Next, the temperature scan from this temperature up to 41 °C was added (Fig. 5). Because one of the studied phenomena was the specimen transparency effect, each time the temperature response was taken from both the graphite coated part of the upper specimen surface and from the area intentionally left free from the coating. For the temperature data collection rectangular areas were applied as shown in Fig. 6.d (see also Figs 5 and 7.d). The signals were compared with the temperature histories taken from a semi-circled segment line around the PTFE and tetragonal line around the PMMA specimen respectively. The a_ψ and a_φ were obtained by solving Eqs (15). The final results are shown in Fig. 11. The data from the steady offset part of experiments is also listed in Table 4. They prove correctness of the applied procedure and show the translucence effect. It should be mentioned that routines of a specimen layering are applied also when investigating TD by any other method involving radiation (see eg. Blumm at all., 2010; Kim & Kim, 2009; Panas at all., 2003; Maglić at all., 1984). The scanning measurements have revealed increasing data scattering that can be attributed to raised convection (Fig. 11.b). This is why wide temperature range measurements of slab specimens become more difficult. Regarding such investigations it should be

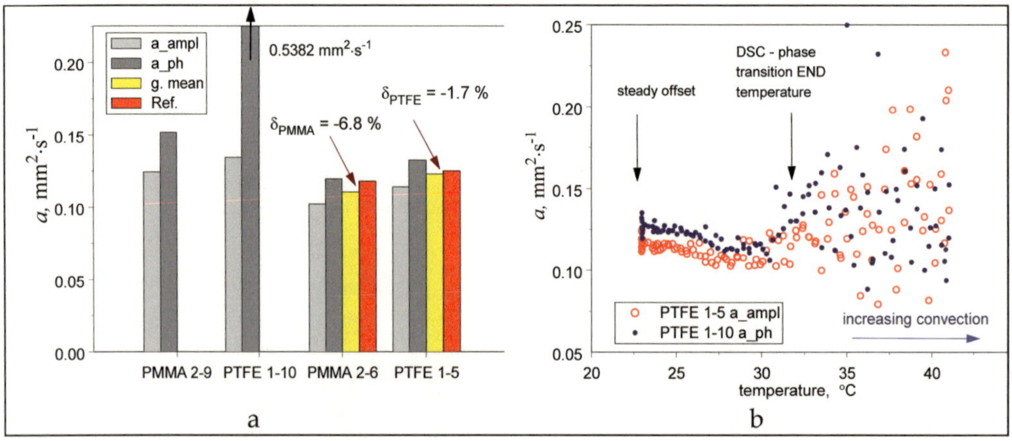

Fig. 11. Results of PMMA and PTFE investigation: a - amplitude and phase apparent TD values from uncoated (2-9 and 1-10) and graphite coated areas (2-6 and 1-5) and their respective geometric means (Eq. 29) compared with the reference data from Salmon & Tye, 2005 and Blum at all., 2010; b – PTFE apparent TD dependence on the temperature (scanning step results; DSC END temp. – from Panas at all., 2003)

Data on:	PMMA				PTFE			
	uncoated	coated	g. mean	Salmon & Tye, 2005	uncoated	coated	g. mean	Blumm at all., 2010
a_ψ, mm²·s⁻¹	0.1243	0.1021	0.1104	0.1179	0.1343	0.1140	0.1229	0.1250*
a_φ, mm²·s⁻¹	0.1518	0.1195			0.5382	0.1325		

Table 4. Results of TD studies of PMMA and PTFE at 22.5 °C compared with literature data (*) - the result was interpolated from results by Blumm at all., 2010)

underlined that the other basic restrictions originate from the IR camera calibration. Nevertheless, the obtained characteristics enable to reveal the phase change effect that is in agreement with previously published data (Blumm at all., 2010; Panas at all., 2003).

4.5 Hard tooth structures – Temperature oscillation treatment

The hard tooth tissue investigation is a crucial one regarding the fact that the described IR "slab" technology is dedicated mostly to such problems. The measurements were performed on test specimens that had been previously studied both applying the IR camera (Panas at all., 2007) and thermocouples. In present investigation a close-up view (Fig. 7.e, f) recordings were taken at 10 Hz sampling rate for specimen A and 2 Hz sampling rate for specimen B1/B2 (see Fig. 6.c,d). The temperature oscillation period was 20 s and steady offsets were 26.7 °C and 31.2 °C for specimens A and B respectively. The temperature signals taken from control sections (Fig. 7. E,f) were processed the same way as discussed in the previous section. The results are shown in Fig. 12.

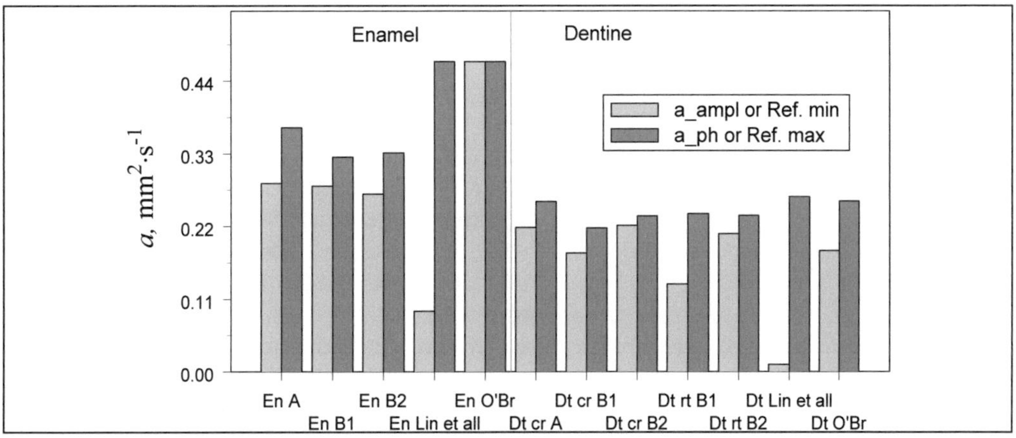

Fig. 12. Results from hard tooth structure TD studies: En indicates enamel, Dt cr - crown dentine, Dt rt - root dentine. The reference min/max data come from O'Brien, 1997 and Lin at all., 2010.a.

Analysing the displayed data one can notice that general relation between the enamel and dentine TD reported in the literature is properly expressed. The obtained ratio of enamel to dentine diffusivity is about 1.5. Moreover, the dentine results agree with most of the previously published data (comp. O'Brien, 1997; Panas at all, 2003, Lin at all., 2010a, 2010b). Considering the enamel data reported by Lin at all., 2010b and by O'Brien, 1997 the present enamel results might be regarded as underestimated. To express the differences more clearly geometric means (Eq. 29) from individual results were calculated first and next the data were averaged over similar characteristic tooth structures. These results are displayed in Table. 5.

In general, the problem of hard tooth tissue thermal properties is still open (comp. eg. Lin at all, 2010a). Taking this into account the obtained results can be regarded as acceptable. In spite of the fact that three different specimens from two different teeth were investigated, the obtained data are in mutual agreement. For that reason we can assume that the IR supported methodology of oscillating heating investigation of such objects is correct.

	Thermal diffusivity, $mm^{2}\cdot s^{-1}$		
	Enamel	Dentine (crown)	Dentine (root)
$a_{A,B1,B2}$ (mean)	0.308	0.221	0.200
O'Brien at all., 1997	0.469	0.183 ÷ 0.258	
Lin at all., 2010b	0.408 (±0.0178)	0.201 (±0.05)	

Table 5. Results of TD studies of PMMA and PTFE at 22.5 °C compared with literature data (*) - the result was interpolated from results by Blumm at all., 2010)

4.6 Hard tooth structures – Step heating experiments

There are few reports on successful implementation of the step heating procedure in quantitative analyses of thermophysical properties of thin non-homogeneous slabs. Such a paper has been recently published by Lin at all., 2010. It presents the results of hard tooth tissue TD measurements and these data have been taken as one of the references in the previous section. However, to obtain quantitative results the researchers were forced to apply a special data processing. More often the analyses are limited to qualitative studies (Panas at all., 2007; 2003a). It is because the step heating boundary condition is difficult to apply. As an illustration of the problem the results of investigation described by Panas at all., 2007 can be indicated. The study was conducted with almost the same arrangement of the system as depicted in Fig. 4 except that Peltier elements had not been present. Tooth specimens B1 and B2 were heated through the copper plate by a sudden flow of water beneath. The temperature histories were taken from points located as indicated in Fig. 13.a. The obtained relative temperature differences $\Delta t = t_{Cu} - t_i$, $i = 1, \ldots, 5$ are shown in Fig. 13.b. As was expected the qualitative differences are easy to recognize. It does not concern quantitative analyses.

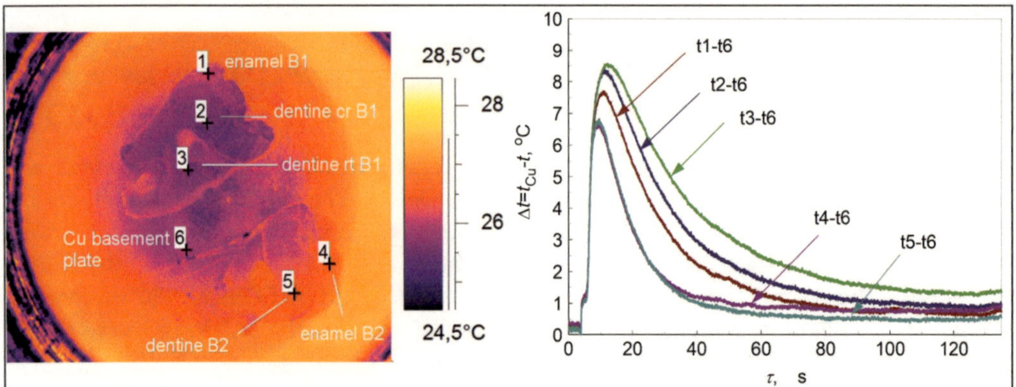

a b

Fig. 13. The specimens B1 and B2 under monotonic heating (a; Panas at all., 2007)) and the temperature changes with reference to the copper plate temperature (b)

To examine the problem let us consider the model of the temperature evolution given by Eq. (20). Assuming perfect surface conductance ($Bi = \infty$) we get from Eqs. (20)-(22):

$$\theta(0,\tau)-\theta(L,\tau)=\left[1-\left(\cos\frac{L}{\sqrt{a\tau_{ch}}}\right)^{-1}\right]\exp\left(-\frac{\tau}{\tau_{ch}}\right)+\sum_{n=1}^{\infty}\frac{2(-1)^{n+1}}{1-a\tau_{ch}\pi^2n^2L^{-2}}\exp\left(-\pi^2n^2\frac{a\tau}{L^2}\right). \quad (30)$$

As we can see, the procedure of single exponential approximation (24) do not comply with the model formulae (30). The "amplitude" of the first exponential term is depended on a and this makes any modification more difficult. Application of Eq. (24) is limited to qualitative analyses (Panas at all., 2007). Only when $\tau_{ch} \to 0$ the TD identification can be done more easily. This is not the case regarding the discussed experimental results.

5. Numerical validation of TD investigation procedures

In comparison with standard procedures the IR imaging techniques are usually more demanding when applied to thermophysical property quantitative investigation. It is because of their increased sensitivity to environmental influences. The environment is also a source of major incompatibilities between the applied mathematical model and real experimental conditions. It was proved that analytical approach in the studies where such discrepancies appear is ineffective in most of the cases. For that reason, the investigation procedures are tested, validated and corrected in the course of numerical simulations. This is also the case in the studies described above. Because a detailed discussion on numerical modelling issue is outside the scope of this work, only a short review of selected results will be provided here.

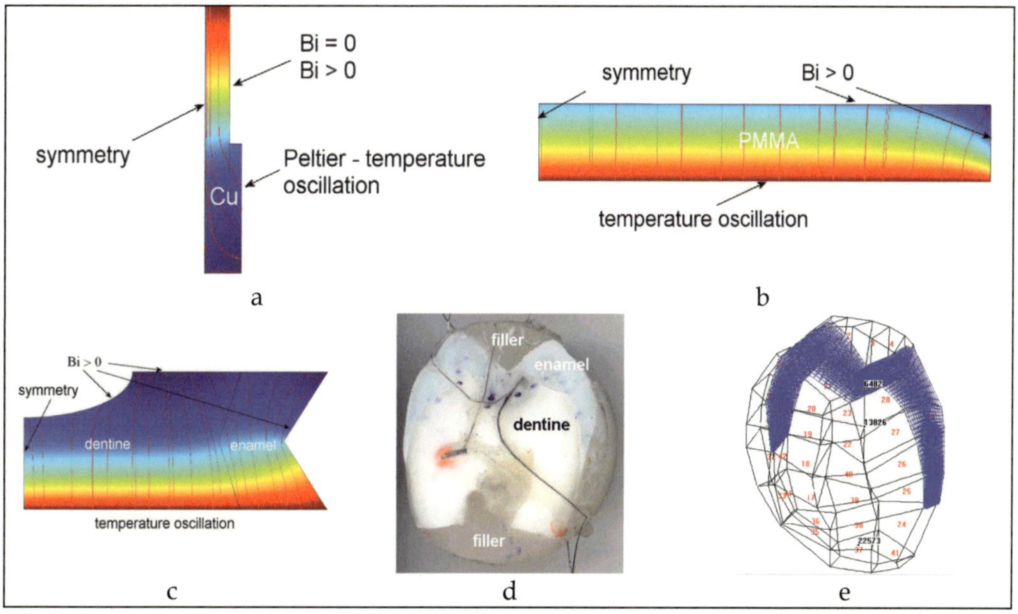

Fig. 14. Illustration of selected analysed numerical models: a - model for investigation of non-unidirectional heat transfer in Cu specimen, b - studies of convective heat losses from a slab specimen, c - 2D model for investigation of effects of non-homogeneity of a tooth specimen under periodic heating, d and e - the modelled structure and 3D model for investigation of laser flash and monotonic heating experiments of hard tooth tissues (Żmuda at all., 2005).

The numerical analyses were focused mostly on such problems as non-unidirectional heat transfer (Fig.14a), convection heat losses that violate assumptions of a slab model (Fig. 14.b), irregularity and non-homogeneity of the investigated tooth structure (Fig. 14.c) and 3D heat transfer effects including incompatibility of the temperature excitation (Fig. 14.d and e; see also Żmuda at all., 2005). In all the instances the elaborated models were also tested for the scanning mode application (comp. Panas & Nowakowski, 2009). The major outcome of these studies is the proof of correctness of the developed procedures for experimental TD investigation. The analyses proved good metrological conditioning of the oscillation technique. This concerns also the scanning mode operation.

Moving to the details, effects of 3D heat transfer in a laterally heated Cu specimen were confirmed (see red streamlines marking total heat flux path in Fig. 14.a). It was observed that both convective and conductive heat losses from side surfaces of slab specimens result in underestimation of amplitude and overestimation of phase results. This is similar to Ångström's model for semi-infinite rod. In the study of hard tooth tissue non-homogeneity effects, significant underestimation of the enamel results was revealed. This concerns not only the "thermal wave" measurements but also pulse heating experiments and monotonic heating investigations (Żmuda at all., 2005). The identified TD values were about 20% lower than that assumed in the model. The simulations showed that there is an influence of the control area location on the obtained enamel parameters. Because of this outcome, IR measurements seem to be more advantageous than traditional techniques that exclude monitoring of spatial temperature distribution.

It should also be mentioned that the numerical simulation technique has been applied for validation of the methodology for investigation of metallic specimens covered with a graphite layer. This concerns Cu and Ni3Al specimen studies. In both cases the effect of the covering has been proved to be negligible. However, in order to get more universal results a more thorough analyses needs to be performed like those concerning laser flash technology (comp. Cernuschi at al., 2002, S.K. Kim & Y.-J. Kim, 2009).

In general, a numerical modelling technique has not only proved to be helpful but also indispensable for discussed tasks. The only problem is that complicated objects need to be treated individually.

6. Conclusion

With its possibility of non-intrusive and simultaneous examination of both the temperature changes in time and temperature spatial distribution the IR imaging technology creates unique possibilities for complex thermal investigations. For that reason it is widely applied eg. in non destructive testing. However, application of IR imaging is restricted mostly to qualitative analyses because of its limitations. This applies also to thermophysical property investigation. In that domain the major disadvantages are related to: difficulties in complying with certain analytical model assumptions; narrow temperature range of operation when compared with standard techniques of TP studies (comp. eg. Maglić at all., 1984); unreliable temperature measurements due to the surface emissivity changes etc. On the other hand, the standard procedures of thermophysical property investigation are demanding for the investigated specimen structures and forms. This concerns mostly heat transfer parameters including the thermal diffusivity. In such a situation the IR imaging technology seems to be indispensable especially when materials possessing unusual properties, exhibiting complicated structure or non-homogeneous composition are tested. As it has been shown in the present study reliable quantitative results of the thermal diffusivity investigation can be obtained providing that the IR technique is combined with a properly selected thermal treatment methodology. Regarding the problem of the temperature field excitation, the temperature oscillation proved to be well metrologically conditioned. This enabled accommodating the classical and modified Ångström's procedures (Ångström, 1961; Bodzenta at all., 2007) to the scanning mode operation (Panas & Nowakowski, 2009). These techniques combined with IR measurements of the spatial and temporal temperature changes enable successful investigations of the thermal diffusivity of "difficult" specimens. Measurements of pyrolitic graphite, nanostructural Ni_3Al intermetallic and hard tooth tissue structures illustrate good performance of that methodology. With no doubt, measurements performed on biological specimens are the most demanding. Because of that the results of these studies should be additionally verified. Numerical simulation proves to be the most effective technique for this task.

7. Acknowledgment

The experimental instrumentation of the research was completed with a support from UE/state POIG.02.02.00-14-022/09 project.

8. References

Ångström, A. J. (1861). Neue Methode, das Warmeleitungsvermogen der Korper zu Bestimmen. *Annalen der Physic und Chemie*, Vol. 114, (1861), pp. 513-530

Belling, J.M. & Unsworth, J. (1987). Modified Ångström's method for measurement of thermal diffusivity of materials with low conductivity. *Review of Scientific Instruments,* Vol.58, No.6, (June 1987), pp. 997-1002, ISSN 0034-6748

Bodzenta, J.; Burak, B.; Nowak, M.; Pyka, M.; Szałajko, M. & Tanasiewicz, M. (2006). Measurement of the thermal diffusivity of dental filling materials using modified Ångström's method, *Dental Materials,* Vol.22, No. , (July 2006), pp. 617–621, ISSN 0 109-5641

Bison, P.G.; Marinetti, S.; Mazzoldi, A.; Grinzato, E. & Bressan, C. (2002). Crosscomparison of thermal diffusivity measurements by thermal methods. *Infrared Physics and Technology*, Vol.43, No.3–5 (June 2002), pp.127–132, ISSN 1350-4490

Blumm, J.; Lindemann, A., Meyer, M. & Strasser, C. (2010). Characterization of PTFE Using Advanced Thermal Analysis Techniques. *International Journal of Thermophysics,* Vol.31, No.1, (October 2010), pp. 1919-1927, ISSN 0198-925X

Carlaw, H. S. & Jaeger, J. C. (2003). *Conduction of Heat in Solids* (2nd Edition), Oxford Univ. Press, ISBN 978-0-19-853368-9, London, Great Britain

Cernuschi, F., Lorenzoni, L.; Bianchi, P. Figari, A. (2002). The effects of sample surface treatments on laser flash thermal diffusivity measurements. *Infrared Physics and Technology*, Vol.43, No.3–5 (June 2002), pp.133–138, ISSN 1350-4490

Heusch, C.A.; Moser, H.-G. & Kholodenko, A. (2002). Direct measurements of the conductivity of various pyrolytic graphite samples (PG, TPG) used as thermal dissipation agents in detector applications. *Nuclear Instruments and Methods in Physics Research A*, Vol. 480, No.2-3 (March 2002), pp. 463-469, ISSN: 0168-9002

Kim, S.-K. & Kim, Y.-J. (2009). Determination of apparent thickness of graphite coating in flash method. *Thermochimica Acta,* Vol.468, No.1-2, (February 5, 2008), pp. 6-9, ISSN: 0040-6031

Lin, M.; Xu, F.; Lu, T.J. & Bai, B.F. (2010a). A review of heat transfer in human tooth—Experimental characterization and mathematical modeling. *Dental Materials,* Vol.26, No.6, (June 2010), pp. 501-513, ISSN: 0109-5641

Lin, M.; Liu, Q.D.; Kim, T.; Xu, F.; Bai, B.F. & Lu, T.J. (2010b). A new method for characterization of thermal properties of human enamel and dentine: Influence of microstructure. *Infrared Physics and Technology,* Vol.53, No.6, (November 2010), pp. 457-463, ISSN 1350-4490

Lykov, A.V. (1967). *Tieoria tieploprovodnosti,* Vysshaya Shkola, UDK-536.2, Moskva, USSR

Maglić, K. D.; Cezairliyan, A. & Peletsky, V. E. (Eds.). (1984). *Compendium of Thermophysical Property Measurement Methods. Volume 1: Survey of Measurement Techniques.* Plenum Press, ISBN 0-306-41424-4, New York, USA

Maglić, K. D.; Cezairliyan, A. & Peletsky, V. E. (Eds.). (1992). *Compendium of Thermophysical Property Measurement Methods. Volume 2: Recommended measurement Techniques and Practices.* Plenum Press, ISBN 0-306-43854-2, New York, USA

Maglić, K. D. & Milošević, N. D. (2004). Thermal Diffusivity Measurements of Thermographite. *International Journal of Thermophysics,* Vol.25, No.1, (January, 2004), pp. 237 - 247, ISSN 0198-925X

Miettinen, L.; Kekäläinen, P.; Merikoski, J.; Myllys, M. & Timonen, J. (2008). In-plane Thermal Diffusivity Measurement of Thin Samples Using a Transient Fin Model and Infrared Thermography. *International Journal of Thermophysics,* Vol.29, No.4, (August, 2008), pp. 1422 - 1438, ISSN 0198-925X

Muscio, A.; Bison, P.G.; Marinetti, S. & Grinzato, E. (2004). Thermal diffusivity in slabs using harmonic and one-dimentional propagation of thermal waves. *International Journal of Thermal Sciences,* Vol. 43, No.5 , (May 2004), pp. 453-463, ISSN 1290-0729

Nakamura, H. (2009). Frequency response and spatial resolution of a thin foil for heat transfer measurements using infrared thermography. *International Journal of Heat and Mass Transfer,* Vol.52, No.21-22, (October 2009), pp. 5040-5045, ISSN 0017-9310

O'Brien, W. J. (1997). *Dental Materials and Their Selection,* Quintessence Publ., ISBN 0-86715-297-4, Chicago, USA

Özişik, M.N. and Orlando, H.R.B. (2000). *Inverse Heat Transfer. Fundamentals and Application,* Taylor & Francis, ISBN 1-56032-838-X, New York, USA

Panas, A. J.; Żmuda, S.; Terpiłowski, J. & Preiskorn, M. (2003a). Investigation of the thermal diffusivity of human tooth hard tissue. *International Journal of Thermophysics,* Vol.24, No.3, (May 2003), pp. 837-848, ISSN 0198-925X

Panas, A.J.; Cudziło, S.; & Terpiłowski, J. (2003b). Investigation of thermophysical properties of metal-polytetrafluoroethylene pyrotechnic compositions. *High Temperatures - High Pressures,* Vol. 34, No.6 , (December 2003), pp. 691-698, ISSN 0018-1544

Panas, A.J.; Preiskorn, M.; Dąbrowski, M. & Żmuda, S. (2007). Validation of hard tooth tissue thermal diffusivity measurements applying an infrared camera. *Infrared Physics and Technology*, Vol.49, No.3, (January 2007), pp. 302-305, ISSN 1350-4490

Panas, A.J. & Nowakowski, M. (2009). Numerical validation of the scanning mode procedure of thermal diffusivity investigation applying temperature oscillation, *Proceedings of Thermophysics 2009*, pp.252-259, ISBN 978-80-214-3986-3, Valtice, 29th÷30th October 2009

Panas, A.J. (2010). Comparative-complementary investigations of thermophysical properties - high thermal resolution procedures in practice, *Proceedings of Thermophysics 2010*, pp.218-235, ISBN 978-80-214-4166-8, Valtice, 3nd÷5th November 2010

Perkowski, Z. (2011). A thermal diffusivity determination method using thermography: Theoretical background and verification. *International Journal of Heat and Mass Transfer*, Vol.54, No.9-10, (April 2011), pp. 2126-2135, ISSN 0017-9310

Phylippov, L.P. (1984). Temperature Wave Techniques, In: *Compendium of Thermophysical Property Measurement Methods. Volume 1: Survey of Measurement Techniques.* Plenum Press, K. D. Maglić, A. Cezairliyan & V. E. Peletsky (Eds.), ISBN 0-306-41424-4, New York, USA

Platunov, E.S. (1992). Instruments for Measuring Thermal Conductivity, Thermal Difusivity, and Specific Heat Under Monotonic Heating, In: *Compendium of Thermophysical Property Measurement Methods, Volume 2: Recommended measurement Techniques and Practices.* Plenum Press, K. D. Maglić, A. Cezairliyan & V. E. Peletsky (Eds.), pp. 337-365, Plenum Press, ISBN 0-306-41424-4, New York, USA

Ražnievič, K. (1966). *Thermal tables and charts*, (In Polish). WNT, Warsaw, Poland, (English edition: K. Raznjevic (1976). *Handbook of Thermodynamic Tables and Charts*, McGraw-Hill, Washington, ISBN 978-0070512702, Washington-New York, USA)

Rémy, B.; Degiovanni, A. & Maillet, D. (2005). Measurement of the In-plane Thermal Diffusivity of Materials by Infrared Thermography. *International Journal of Thermophysics*, Vol.26, No.2, (March 2005), pp. 493 - 505, ISSN 0198-925X

Sakagami, T. & Kubo, S. (2002). Applications of pulse heating thermography and lock-in thermography to quantitative nondestructive evaluations. *Infrared Physics and Technology*, Vol,.43, No.3-5, (June 2002), pp. 211-218, ISSN 0017-9310

Tye, R. P. & Salmon, D. R. (2005). Thermal Conductivity Certified Reference Materials: Pyrex 7740 and Polymethylmethacrylate. In: *Thermal Conductivity 26 / Thermal Expansion 14*, R. B. Dinwiddie, R. Mannello, (Eds.), 291-298, DEStech Publications, Inc., ISBN 1-932078-36-3, Lancaster PA, USA, pp. 437-451

Volokhov, G.M & Kasperovich, A.S. (1984). Monotonic Heating Regime Methods for the Measurement of Thermal Diffusivity, In: *Compendium of Thermophysical Property Measurement Methods,* K. D. Maglić, A. Cezairliyan & V. E. Peletsky (Eds.), pp. 337-365, Plenum Press, ISBN 0-306-41424-4, New York, USA

Wang, Z.I. & Tung, D.W. (2011). Experimental Reconstruction of Thermal Parameters in CNT Array Multilayer Structure. *International Journal of Thermophysics*, Vol.32, No.5, (May 2011), pp. 1013-1024, ISSN 0198-925X

Zhang, S.; Vinson, M.; Beshenich, P. & Montesano, M. (2006). Evaluation and finite element modelling for new type of thermal material annealed pyrolytic graphite

(APG). *Thermochimica Acta*, Vol.442, No.1-2, (March 15, 2006), pp. 6-9, ISSN: 0040-6031

Żmuda, S.; Panas, A. J.; Sypek, J. & Preiskorn, M. (2005). Validation of thermal diffusivity measurements of hard tissue specimens by finite element analysis, In: *Thermal Conductivity 27/ Thermal Expansion 15,* H. Wang , W. Porter (Eds.), pp. 113-120, DEStech Publications, Inc., ISBN 1-932078-34-7, Lancaster PA, USA

Part 2

Life Sciences

5

Infrared Thermography – Applications in Poultry Biological Research

S. Yahav and M. Giloh
Institute of Animal Science,
ARO the Volcani Center, Bet-Dagan
Israel

1. Introduction

Infrared (IR) thermal imaging technology has undergone major development during the past decade. Since 2000 this technology has been developed rapidly, through use of automated systems which have become widely available and cheaper as the technology developed (FLIR Systems, 2004).

Infrared thermal imaging cameras (IRTIC) measure the amount of near-IR radiation - characterized by wavelength of 8 to 12 µm - that is emitted from a surface, and convert it to a radiative temperature reading, according to the Stefan-Boltzmann equation:

$$R = \varepsilon\sigma T^4$$

where ε is the emissivity of the surface, which is defined as the ability of a surface to emit and absorb radiation, and which, for biological surface tissues, ranges between 0.94 and 1.0 (Hammel, 1956; Monteith & Unsworth, 1990); σ is the Stefan-Boltzmann constant (5.67×10^{-8} $Wm^{-2}K^{-4}$); and T is the surface absolute temperature in kelvins (K).

The images that visualize the temperature can be displayed in gray or colors (Figure 1), and can be analyzed as spots (1 spot = 1 pixel) or as user-defined Regions of Interest (ROI), by means of a special computer program.

The IRTI technique plays a major role in biological research applications such as: thermal physiology in mammals (Klir & Heath, 1992; Klir et al., 1990; Mohler & Heath, 1988; Philips & Heath, 1992) and birds (Philips & Sanborn, 1994; Stewart et al., 2005; Yahav 2009; Yahav et al., 1998, 2004); fever diagnosis in homeotherm birds and mammals, including humans (Chiu et al., 2005; Nguyen et al., 2010; Teunissen & Daanen, 2011); cancer diagnosis (Cetingul & Herman, 2010; Herman & Cetingul, 2011; Kontos et al., 2011); and animal population counts.

The major advantage of IR is that it is a non-invasive and contact-free method of measuring surface temperature, at either short or relatively long distance, depending on the goal of the thermal study. For any thermal study: a thermal resolution of 0.1°C is recommended; and spatial resolution must be sufficient and appropriate for the size of the studied object and its distance from the IRTI camera. In general, a spatial resolution of 320 × 240 pixels – as

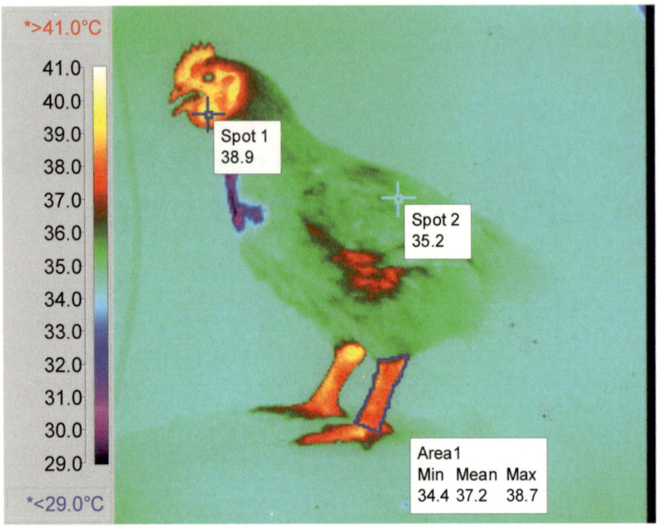

Fig. 1. Thermal image of a broiler chicken exposed to 35°C. The blue outline on the leg defines an ROI that comprises a section of surface area for which minimum, average and maximum surface temperatures were determined. Spots 1 and 2 exhibit the surface temperature in each spot. Thermal images and their analysis were obtained with the Thermacam P 2.8 SR - 1 program (Yahav et al., 2005).

determined by the characteristics of the detector array - is used and is found sufficient. Use of a lens with a 24-degree field of view, with the camera placed 60–80 cm from the object resulted in images in which each pixel corresponded to a rectangular detail of 0.9–1.5 mm^2 (see also Section 4.3.4). The IRTI provides accurate measurements of the surface temperature of the whole object; measurements that are undoubtedly more precise and accurate than those provided by thermocouples (Mohler & Heath, 1988).

For birds the actual temperature measured by the IRTI represents the temperature beneath the outer physical surface of the insulating layer. The matrix nature of plumage - which in general lasts for up to 3 weeks on domestic fowl - and of the subsequent feathers causes radiation to be exchanged at varied depths within the outer layer, with the result that the radiative temperature is usually higher than the temperature at the physical surface (McCafferty et al., 2011). Therefore, the IRTI measures the temperature of the feathers/plumage several millimeters below the surface. However, several regions, such as the face, wattle, comb, legs, beak, and unfeathered areas below the wings radiate directly, and their measured temperatures will be those at the surface (Figure 1).

Surface temperature is significantly correlated to the vasomotor responses of the bird. It is well documented that the vasomotor response - vasodilatation or vasoconstriction (Figure 8, for details see Section 4.3.1) is affected by the bird's body temperature (T_b): hyperthermic or hypothermic, respectively (Yahav, 2009).

Birds are endotherms, i.e., they are able to maintain their T_b within a narrow range. In endothermic animals (birds and mammals) T_b is physiologically the most closely controlled parameter of the body, therefore the thermoregulatory system in these animals operates at a very high gain in order to hold T_b within a relatively narrow range, despite moderate to extreme changes in environmental conditions. The ability to maintain a stable T_b depends on the mechanisms that control heat production and heat loss; mechanisms that changed in the course of evolution, to enable endothermia to replace ectothermia (Silva, 2006). The evolutionary changes from ectothermia to endothermia were achieved because the developmental regulatory mechanisms maintained a balance between heat production and heat loss (Equation 1). Both mechanisms, especially heat production, are probably older than endothermy, but both are permanently activated and regulated by both neuronal and hormonal signals (Morrison et al., 2008; M.P. Richards & Proszkowiec-Weglarz, 2008; Silva, 2006).

Collectively, heat transfer modeling has been used to understand thermoregulation in endotherms in general. These models take into consideration a constant T_b for an animal not performing external work and are represented by the following heat-balance equation, based on the first law of thermodynamics:

$$S = H - E \pm R \pm C \pm K \qquad (1)$$

where S is the bodily heat gain or loss that must be balanced by: H metabolic heat production; E evaporative heat loss; R radiative heat gain or loss; C convective heat loss or gain; K conductive heat loss or gain. Body temperature will remain unchanged when S is zero, i.e., heat gain matches heat loss. If more heat is produced and gained than lost, then S is positive and T_b will rise and vice versa (Dawson & Whittow, 2000).

Within equation (1), thermogenesis (heat production - H) can be divided into obligatory and facultative thermogenesis (Silva, 2006). Obligatory thermogenesis refers to the energy required to maintain T_b as long as the ambient temperature (T_a) lies in the thermoneutral zone, the range within which the body is in thermal equilibrium with the environment and produces energy at a level termed the resting metabolic rate (RMR) (Gordon, 1993). Facultative thermogenesis refers to stimulated energy production that is required when T_a deviates below or, to some extent, above the thermoneutral zone. Heat loss - especially with regard to birds, in the present context- is dissipated through respiratory (Marder & Arad, 1989; S.A. Richards, 1968, 1970, 1976; Seymour, 1972) and cutaneous evaporative mechanisms (E) (Ophir et al., 2002; Webster & King, 1987), and sensible heat loss via radiation (R), convection (C) (Yahav et al., 2005) and conduction (K) (Wolfenson et al., 2001).

Because domestic fowl are highly productive their biology causes severe difficulties in coping with changes in the environment, especially with regard to T_a and ventilation. These difficulties, in turn, may lead to severe difficulties in maintaining a dynamic steady state, which can impair productivity. To avoid such deleterious effects, the bird activates evaporative and sensible heat loss mechanisms. However, heat loss via panting is accompanied by loss of body water content, therefore dehydration will reduce heat loss via this pathway, as well as that through extensive passive cutaneous evaporation. Panting is also associated with respiratory alkalosis, which may affect the evaporation (Yahav et al., 1995). An increase in sensible heat loss may reduce the intensity and consequences of evaporative heat loss.

During recent years alterations in incubation temperature (termed "thermal manipulation") have been used to enhance thermotolerance acquisition in the post-hatch fowl (Druyan et al., 2011; reviewed by Yahav et al., 2009). Briefly, thermal manipulation during incubation aims to change the threshold response of the preoptic anterior hypothalamus (PO/AH) - the principal site in the central nervous system where heat production and heat loss are controlled - in order to elicit appropriate thermoregulatory responses via control of physiological, endocrinological, and behavioral responses and, thereby, to keep the core T_b relatively constant (Boulant, 1996).

To understand the contribution of sensible heat loss to the ability of the domestic fowl to balance the dynamic steady state of its body energy content an IRTI camera had to be used to monitor its body surface temperature, and a special numerical model had to be developed to accurately calculate heat loss by radiation and convection. Furthermore, this technique is used during incubation to determine the effect of incubation temperature on eggshell temperature, and to enhance the effects of these changes on post-hatch thermotolerance acquisition (Piestun et al., 2008, 2009; Shinder et al., 2009).

This Chapter will focus on:

a. The specification of the IRTI camera used for sensible heat loss evaluations; and
b. The sensible heat loss models for domestic fowl (pre- and post-hatch), and their applications.

2. The IRTI camera specifications – Thermal accuracy and calibration

Most infrared cameras that are used for temperature measurements have a guaranteed thermal resolution of 0.1-0.2°C and an absolute accuracy of ± 1-2°C, or 1-2% of the absolute temperature (kelvins, K). This accuracy is not sufficient in biological applications, where

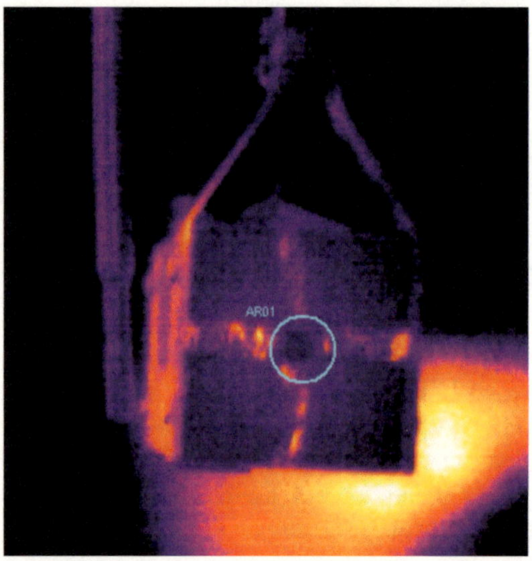

Fig. 2. Infrared thermography of the experimental setup for studying fluctuation of measurements. On the left side, the alcohol thermometer is visible. The blackbody is shown as a box in which the cylinder is confined. The hole of the blackbody cavity appears as a dark area, surrounded by the limits of the ROI (Region of Interest), as defined by the camera software. An electronic temperature probe which measured the actual temperature of the blackbody was inserted into the rear of the cylinder and is invisible to the camera.

comparisons between different thermograms are needed, and the expected temperature changes are, in most cases, smaller than ± 2°C. In order to compensate for differences in the calibration among different thermograms, a common reference point with a known temperature and emissivity must be used. In the studies presented below in this Chapter a simple black body was used (Figure 2); it comprised a closed hollow cylinder with a circular aperture of area about 1 cm^2 in its lid. A high-precision temperature probe connected to a digital thermometer was inserted into the blackbody, and the exact temperature was recorded manually for each thermogram. The blackbody temperature was compared with the camera temperature reading, and the difference between the readings was used to correct the camera readings for each spot in the thermogram. The system was tested also in a commercial chicken farm by comparing simultaneous readings of three temperature measurement systems (Figure 3): 1) a high-precision alcohol thermometer placed in the air; 2) a blackbody with a temperature probe as described above; 3) an IR camera that registered temperature readings for the aperture of the blackbody.

Figure 3 displays typical results of the relation among different measurement systems. During the first 15 minutes, the temperature in the building slowly increased according to the alcohol thermometer which has a shorter equilibration time than the blackbody. After 15

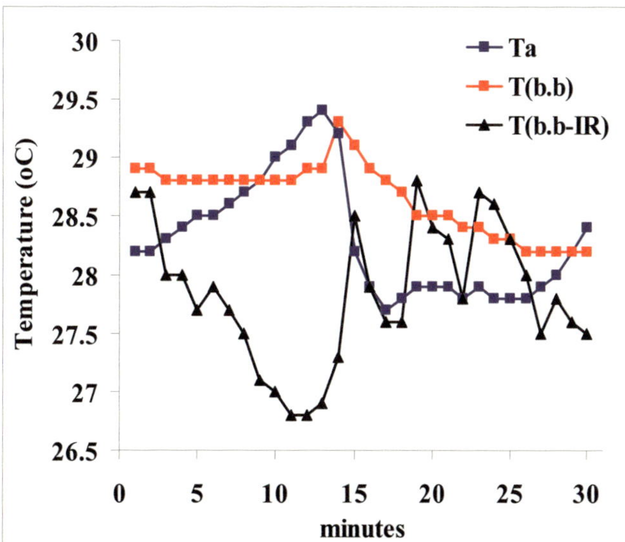

Fig. 3. Comparison between blackbody temperatures measured with a precision digital thermometer, blackbody temperature measured by IR thermography, and ambient temperature measured by a high-precision alcohol thermometer at subsequent instants. A manufacturer-calibrated standard alcohol thermometer was situated 1.5 m above the floor in a commercial poultry house, and the blackbody (described above) was situated immediately adjacent to it. The IR camera was placed 80 cm from the blackbody. The minimum temperature in the hole of the blackbody was measured once a minute, and the digital reading of the blackbody temperature was recorded manually, as was the ambient temperature, as measured by the alcohol thermometer. T(b.b): Digital reading of blackbody temperature; T_a: Ambient temperature, as measured by the alcohol thermometer; T(b.b-IR): Blackbody temperature, as measured by the IR camera.

minutes the ventilation in the farm went on automatically, and the alcohol thermometer displayed an abrupt decline in ambient temperature. The digital thermometer of the blackbody displayed an exponential decline with the same magnitude. During the whole event, before and after the ventilation went on, the absolute blackbody temperature demonstrated both a systematic drift and stochastic fluctuations, displaying errors of up to 2 centigrade. Use of reference temperature measurements implies that each thermogram analysis is based on three measured inputs: the camera reading of the measured spot or the ROI, the camera reading of the blackbody temperature, and the digital reading of the blackbody temperature. In light of the uncertainty of equilibration, the effect of changes in the environment on the camera readings, and the intrinsic fluctuation in each of the inputs, one can hardly expect that using a reference point could eliminate stochastic fluctuations in the measurements. However, systematic drift in the camera calibration can be compensated for accurately.

Figure 4 illustrates the setup for field measurements; it enabled the blackbody to be kept in focus in all thermograms. Figure 5 shows the experimental setup in the laboratory, and Figure 6 shows a typical thermogram produced in the laboratory; the blackbody is clearly distinguishable. The camera reading of the blackbody temperature was obtained by means of an automatic feature that enabled display of the minimal temperature if the blackbody was colder than the surroundings in the region of interest (ROI), or the maximal temperature if it was warmer. However, in some cases the aperture was not distinguishable in the thermogram because the temperature differences in the region were too small, i.e., if the difference between maximum and minimum temperatures in the ROI was less than 0.5°C. In such cases the average temperature of the ROI was used, thus allowing for a

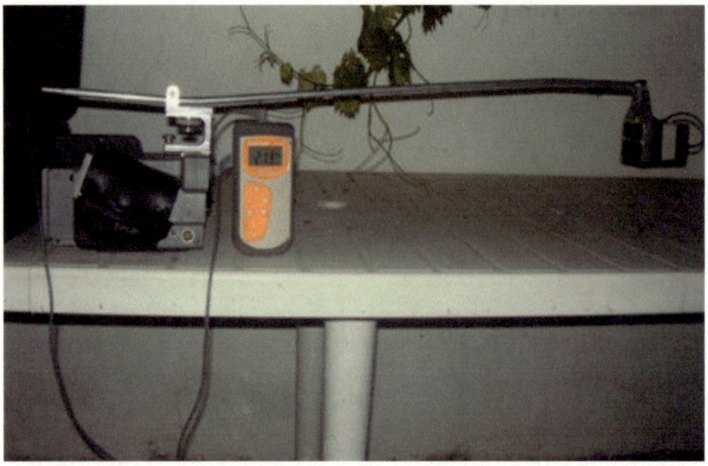

Fig. 4. Setting of field observations. The camera was attached in a permanent position at one end of a special "antenna", and the blackbody was situated at the other end, with the hole facing the camera. The distance between the lens and the blackbody was 80 cm. At the right-hand side of the picture can be seen the wire that connected a temperature probe that was inserted into the rear of the blackbody to the digital thermometer, which is displayed in front of the camera only for this photograph. During experimental measurements this thermometer was carried in a small bag, and the digital blackbody temperature was recorded manually for each thermogram.

Fig. 5. Experimental procedure in the laboratory. The setting was identically arranged and located in each of the four climate-controlled rooms in which birds were exposed to various environmental conditions. The IR camera was permanently placed 80 cm from the blackbody.

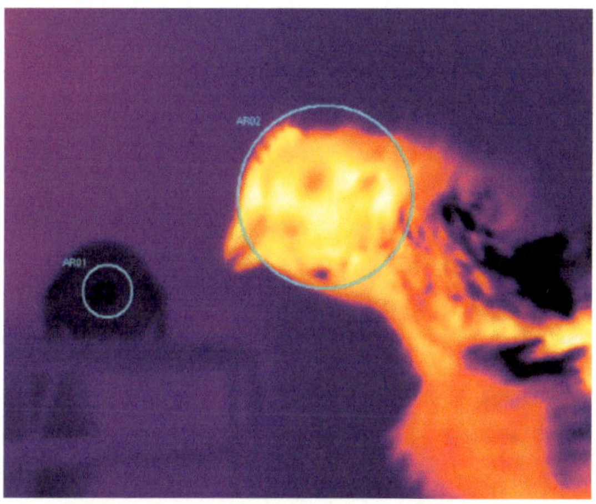

Fig. 6. Typical thermogram obtained in laboratory measurements. Two regions of interest were used for the analysis: one around the hole of the blackbody and the other around the hotter parts of the chicken's face. Camera software was used to read the minimal temperature of the blackbody cavity if it was colder than the surrounding, its maximal temperature if it was hotter than the surrounding, and its average temperature if the cavity was indistinguishable on the thermogram because the temperature differences were too small. For facial temperature measurement, the hottest temperature inside the ROI surrounding the face was recorded.

maximum error of 0.3°C. Since this occurred in a relatively small number of thermograms, the effect on averages of several thermograms was estimated to be not greater than 0.1-0.2°C, which was not crucial in this application.

3. Sensible heat loss from fertile eggs of domestic fowl – The rationale, the model and the applications

3.1 The rationale

During the perinatal period, a time window occurs in which physiological control systems can be imprinted. This imprinting is probably related to neural changes at the microstructural level, i.e., in terms of synaptic plasticity, as well as by environment-induced modifications of gene expression. Perinatal epigenetic temperature adaptation may be used as a tool to adapt poultry embryos or hatchlings to subsequent climatic conditions (Tzschentke, 2007; Tzschentke & Plagemann, 2006). In chickens and other precocial birds, epigenetic temperature adaptation can be induced by changes in incubation temperature during late-stage embryonic development (Collin et al., 2007; Minne & Decuypere, 1984; Piestun et al., 2009; Shinder et al., 2009; Tzschentke & Halle, 2009; Tzschentke & Nichelmann, 1997), as well as by post-hatch thermal conditioning (Yahav & Hurwitz, 1996; Shinder et al., 2002). To calculate heat loss by radiation and convection from fertile eggs a specific model was developed.

3.2 The model

Sensible heat loss from fertile eggs can be calculated according to van Brecht et al. (2005) and Shinder et al. (2007), on the assumption that heat exchange occurs via the whole eggshell surface. The eggshell surface area is determined from the length and breadth of the egg, which, for the purpose of the following model were measured with calipers (Mitutoyo-Ser. No 51015023; precision, 0.02 mm; Mitutoyo Corporation, Kanagawa, Japan). The egg was assigned a characteristic dimension, i.e., the diameter of a sphere with the same surface area (van Brecht et al., 2005).

The very low air velocity (less than 0.3 m s^{-1}) to which the eggs were exposed necessitated the use of a theoretical heat transfer model based on free convection and radiation. The spherical-egg model, incorporating mean values of available or especially derived heat-transfer correlations, was used to calculate the sensible heat loss (Shinder et al., 2007).

The exponential dynamics of cooling following an abrupt decrease in T_a (Tazawa & Nakagawa, 1985; Tazawa et al., 2001) was fitted to the exponential equation:

$$T_{egg}(t) = T_1 + A \exp(-k\, t) \tag{2}$$

where T_{egg} is the eggshell temperature as measured by IRTI; T_1 is the extrapolated asymptotic temperature as $t \rightarrow \infty$, which may differ from T_a during cold exposure; $T_1 + A$ is the extrapolated temperature at the moment of onset of the cold exposure ($t = 0$); The decay-rate or cooling-rate coefficient, k is the reciprocal of the time constant, t_1, and is related to the thermal half-time, $t_{1/2}$ (min) according to:

$$k = 0.693/\, t_{1/2} \tag{3}$$

The estimated total accumulated heat loss (measured in units of energy) from the egg during its cold exposure may be a physiologically relevant measure of the intensity of the cold exposure, and it can be calculated from T_{egg}, as measured by IR thermography. Since the total heat content of the egg is proportional to T_{egg}, the instantaneous value of SHL, q (watt) is proportional to dT_{egg}/dt, which has the same decay rate, k, as T_{egg}. After the abrupt reduction of T_a, q can, therefore, be fitted into an exponential decay curve described by:

$$q = q_0 + q_1 \bullet \exp(-kt) \tag{4}$$

where q_0 is the extrapolated value at large t, when the living embryo is in quasi-equilibrium with the environment. The extrapolated heat loss at t = 0 is $q_0 + q_1$. The decay rate k [min^{-1}] is as described above. The initial heat loss per degree of temperature change, in watt/°C was designated as G:

$$G = (q_0 + q_1)/\Delta T \tag{5}$$

where ΔT is the initial difference between T_{egg} and T_a. To calculate G, the incubation temperature was used instead of T_{egg}, because the differences between the two parameters could be neglected within the accuracy of the overall calculation.

Disregarding the asymptotic "plateau" q_0, which represents the metabolic heat production of the embryo, the total sensible heat loss (Q) caused by a cold exposure for Δt min will be:

$$Q = \int_0^{\Delta t} (q - q_0) \bullet dt = (q_1/k) \bullet (1 - \exp\{-k \bullet \Delta t\}) \bullet 60/4.19 \tag{6}$$

where Q = total accumulated heat loss (Calories); q = instantaneous rate of heat loss (watt); k = decay rate (min^{-1}).

The parameter Q in Equation (6) represents a quantitative estimate of the total physiological impact of the cold exposure, and it can serve as a suitable indicator for comparing different temperature manipulation strategies.

3.3 The applications

Short-term cold exposure during the late phase of embryogenesis, when embryos switch from ecto- to endothermic behavior, was found to induce an enhanced thermogenesis pattern in the hatched chicks (Minne & Decuypere, 1984; Nichelmann, 2004).

In order to study the effect of reducing incubation temperature during the final phase of embryogenesis on thermotolerance acquisition of post-hatch chicks, embryos were exposed to 15°C for 30 or 60 min on embryogenesis days 18 and 19. IRTI was used to measure the eggshell surface temperature, to determine sensible heat loss from the egg during 60 min of cold exposure (Figure 7), in order to determine its effect on post-hatch parameters (Shinder et al., 2009).

Figure 7 depicts the calculated time variation of SHL from the eggs, obtained by fitting the data to an exponential-decay regression curve according to Equation (4). The calculated q_0 was 0.55 ± 0.14 W; q_1 was 1.81 ± 0.11 W; and k was 0.032 ± 0.006 min^{-1}. Eggs exposed to 15°C for 60 min exhibited decreases in temperature and in SHL with a high degree of correlation with the regression curve (R^2 = 0.98). The calculated G was 0.1 ± 0.01 W/°C (Equation 4). The total SHL from eggs, according to Equation 6, with the extreme values of q_1 and k were 512 ± 66 and 718 ± 126 cal for cold exposures of 30 and 60 min, respectively. These findings are of great importance in the application of cooling of fertile eggs during incubation.

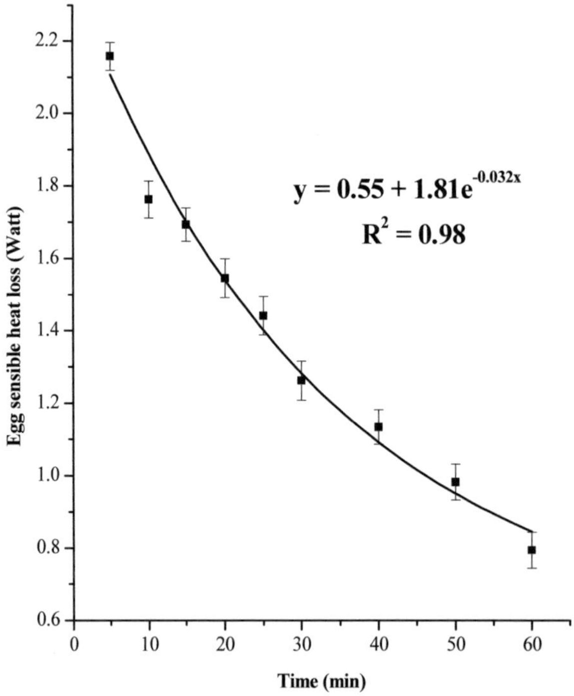

Fig. 7. Sensible heat loss from eggs by convection and radiation during 60 min of cold exposure (15°C) on d 19 of incubation (E19), fitted to an exponential regression curve (Mean ± SE; $n = 8$; $P < 0.05$) (According to Shinder et al., 2009).

4. Convective and radiative heat transfer from fowls: Heat transfer relationships – The rationale, the model and the applications

4.1 The rationale

It has been assumed that sensible heat loss does not play an important role in the domestic fowl when T_a is above the upper boundary of the thermoneutral zone (for review see Hillman et al., 1985). This assumption was based on: a. the small difference between the body surface temperature (T_s) and T_a; and b. the fact that in fully feathered birds only limited areas are unfeathered, i.e., legs, head, wattle and comb. However, use of the IRTI camera for non-invasive surface-temperature measurements has shown that sensible heat loss can amount to approximately 45% of the total heat loss under specific environmental conditions (Yahav et al., 2004) and thus can reduce the deleterious effects of panting.

4.2 The model (according to Yahav et al., 2005)

To calculate heat transfer from the fowl, each part of the surface is represented by a corresponding geometrical shape. For each part, radiative and convective heat transfers are estimated by means of available and especially developed heat transfer relationships. Below is a short introduction on convective and radiative heat transfer, followed by a detailed discussion on the relationships used.

4.2.1 Convective heat transfer

Heat is transferred by convection when a body at a given temperature is in contact with air at another temperature. The convective heat flux, q_c, depends on the temperature difference ΔT, between the body and the air, the area of contact A, and the heat-transfer coefficient h:

$$q_c = hA\Delta T \tag{7}$$

The average heat transfer coefficient, h, depends on the geometry of the body, the physical properties of the air and the flow regime. The major difficulty in calculating q_c stems from the strong dependence of h on the flow regime.

The heat transfer coefficient, h, is expressed through the nondimensional group known as the Nusselt number:

$$Nu = \frac{hD}{k} \tag{8}$$

where D is the body length scale, e.g., diameter in the case of a sphere or a cylinder, and k is the thermal conductivity of the air. Heat-transfer relationships given in the literature relate the Nusselt number to two other nondimensional groups. The first is the Reynolds number:

$$Re = \frac{UD}{\nu} \tag{9}$$

where U is the air velocity and ν is the kinematic viscosity. The second is the Prandtl number:

$$Pr = \frac{\nu}{\kappa} \tag{10}$$

where κ is the thermal diffusivity of the air. The three groups are related, in general, by:

$$Nu = f(Pr, Re) \tag{11}$$

where the function f is specific to each geometry and flow regime.

Below we present, for each part of the fowl surface, the appropriate relationship in the form of Eq. (11). This enables calculation of h and heat transfer for each part by using Eq. (7) and thence the total convective heat transfer from the bird.

4.2.2 Radiative heat transfer

Heat transfer by radiation occurs through electromagnetic radiation from one surface to another because of the temperature difference between them. The rate of radiative heat transfer between two surfaces depends on their temperatures, the view area, and the respective surface emissivities.

Radiative heat transfer between the fowl and its environment and among adjacent organs, e.g., the two legs, of the fowl itself occurs if temperatures are different. The view area changes frequently because of movement of the bird. In our model it is assumed that radiative heat transfer takes place only between the fowl and its environment, and we neglect radiation among the bird's surface parts. We also treat the environment as a large surface at uniform temperature that surrounds the relatively small bird.

Radiative heat flux from (or to) the animal is estimated as:

$$q_r = \varepsilon_1 \sigma A_1 (T_1^4 - T_2^4) \tag{12}$$

where subscript *r* stands for radiation, indices 1 and 2 represent, respectively, the body surface and the environment, ε (= 0.96) is the emissivity of a biological tissue, σ is the Stefan-Boltzmann constant (= 5.669×10^{-8} $Wm^{-2}K^{-4}$), *A* is the surface area and *T* is the absolute temperature.

4.2.3 Convective heat transfer from different parts of the bird surface

The present simple model is based on the following two assumptions:

- The bird is at rest, facing the upstream airflow direction. In practice the bird is frequently in motion, but this is not considered in the model.
- Heat transfer is by forced convection because the wind speed is rarely low enough for free convection to occur. This was validated for the air speeds associated with the present study.

The IR thermal imaging system measured the surface temperature of each surface part as well as the ambient temperature. The areas of the surface parts also were estimated from the thermal images. As mentioned above, to estimate the heat transfer by convection, the coefficient *h* was estimated for each surface part by using the following relationships:

The comb was simulated as a rectangular flat plate exposed to uniform flow parallel to its long side. The corresponding heat-transfer relationship was that for forced convection from a flat plate. According to the experimental air speed and dimensions of the comb the flow was laminar, therefore (Holman, 1989):

$$Nu = 0.664 Re^{1/2} Pr^{1/3} \tag{13}$$

where the length scale of the nondimensional parameters Re and Nu is the comb length. The heat-transfer coefficient is calculated by using Eqs. (8) and (13), and the convective heat transfer is estimated by using Eq. (7) where *A* is the total area of the two sides of the comb, taken as two equal rectangular surfaces.

Each wattle is modeled as a circular flat plate immersed in a uniform flow parallel to its surface. Our literature search for a relationship for such a configuration was unsuccessful. Therefore the required relationship was derived from that for a rectangular flat plate (Eq. (13) above). In particular, the surface of the circular plate was simulated as an infinite number of narrow rectangular plates, each of different length. For each thin plate Eq. (13) was applied and the average heat transfer coefficient over the circular plate was obtained by integration:

$$Nu = 0.7389 Re^{1/2} Pr^{1/3} \tag{14}$$

The characteristic length scale was the wattle diameter and the area was the total for four sides of the two wattles.

The bird's face is modeled the same way as the wattles, i.e., as a flat circular surface exposed to a parallel uniform airflow. Thus Eq. (14) is applied, with the effective area being the two sides of the face. In some adult birds it was evident that each face surface was closer to a semicircle than a full one: in these cases, the same relationship was used (i.e., Eq. (14)) with the adjusted heat transfer area applied in Eq. (7).

The leg is modeled as a circular cylinder in uniform cross flow. Many empirical heat transfer relationships for this configuration can be found in the literature. In the one chosen here, the relationship depends on the value of Re (Holman, 1989):

$$\mathrm{Nu} = (0.43 + 0.5\mathrm{Re}^{1/2})\mathrm{Pr}^{0.38} \qquad (\text{for } 1 < \mathrm{Re} < 1000) \qquad (15)$$

$$\mathrm{Nu} = 0.25\mathrm{Re}^{0.6}\,\mathrm{Pr}^{0.38} \qquad (\text{for } 1000 < \mathrm{Re} < 200{,}000) \qquad (16)$$

The length scale in the expressions for Nu and Re is the diameter of the extremity and the area is its surface area.

The toes are also modeled as circular cylinders in uniform cross flow. We assume that for all toes the flow is perpendicular to the cylinder axis. Therefore Eq. (15) or (16) is applied but, because of the small diameter of the toes, and the consequently small Reynolds number, Eq. (15) was usually used. In calculating the heat flux, eight equal-area toes were assumed, although in practice the toes' areas are not identical.

The bird's neck also is modeled as a cylinder in a uniform cross flow. The heat-transfer coefficient is calculated by using either Eq. (15) or Eq. (16), depending on the value of Re.

The body of the fowl is modeled as a sphere immersed in a uniform flow. We assume that the limbs are always in contact with the body, although sometimes they are not; the effect of separated wings is discussed below. The relationship for air at room temperature (Pr = 0.71) is given by Holman (1989):

$$\mathrm{Nu} = 2 + (0.25 + 3\times10^{-4}\,\mathrm{Re}^{1.6})^{1/2} \qquad (\text{for } 100 < \mathrm{Re} < 300{,}000) \qquad (17)$$

The bird separates its **wings** from its body under extremely warm conditions, thus forming two channels through which airflow enhances cooling. Assuming that the two channels are rectangular, we apply the Petukhov equation (Ozisik, 1985):

$$\mathrm{Nu} = \frac{\mathrm{Re}\,\mathrm{Pr}}{8X}f \qquad (\text{for } 10^4 < \mathrm{Re} < 5 \times 10^6) \qquad (18)$$

where

$$X = 1.07 + 12.7(\mathrm{Pr}^{2/3} - 1)\left(\frac{f}{8}\right)^{1/2} \qquad (19)$$

and f is the friction factor (= 0.07) which can be evaluated from the Moody diagram. The Reynolds number is based on the hydraulic diameter, $D_h = 4A_c / P$, in which A_c is the cross-sectional area of the duct and P is its perimeter. The airspeed substituted in Re was half the ambient air speed surrounding the bird. This roughly accounted for the reduction in air speed caused by the high frictional surface of the wing. The resulting Re $\approx 3 \times 10^3$, is smaller than the lower validity limit of Eq. (12). Nevertheless this equation was used in the model to provide a rough estimate of the convective heat flux through the wing channels.

4.2.4 Radiative heat transfer from different parts of the body surface

In using the above assumptions, each part of the body surface is considered to be a small body surrounded by a large, uniform-temperature environment. The radiative heat flux is calculated by using Eq. (12) with A_1 as the surface area of the body part. It is assumed that there is no radiative heat transfer associated with the channels formed by the separated legs.

4.3 The applications

The vasomotor response of the peripheral blood system that result from changes in environmental conditions (temperature, ventilation, relative humidity) can be evaluated non-invasively by IRTI. The vasomotor response immediately affects skin temperature, which increases or decreases as a result of vasodilation or vasoconstriction, respectively. In many biological studies the effects of various treatments on vasomotor responses is of great importance, both scientifically and practically. These two aspects are discussed below.

4.3.1 Evaluating vasomotor response in domestic fowl: Vasoconstriction vs vasodilation

The epigenetic temperature adaptation approach is based on the hypothesis that thermal manipulations during embryogenesis may affect the PO/AH threshold for heat loss, i.e., they may affect heat loss by radiation and convection; in other words, they may change the vasomotor response.

Based on this hypothesis, Piestun et al. (2008, 2009), elevated the incubation temperature between embryogenesis (E) days E7 to E16 for 12 h/d (12H) or continuously (24H) to establish a better post-hatch vasodilatation response of the peripheral blood system.

This phenomenon could be evaluated by measuring the temperatures of the chicken surface and of its surroundings (Figure 8). This information was used to calculate sensible heat loss and indirectly to determine the vasomotor efficacy.

Figure 8 illustrates the efficacy of the vasomotor response of those chickens that had been thermally manipulated during embryogenesis. Under both high and relatively low T_a conditions the manipulated broilers exhibited significantly better vasodilatation (35°C) or vasoconstriction (23°C) than controls, as highlighted by the significantly higher and lower heat losses at the low and high T_a, respectively.

4.3.2 Correlation between body temperature and facial surface temperature

The commonly used ways to measure T_b are invasive and time consuming and, therefore, may be traumatic, affecting T_b within a few minutes (Cabanac & Aizawa, 2000; Cabanac & Guillemette, 2001), but the adoption of IRTI in biological sciences has enabled non-invasive, non-contact measurement of surface temperature.

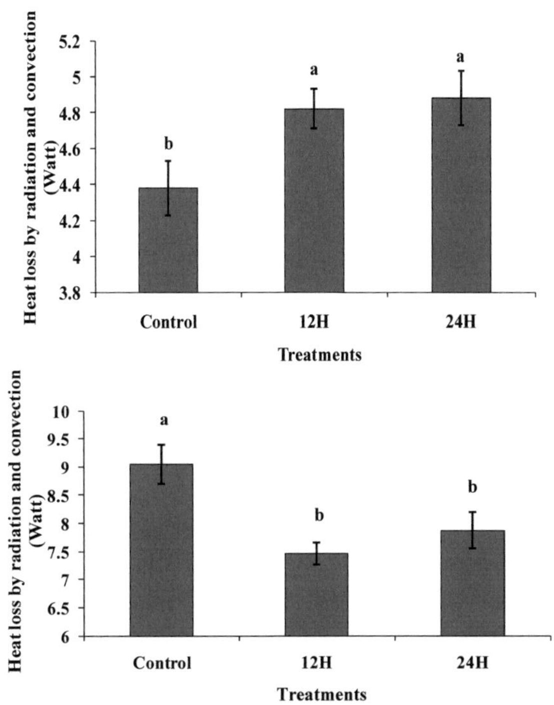

Fig. 8. The effects of two different ambient temperatures (35°C - upper; 23°C - lower) on heat loss by radiation and convection from broiler chickens that had been thermally manipulated during embryogenesis (embryogenesis days 7 to 16) for 12 h/d -12H - or continuously - 24H - in order to change their vasomotor responses.

However, use of IRT measurement of skin surface temperature in monitoring the thermal status of chickens in a commercial flock necessitates selection of a specific surface site, and determination of the exact correlation of its temperature with T_b under various environmental conditions. A reliable correlation between facial surface temperature and T_b could shed light on the thermal-physiological status of the organism. This application, in which facial surface temperature measured by IRTI can serve as an important indicator of the thermal-physiological status of the fowl, could dramatically improve poultry house environmental control by adapting it to the needs of the client - the domestic fowl.

Thermographic imaging in poultry breeding can be relevant only if a satisfactory correlation exists between the chickens' body temperature and facial surface temperature. Our studies demonstrated that such a correlation can indeed be found. In the laboratory, where the climatic conditions were relatively similar for all the chickens, individual body temperatures correlated well with individual facial surface temperatures for chickens that were exposed to acute environmental heat conditions, both with and without ventilation, and to persistent heat without ventilation (Figure 9). Moreover, if a distinction was made between acute and persistent heat exposure, body and facial temperatures correlated remarkably better than ambient temperature with hormonal stress symptoms (Figure 10).

The hormonal analysis was crucial for correlating temperature data with the changes in the physiological status of the chicken that resulted from exposure to acute or persistent heat.

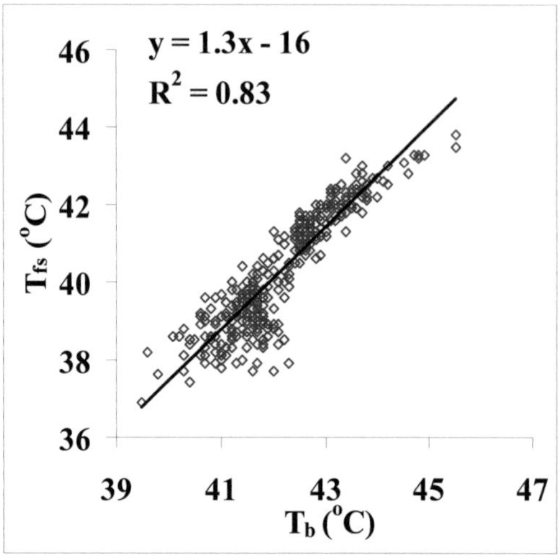

Fig. 9. Correlation of individual Facial Temperature (T_{fs}) with Body Temperature (T_b). Chickens were reared under controlled conditions, and on each of days 8, 15, 22, 29 and 36 a group was randomly chosen for exposure to three different heat treatments: acute heat only; acute heat together with ventilation; and persistent heat. In each treatment 27 chickens were used: 18 chickens in each treatment were used for measuring the response of hormone concentrations to the treatments; the remaining nine were used to monitor the responses of body and facial temperatures. The data presented in the graph are for all measurements of individual chickens in all treatments and of all ages. The linear regression curve is a least-squares curve calculated by Excel.

Four major hormones were selected: thyroxin (T_4), triiodothyronine (T_3), corticosterone and arginine vasotocine (AVT). The first two of these relate directly to the metabolic status of the chicken, whereas corticosterone is a well known indicator for stress level, and AVT for hydration status.

The correlations of the plasma concentrations of T_4 and T_3 with body, facial surface and ambient temperature during persistent exposure to heat are presented in Figure 10, as examples. Table 1 summarizes the statistical analysis and the R^2 values for correlations between T_4, T_3, corticosterone and AVT concentrations and the various temperature parameters - body, facial and ambient. In acute treatments, with or without ventilation, the corticosterone concentration displayed a statistically significant positive correlation and triiodothyronine a significant negative correlation with body and facial temperature ($p < 0.05$), whereas the regression with ambient temperature was not significant. The responses of these hormones were found to be consistent with previously described physiological reactions to acute heat stress (Darras et al., 1996; Iqbal et al., 1990).

During exposure to persistent heat, significant negative regressions were found for the correlations of T_4 and T_3 with body and facial temperatures. Only T_3 displayed a significant regression with ambient temperature also, although this regression was somewhat weaker (R^2 lower) than that with physiological temperatures. AVT showed a strong negative correlation with facial temperature, a weaker correlation with ambient temperature, and no correlation with body temperature.

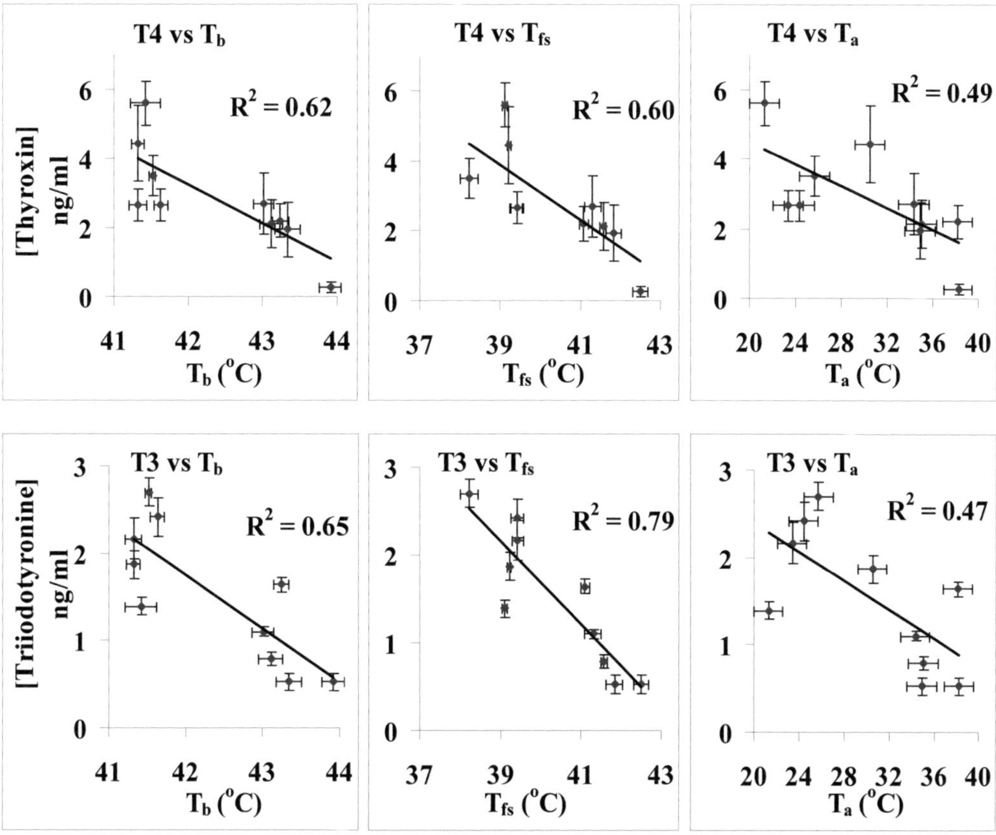

Fig. 10. Correlation of blood plasma concentrations of thyroid hormones - Thyroxine (T_4) and Triiodothyronine (T_3) - with three different temperature parameters during persistent exposure to heat. Data are from the experiment described in the legend to Figure 8. Each data point represents the average for one group of chickens aged between 7 and 35 days, during one measurement ($n = 9$). T_4, Thyroxin; T_3, Triiodothyronine; T_b, Body temperature; T_{fs}, Facial Surface temperature as measured with IRTI; T_a, Ambient temperature. The linear regression curves are least-squares curves calculated by Excel. Error bars are ± s.e.

One can conclude from the above results that in the laboratory setting facial temperature, as measured by thermographic imaging was more closely related than the ambient temperature to parameters that indicate the birds' thermal and stress status. It further can be concluded that the high positive correlation between facial and body temperature can be exploited commercially to detect thermal stress in the birds without use of invasive measurements.

4.3.3 Field observations

In commercial poultry houses it is well-known that environmental conditions are far from homogenous, even when a building is equipped with a climate-control system. It was therefore not surprising that in field studies the variation in the correlation between T_b and facial surface temperature was larger than in the laboratory. However, even for individual

		Chronic heat			Acute heat		
		T_b	T_{fs}	T_a	T_b	T_{fs}	T_a
Corticosterone	R^2	0.01	0.00	0.07	0.33	0.29	0.12
	p	0.79	0.97	0.49	**0.03**	**0.04**	0.21
Thyroxin	R^2	**0.62**	**0.60**	0.49	0.05	0.05	0.07
	p	**0.02**	**0.02**	0.06	0.34	0.35	0.26
Triiodothyronine	R^2	**0.65**	**0.79**	0.47	0.34	0.40	0.17
	p	**0.00**	**0.00**	**0.03**	**0.02**	**0.01**	0.13
Arginine vasotocin	R^2	0.36	**0.51**	0.16	0.01	0.05	0.02
	p	0.06	**0.02**	**0.03**	0.72	0.43	0.65

R^2 values larger than 0.50 and *p* values smaller than 0.05 are printed in bold.

Table 1. R^2 values, and significance probability (*p*) for the correlations of average hormone concentrations with body temperature (T_b), facial surface temperature (T_{fs}), and ambient temperature (T_a).

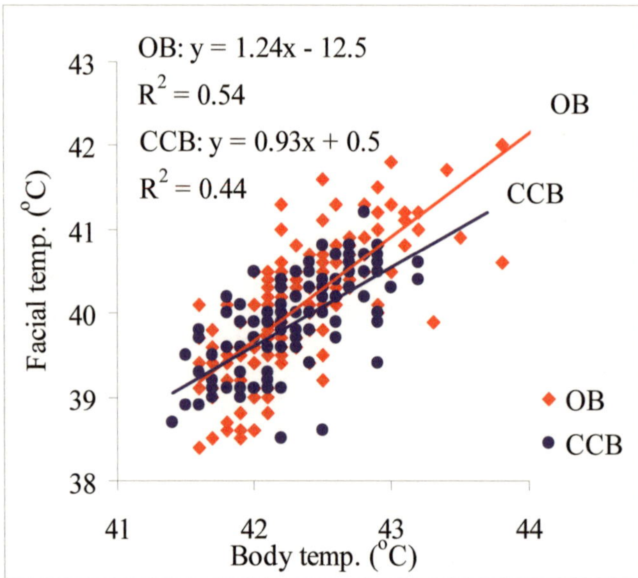

Fig. 11. Correlation between individual body and facial temperatures in two commercial buildings. Two commercial poultry flocks in Kibbutz Lavi were monitored during 36 h. The flocks were in two different buildings: a closed climate-controlled building (CCB) and an open building (OB). Measurements were performed in "stations" in which 50–70 randomly selected chickens were confined in a detachable cage, and were then immediately caught one by one, and their body and facial temperatures were measured simultaneously. When behavioral stress became evident the procedure was interrupted, generally after 10–20 chickens had been measured. Three measuring sessions were performed: starting at midday when the chickens were 30 days old, the following midnight, and the following midday. In each measuring session three to five stations were surveyed in each flock.

chickens a certain correlation was found in the field measurements (Figure 11). The correlation between the two temperatures (face and body) was clearly revealed when averages of facial temperature for each "measurements station" were plotted against the corresponding average of T_b (Figure 12). Although an average of only 15 chickens were recorded for body and facial temperatures in each station, a strong linear regression could be drawn between the averages.

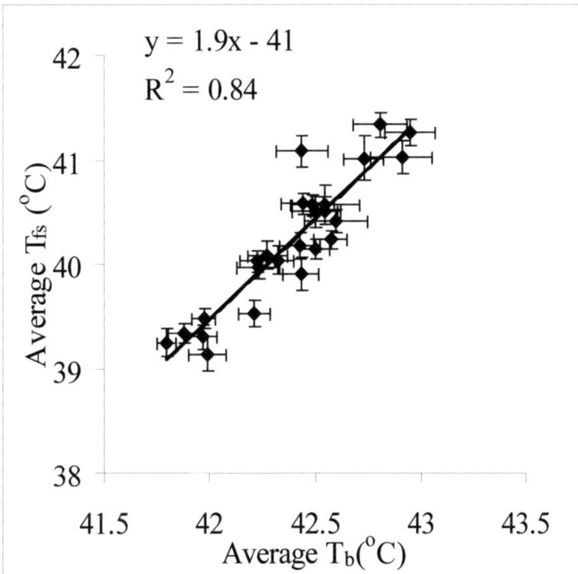

Fig. 12. Correlation between sub-population averages of body and facial temperatures in two commercial buildings. The experiment is described in the legend to Figure 11. Data are averages of body and facial temperatures that were calculated for each measuring station under identical environmental conditions. Regression for both of the buildings together is displayed. Error bars are ± s.e.

4.3.4 The possibility of reducing costs by using low spatial resolution

The algorithm discussed in this Chapter - for calculating the facial temperature of chickens from the thermal data of a thermographic image - was based on locating the maximal temperature within a manually defined circular region of interest (ROI) that comprised the relatively unfeathered area around the ear and eye on the chicken's face. This temperature - termed the "hottest spot temperature" - was preferred over the calculated average temperature of the chicken's face because of ambiguity in defining the face. However, the notion of a "spot" is misleading because of the limited spatial resolution of the images. The camera was equipped with a sensor array of 320 × 240 pixels. Using a lens with a 24-degree field of view, and placing the camera 60–80 cm from the object (chicken) resulted in images in which each pixel corresponded to a rectangular area of 0.9–1.5 mm^2. The "hottest spot" is therefore equivalent to the one among an array of such areas that had the highest average temperature.

Analysis of the economic feasibility of hypothetical commercial applications of IRTI in poultry breeding clearly showed that including IR cameras that cost tens of thousands of

American dollars in a flock-monitoring system was not a realistic option. It was therefore clear that for commercial applications one must compromise in spatial resolution, which today is one of the main factors that determine the prices of cameras. The question is, therefore: What is the smallest detail that has to be identified in a thermographic image in order to maintain the strong correlation between "hottest spot" temperature and body temperature? In order to check the sensitivity of this correlation to spatial resolution we re-analyzed a few experiments, simulating the "hottest spot" algorithm for the same images with lower spatial resolution. An array of rectangular ROIs was created in the image, to represent part of a (44 × 40)-pixel array, within the (320 × 240)-pixel field of view of the camera. In this configuration each pixel corresponded to about 34 mm^2 for an object located 60 cm from the camera.

The average temperature of each region of interest was listed, and the highest average was chosen as a "low-resolution hottest spot" temperature. Facial temperatures determined in this low-resolution algorithm and those determined with the high-resolution algorithm were then plotted against body temperature (Figure 13). As expected, low-resolution "hottest spot" temperatures were somewhat lower than high-resolution "hot spot" temperatures because, in the low-resolution configuration peak temperatures were blurred by averaging over a larger area.

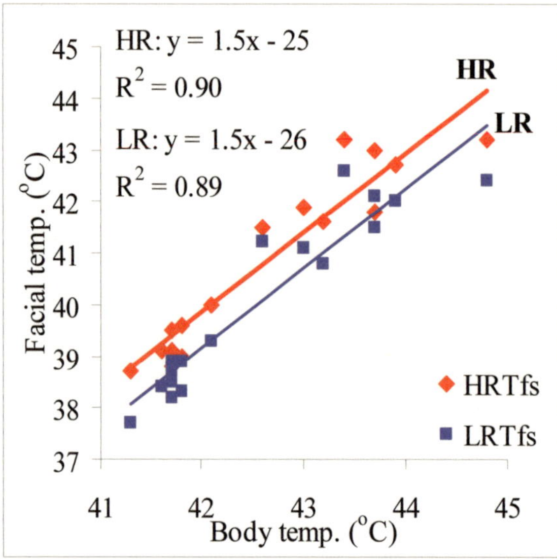

Fig. 13. Typical experimental correlation between facial temperature and body temperature, calculated with a high-resolution algorithm (320 × 240 pixels) and a low-resolution algorithm (40 × 44 pixels). HR, High resolution; LR, Low resolution; Tfs, facial temperature. Regression lines are least squares fit linear regressions computed with Excel.

However, surprisingly, the regression (R^2) between T_b and facial surface temperature was not sensitive to spatial resolution. In the example shown in Figure 13 R^2 values were practically identical: 0.90 for high resolution and 0.89 for low resolution. Three other experiments were analyzed in this cumbersome way, with similar results (not shown). One can conclude that future commercial applications of IRTI in poultry breeding will be able to use low-resolution equipment, thereby reducing costs significantly without impairing the accuracy of the results.

5. Overall conclusions

- The IRTI camera is an efficient and accurate means of measuring the body surface temperature of domestic fowl as a basis for calculating heat loss by radiation and convection and for evaluating the vasomotor response of fowl subjected to varied environmental conditions.
- This technique enables determination of the rate of heat loss from a fertile egg during thermal manipulations that impart epigenetic temperature adaptation to the manipulated embryo.
- Use of The IRTI technique facilitates the correlation between body temperature and facial surface temperature to assess the physiological status of the chicken under laboratory or commercial environmental conditions.
- The possibility to assess changes in the thermal status of a commercial flock through non-contact measurements can potentially be utilized for improving climate-control systems and spotting situations of acute thermal stress.
- Future commercial applications of this technique could, therefore, contribute to improving both performance and welfare of the flock, and thereby enable the poultry industry to meet future challenges posed by global climate change and the need for cheap sources of meat for the world's growing population.

6. Acknowledgements

Thanks are due to: Dmitry Shinder, The Volcani Center for technical assistance; Jens Vogt, Infratec, Dresden, Germany, for advice and guidance during the initial stages of the project; the teams of RDT and Asio Vision, Tel-Aviv, Israel, for generously lending equipment when needed; and the poultry breeding team in Kibbutz Lavi, Israel, for assisting with observations in the commercial farm.

7. References

Boulant, J. A. (1996). Hypothalamic neurons regulating body temperature, In: *Handbook of Physiology*. Section 4: Environmental physiology. Fregly, M. J. & Blatteis, C. M. Eds.), pp. 105-126. APS, Oxford University Press, New York.

Cabanac, M. & Aizawa, S. (2000). Fever and tachycardia in a bird (*Gallus domesticus*) after simple handling. *Physiol. Behav.*, Vol. 69, pp. 541–545.

Cabanac, A. J. & Guillemette, M. (2001). Temperature and heart rate as stress indicators of handled common eider. *Physiol. Behav.*, Vol. 74, pp. 475– 479.

Cetingul, M. P. & Herman, C. (2010). A heat transfer model of skin tissue for the detection of lesions: sensitivity analysis. *Phys. Med. Biol.*, Vol. 55, pp. 5933–5951.

Chiu, W. T., Lin, P. W., Chiou, H. Y., Lee, W. S., Lee, C. N., Yang, Y. Y., Lee, H. M., Hsieh, M. S., Hu, C. J., Ho, Y. S., Deng, W. P. & Hsu, C. Y. (2005). Infrared thermography to mass-screen suspected SARS patients with fever. *Asia Pac. J. Public Health*, Vol. 17, pp. 26-28.

Collin, A., Berri, C., Tesseraud, S., Rodón, F. E., Skiba-Cassy, S., Crochet, S., Duclos, M. J., Rideau, N., Tona, K., Buyse, J., Bruggeman, V., Decuypere, E., Picard, M. & Yahav, S. (2007). Effects of thermal manipulation during early and late embryogenesis on thermotolerance and breast muscle characteristics in broiler chickens. *Poultry Science,* Vol. 86, pp. 795–800.

Darras, V. M., Kotanen, S. P., Geris, K. L., Berghman, L. R. & Kühn, E. R. (1996). Plasma thyroid hormone levels and iodothyronine deiodinase activity following an acute glucocorticoid challenge in embryonic compared with posthatch chickens. *Gen. Comp. Endocrinol.,* Vol. 104, pp. 203–212.

Dawson, W. R. & Whittow, G. C. (2000). Regulation in body temperature, In: *Sturkie Avian Physiology.* Whittow, G. C. (Ed.). pp. 343-390. Academic Press, San Diego, CA, USA.

Druyan, S., Piestun, Y. & Yahav, S. (2011). Heat stress in domestic fowl - genetic and physiological aspects, In: *Body Temperature Control.* Cisneros, A. B. & Goins, B. L. (Eds.) Nova Science Publishers Inc. New York (in press).

FLIR Systems (2004). *ThermaCAM P65 Operators Manual.* FLIR Systems Publications Vol. 1, pp. 557-954.

Gordon, C. J. (1993). *Temperature Regulation in Laboratory Rodents,* New York, Cambridge University Press.

Hammel, H. T. (1956). Infrared emissivities of some arctic fauna. *J. Mammal.,* Vol. 37, pp. 375–378.

Herman, C. & Cetingul, M. P. (2011). Quantitative visualization of skin cancer using dynamic thermal imaging. *J. Vis. Exp.,* Vol. 5, p. 2679 (abstract).

Hillman, P. E., Scott, N. R. & Van Tienhoven, A. (1985). Physiological responses and adaptations to hot and cold environments, In: Stress *Physiology in Livestock* Yousef, M. K. (Ed.), pp. 27-71, CRC Press, Boca Raton, FL.

Holman, J. P. (1989). *Heat Transfer,* pp. 231, 292, 295. McGraw-Hill, Singapore.

Iqbal, A., Decuypere, I. E., Abd El Azim, A. & Kühn, E. R. (1990). Pre- and post-hatch high temperature exposure affects the thyroid hormones and corticosterone response to acute heat stress in growing chicken (*Gallus domesticus*). *J. Therm. Biol.,* Vol. 15, No. 2, pp. 149–153.

Klir, J. J. & Heath, J. E. (1992). An infrared thermographic study of surface temperature in relation to thermal stress in three species of foxes: the red fox (*Vulpes vulpes*), arctic fox (*Alopex lagopus*), and kit fox (*Vulpes macrotis*). *Physiol. Zool.,* Vol. 65, pp. 1011–1021.

Klir, J. J., Heath, J. E. & Benanni, N. (1990). An infrared thermographic study of surface temperature in relation to thermal stress in the Mongolian gerbil, *Meriones unguiculatus. Comp. Biochem. Physiol.,* Vol. 96A, pp. 141–146.

Kontos, M., Wilson, R. & Fentiman, I. (2011). Digital infrared thermal imaging (DITI) of breast lesions: sensitivity and specificity of detection of primary breast cancer. *Clin. Radiol.,* Vol. 66, pp. 536–539.

Marder, J. & Arad, Z. (1989). Panting and acid-base regulation in heat stressed birds. *Comp. Biochem. Physiol.,* Vol. 94, pp. 395–400.

McCafferty, D. J., Gilbert, C., Paterson, W., Pomeroy, P. P., Thompson, D., Currie, J. I. & Ancel, A. (2011). Estimating metabolic heat loss by combining infrared thermography with biophysical modeling. *Comp. Biochem. Physiol. A.,* Vol. 158, pp. 337–345.

Minne, B. & Decuypere, E. (1984). Effects of late prenatal temperatures on some thermoregulatory aspects in young chickens. *Arch. Exp. Vet.,* Vol. 38, pp. 374–383.

Mohler, F. S. & Heath, J. E. (1988). Comparison of IR thermography and thermocouple measurement of heat loss from rabbit pinna. *Am. J. Physiol.,* Vol. 254, pp. 389–395.

Monteith, J. L. & Unsworth, M. H. (1990). *Principals of Environmental Physics*, 291 pp., Edward Arnold, London.

Morrison, S. F., Nakamura, K. & Madden, C. J. (2008). Central control of thermogenesis in mammals. *Experimental Physiology*, Vol. 93, pp. 773-797.

Nguyen, A. V., Cohen, N. J., Lipman, H., Brown, C. M., Molinari, N. A., Jackson, W. L., Kirking, H., Szymanowski, P., Wilson, T. W., Salhi, B. A., Roberts, R. R., Stryker, D. W. & Fishbein, D. B. (2010). Comparison of 3 infrared thermal detection systems and self-report for mass fever screening. *Emerg. Infect. Dis.*, Vol. 16, No. 11, pp. 1710-1717.

Nichelmann, M. (2004). Activation of thermoregulatory control elements in precocial birds during the prenatal period. *J. Therm. Biol.*, Vol. 29, pp. 621-627.

Ophir, E., Arieli, Y., Marder, J. & Horowitz, M. (2002). Cutaneous blood flow in the pigeon *Columba livia*: its possible relevance to cutaneous water evaporation. *J. Exp. Biol.*, Vol. 205, pp. 2627-2636.

Ozisik, M. N. (1985). Heat Transfer - A Basic Approach. McGraw-Hill, Singapore.

Phillips, P. K. & Heath, J. E. (1992). Heat loss by the pinna of the African elephant (*Loxodonta africana*). *Comp. Biochem. Physiol.*, Vol. 101A, pp. 693-699.

Phillips, P. K. & Sanborn, A. F. (1994). An infrared, thermographic study of surface temperature in three ratites: ostrich, emu and double-wattled cassowary. *J. Therm. Biol.*, Vol. 19, pp. 423-430.

Piestun, Y., Shinder, D., Ruzal, M., Halevy, O., Brake, J. & Yahav, S. (2008). Thermal manipulations during broiler embryogenesis: effect on the acquisition of thermotolerance. *Poultry Science*, Vol. 87, pp. 1516-1525.

Piestun, Y., Halevy, O. & Yahav, S. (2009). Thermal manipulations of broiler embryos - the effect on thermoregulation and development during embryogenesis. *Poultry Science*, Vol. 88, pp. 2677-2688.

Richards, M. P. & Proszkowiec-Weglarz, M. (2007). Mechanisms regulating feed intake, energy expenditure, and body weight in poultry. *Poultry Science*, Vol. 86, pp. 1478-1490.

Richards, S. A. (1968). Vagal control of thermal panting in mammals and birds. *J. Physiol.*, Vol. 199, pp. 89-101.

Richards, S. A. (1970). The biology and comparative physiology of thermal panting. *Biol. Rev. Camb. Philos. Soc.*, Vol. 45, pp. 223-264.

Richards, S. A. (1976). Evaporative water loss in domestic fowls and its partition in relation to ambient temperature. *J. Agric. Sci.*, Vol. 87, pp. 527-532.

Seymour, R. S. (1972). Convective heat transfer in the respiratory systems of panting animals. J. Theor. Biol., Vol. 35, pp. 119-127.

Shinder, D., Luger, D., Rusal, M., Rzepakovsky, V., Bresler V. & Yahav S. (2002). Early age cold conditioning in broiler chickens (Gallus domesticus): thermoregulatory and growth responses. J. Therm. Biol., Vol. 27, pp. 517-523.

Shinder, D., Rusal, M., Tanny, Y., Druyan, D. & Yahav, S. (2007). Thermoregulatory response of chicks (*Gallus domesticus*) to low ambient temperatures at an early age. *Poultry Science*, Vol. 86, pp. 2000-2009.

Shinder, D., Rusal, M., Giloh, M. & Yahav, S. (2009). The effect of repetitive acute cold exposures at the latest phase of embryogenesis of broilers on cold resistance during life span. *Poultry Science*, Vol. 88, pp. 636-646.

Silva, J. E. (2006). Thermogenic mechanisms and their hormonal regulation. *Physiol. Rev.*, Vol. 86, pp. 435-464.

Stewart, M., Webster, J. R., Schaefer, A. L., Cook, N. J. & Scott, S. L. (2005). Infrared thermography as a non-invasive tool to study animal welfare. *Anim. Welf.*, Vol. 14, pp. 319–325.

Tazawa H, & Nakagawa S. (1985). Response of egg temperature, heart rate and blood pressure in the chick embryo to hypothermal stress. *J. Comp. Physiol. B*, Vol. 155, pp. 195–200.

Tazawa H., Moriya, K., Tamura, A., Komoro, T. & Akiyama, R. (2001). Ontogenetic study of thermoregulation in birds. *J. Therm. Biol.*, Vol. 26, pp. 281–286.

Teunissen, L. P. J. & Daanen, H. A. M. (2011). Infrared thermal imaging of the inner canthus of the eye as an estimator of body core temperature. *J. Med. Eng. Technol.*, Vol. 35, No. 3-4, pp. 134–138.

Tzschentke, B. (2007). Attainment of thermoregulation as affected by environmental factors. *Poultry Science*, Vol. 86, pp. 1025–1036.

Tzschentke, B. & Halle, I. (2009). Influence of temperature stimulation during the last 4 days of incubation on secondary sex ratio and later performance in male and female broiler chicks. *British Poultry Science,* Vol. 50, pp. 634–640.

Tzschentke, B., & Nichelmann, M. (1997). Influence of prenatal and postnatal acclimation on nervous and peripheral thermoregulation. *Ann. N. Y. Acad. Sci.*, Vol. 813, pp. 87–94.

Tzschentke, B. & Plagemann, A. (2006). Imprinting and critical periods in early development. *World Poult. Sci. J.*, Vol. 62, pp. 626–638.

van Brecht, A., Hens, H., Lemaire, J. L., Aerts, J. M., Degraeve, P. & Berckmans, D. (2005). Quantification of the heat exchange of chicken eggs. *Poultry Science,* Vol. 84, pp. 353–361.

Webster, M. D. & King, J. R. (1987). Temperature and humidity dynamics of cutaneous and respiratory evaporation in pigeons, *Columba livia. J. Comp. Physiol.*, Vol. 157, pp. 253–260.

Wolfenson, D., Bachrach, D., Maman, M., Graber, Y. & Rozenboim, I. (2001). Evaporative cooling of ventral regions of the skin in heat-stressed laying hens. *Poultry Science,* Vol. 80, pp. 958–964.

Yahav, S. (2009). Alleviating heat stress in domestic fowl - different strategies. *World Poult. Sci. J.*, Vol. 65, pp. 719–732.

Yahav, S., & Hurwitz, S. (1996). Induction of thermotolerance in male broiler chickens by temperature conditioning at an early age. *Poultry Science,* Vol. 75, pp. 402–406.

Yahav, S., Goldfeld, S., Plavnik, I. & Hurwitz, S. (1995). Physiological responses of chickens and turkeys to relative humidity during exposure to high ambient temperature. *J. Therm. Biol.*, Vol. 20, pp. 245–253.

Yahav, S., Luger, D., Cahaner, A., Dotan, M., Rusal, M. & Hurwitz. S. (1998). Thermoregulation in naked neck chickens subjected to different ambient temperatures. *British Poultry Science,* Vol. 39, pp. 133–138.

Yahav, S., Straschnow, A., Luger, D., Shinder, D., Tanny, J. & Cohen, S. (2004). Ventilation, sensible heat loss, broiler energy, and water balance under harsh environmental conditions. *Poultry Science,* Vol. 83, pp. 253–258.

Yahav, S., Shinder, D., Tanny, J. & Cohen, S. (2005). Sensible heat loss: the broiler's paradox. *World Poultry Science J.*, Vol. 61, pp. 419–434.

Yahav, S., Shinder, D., Ruzal, M., Gilo, M. & Piestun, Y. (2009). Controlling body temperature – the opportunities for highly productive domestic fowl, In: *Body Temperature Control.* Cisneros, A. B. &. Goins, B. L (Eds.), pp. 65-98, Nova Science Publishers, New York.

6

Thermographic Applications in Veterinary Medicine

Calogero Stelletta, Matteo Gianesella, Juri Vencato, Enrico Fiore and Massimo Morgante
Department of Animal Medicine, Production and Health, University of Padova Italy

1. Introduction

Thermography is a non-contact, non-invasive technique that detects surface heat emitted as infrared radiation. Because skin temperature reflects the status of tissue metabolism and blood circulation, abnormal thermal patterns can signify areas of superficial inflammation or circulatory impairments.

Veterinary infrared thermography is a term indicating *in vivo* digitally imaging an animal with an infrared camera using computer interpretation of thermal maps. Various trials were performed with different species (horse, pig and cows)[1,2,3,4,5] to assess the validity of thermographic instrument. Infrared thermographic systems are capable of seeing energy emitted by most objects at a temperature above -35°C. Therefore colour or visible light does not interfere with the possible images seen by thermographic system. The maximum heat emitter is considered a black body which have an emissivity of 1 because it adsorbs all radiated heat. The emission factor of skin is approximately 0.93-0.98 depending on coat quantity and length. Heat is the primary sign of inflammation process and different disease processes affect the microcirculation of the skin. Therefore variations of the skin temperature become interesting indicator of such conditions that can range from specific vascular alterations to referred conditions also physiologically. Since skin temperature may be used in order to estimate tissue integrity because it reflects the underlying circulation and tissue metabolism.

Infrared thermography (IRT) uses thermal radiation emitted by objects to visualize and measure their surface temperature. The temperature is detected over wide areas, at a distance and measuring time is fast. It does not require physical contact and, therefore, it is entirely non-invasive. The colours of the images represent different temperatures, highlighting hot and cold spots and showing the map of the thermal distribution of an object or a body surface. Thermal imaging cameras can produce very sharp images of the distribution of body surface temperatures to a precision of 0.08°C. These characteristics allow the IRT to be applied where the temperature of live animals or carcases is difficult to measure under housing conditions or in the situation occurring during commercial slaughter.

Since the 2001 it has been carried out numerous efforts to introduce the thermography in the veterinary clinical practice and below are reported some clinical applications and experimental approaches.

2. Veterinary clinical applications of the thermography

The thermographic applications in veterinary medicine are very numerous considering the difficulties that in some cases due to the characteristics of the patients. In the past time the major application was on the equine diagnostic procedures above all for the lameness. More recently different applications were on bovine medicine and particularly for the mastitis and the welfare evaluation. In animals, body surface temperature is a function of blood flow and metabolic rate of underlying tissues. Thus, the physiological state of underlying cells could potentially be assessed by measuring skin temperatureusing IRT[6]. Infrared thermography and potential veterinary applications for this imaging technique have been described[6,7,8]. These reports mostly described thermographic imaging of spontaneous disease and attempts to correlate images to disease or injury diagnosed by other means. A study in cattle revealed successful utilization of thermography in the detection of localized sepsis in the pinna after contaminated growth stimulant pellets had been administered[7]. Infrared thermography has been used to predict changes in udder temperature[9] and to detect inflammation associated with hot- iron and freeze branding in cattle[10] and bovine viral diarrhea infection in calves[11]. Soles of hooves affected by subclinical laminitis commonly appear soft and warm long before the appearance of yellowish discoloration, lesions, and ulcers[12]. Our experiences were based on different species (bovine, ovine, south american camelids, horse, dog) and with the main objective of diagnostic procedures standardization[13,14,15,16,17,18,19,20]. The most important problem that have to be consider is the specie-specific heat transfer equation which is influenced by numerous factors. The following paragraphs report our experimental approaches for the different species.

3. Experimental approaches to the use of the thermography in veterinary medicine

3.1 Effect of GnRH test on scrotal surface temperature in Alpaca

Treatment with gonatropin realising hormone (GnRH) increases blood concentrations of LH and FSH. LH and GnRH directly act on Leydig cells stimulating them to release testosterone from the testis. Testis have a local regulation due to its structural organization (avascular seminiferous tubules compartment and vascularised interstitial compartment) and to its organization and hormonal control of spermatogenesis. The regulation of the testis functionality is based on general and local information through hyphophyseal LH and GnRH. Increments of interstitial testosterone level following the local action of GnRH is generally faster then hypotalamic-hyphophyseal-gonadal way. High interstitial testosterone level might activate a mechanism of secretion based on the neural local regulation of the blood flow and muscular contraction.

Aim of this work was to evaluate variations in terms of testosteronemia and scrotal surface temperature (ST) during GnRH test. Five adult males (4 Huacaya, 1 Suri) were tested to evaluate their testicular functionality through the GnRH test (administration I.M. of 6-9 mcg of GnRH analogue buserelin/male). Trial 1 (T1): males were completely isolated from

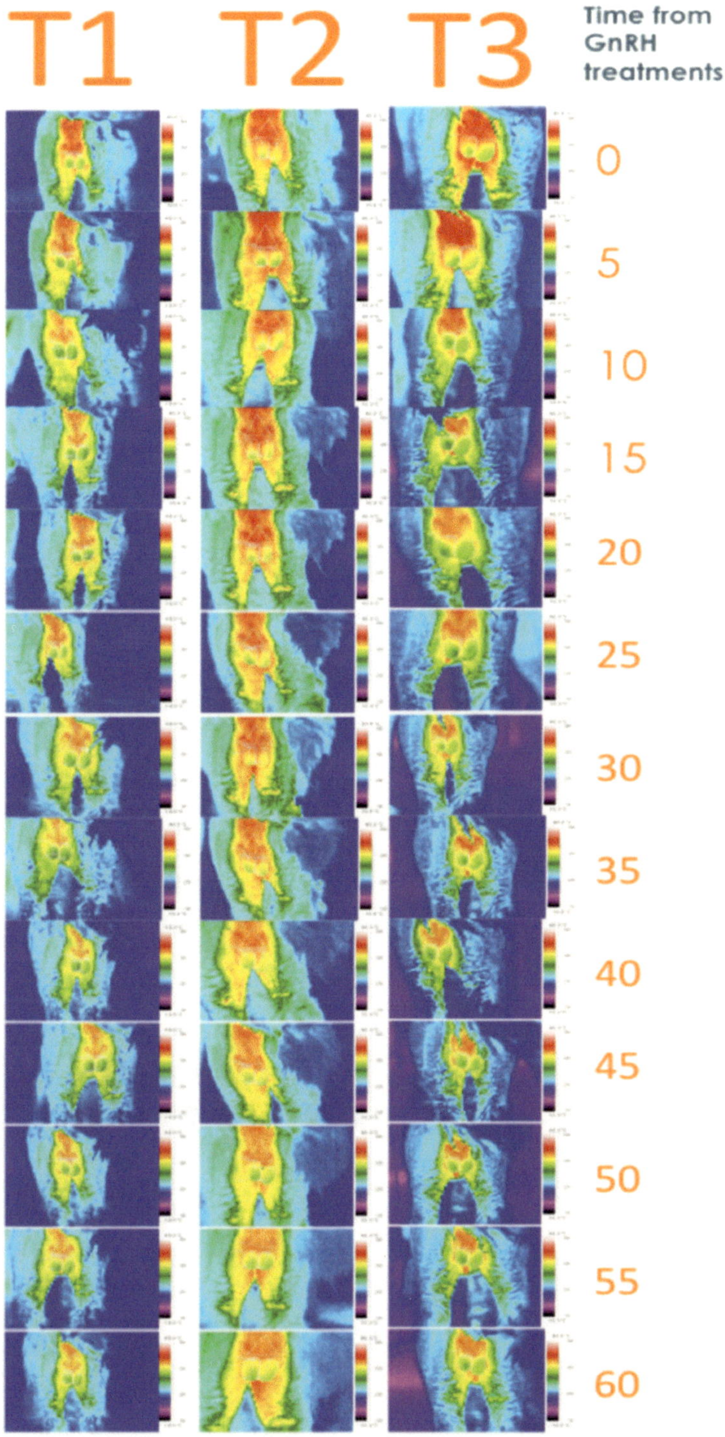

Fig. 1. Example of Alpaca's scrotum thermographic images during three trials. (T1: isolation from female; T2: exposure without mounts; T3: exposure with mounts)

females 2 months before. Trial 2 (T2): males were exposed to females without mounts (3 weeks; two times each week for 15 min). Trial 3 (T3): males were exposed to females with mounts (3 weeks; two times each week, 15-20 minutes). Thermometric measurements of scrotal surface were carried out using an infrared-camera (P25, Flir System) every 5 minutes during each GnRH test (Figure 1.). Thermografic images were analysed using a specific software (ThermaCam Researcher Basic 2.08, Flir System). Testosteronemia was determined using a chemiluminescent method previously validated. Data obtained were analysed through ANOVA for repeated measures, using the GLM procedure of the statistical software SIGMASTAT 2.03, taking into consideration the trial as independent variable while ST and testosteronemia as dependent variables; besides Pearson correlation coefficients were calculated for the variables considered.

GnRH administration influenced testosteronemia and scrotal temperature during the monitoring time (60 minutes). Comparisons among monitoring times are reported in Table 1.

Time after GnRH administration	Trial 1	Trial 2	Trial 3	S.E.M.
0	30.790	30.900	30.270	
5	29.460^{a}	30.970^{b}	29.620^{ac}	
10	29.130^{a}	30.800^{b}	29.350^{ac}	
15	29.080^{a}	30.700^{b}	29.540^{ac}	
20	28.870^{a}	30.490^{b}	29.720^{ac}	
25	28.700^{a}	30.620^{b}	29.420^{c}	
30	28.700^{a}	30.680^{b}	29.740^{c}	0.195
35	28.460^{a}	30.600^{b}	$29.250a^{c}$	
40	28.280^{a}	30.570^{b}	$29.190a^{c}$	
45	28.620^{a}	30.310^{b}	$29.410a^{c}$	
50	28.460^{a}	30.030^{b}	29.250^{b}	
55	28.480^{a}	30.260^{b}	29.290^{ac}	
60	28.940^{a}	30.250^{b}	29.330^{ac}	

Table 1. Variation of scrotal temperature during the three GnRH tests.

Frequency variations of temperature ranges are reported in graphics 1, 2 and 3 for 1st, 2nd and 3rd test respectively. ST decrease during the monitoring period (60 minutes) after GnRH administration with the higher variation before the female exposure.

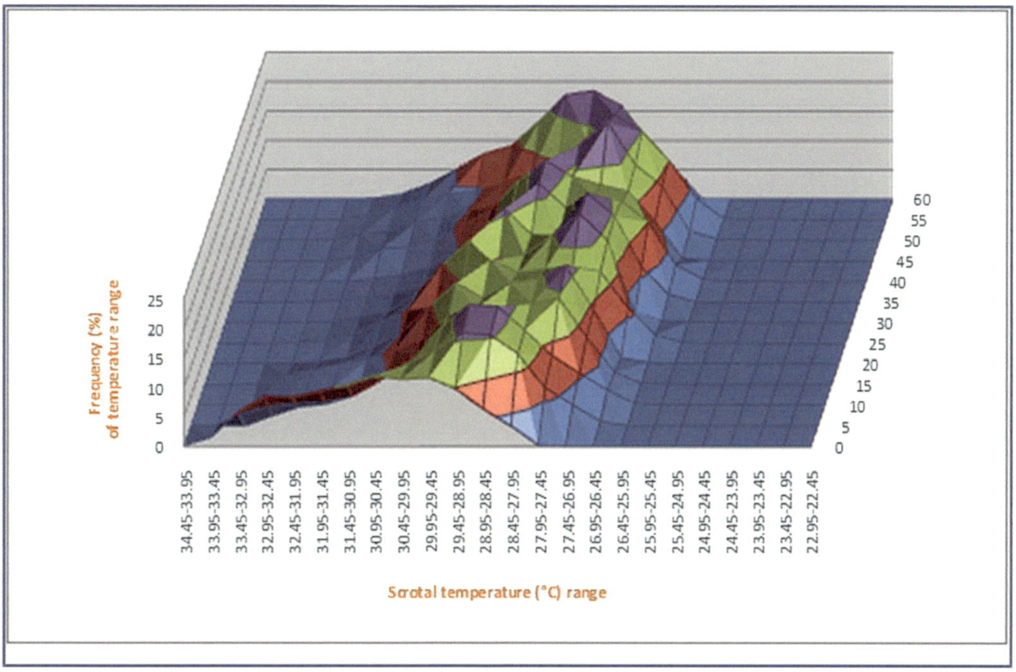

Graphic 1. Distribution of frequencies of scrotal temperature range during GnRH test after isolation from female

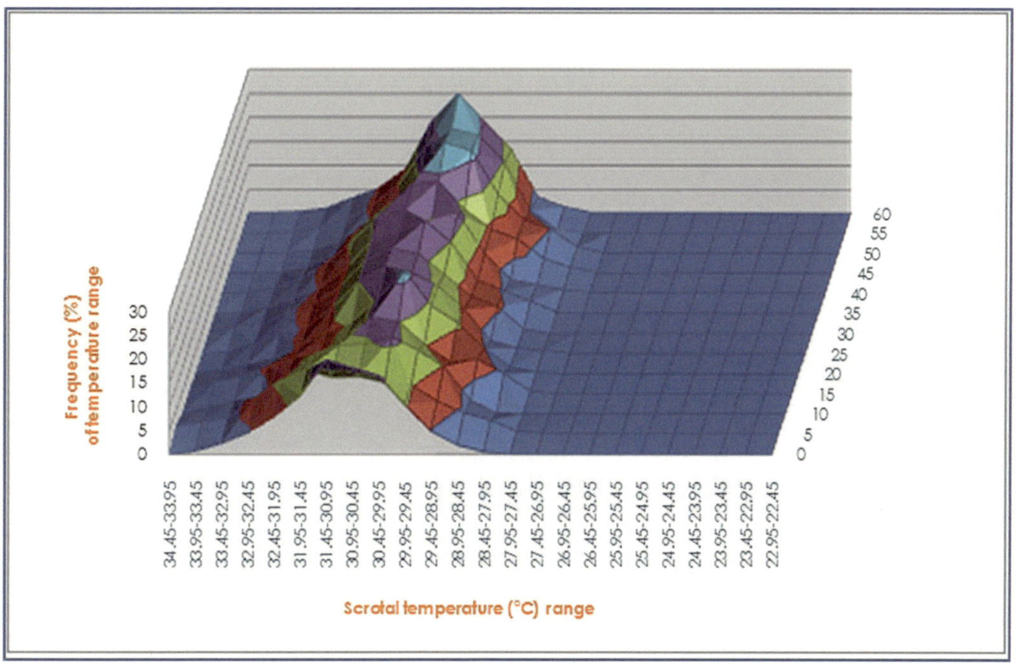

Graphic 2. Distribution of frequencies of scrotal temperature range during GnRH test after female exposure

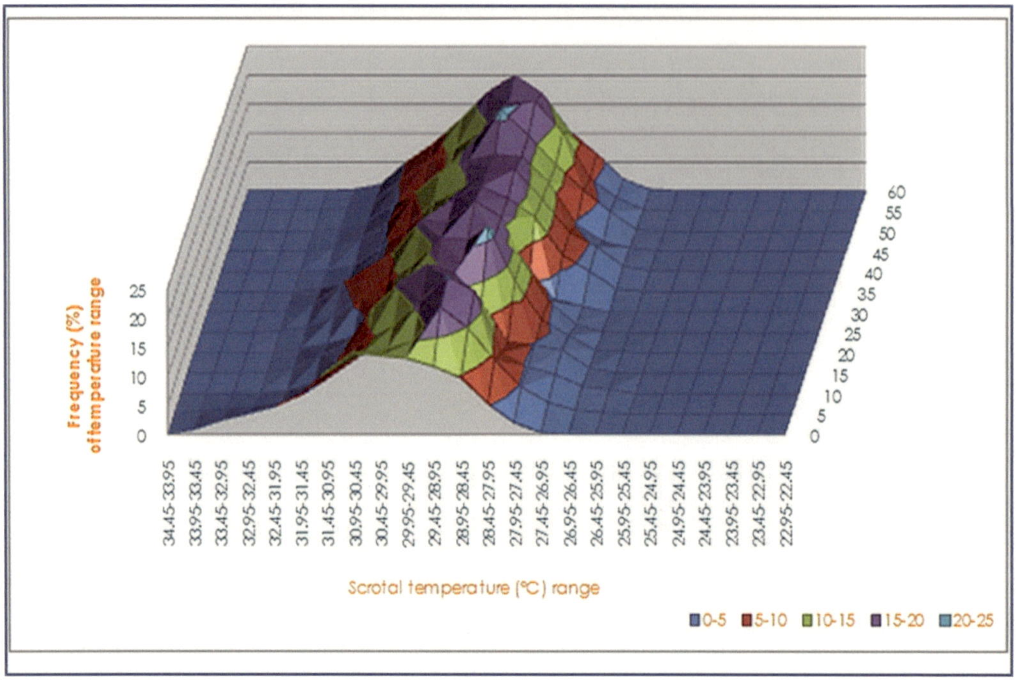

Graphic 3. Distribution of frequencies of scrotal temperature range during GnRH test after mounts

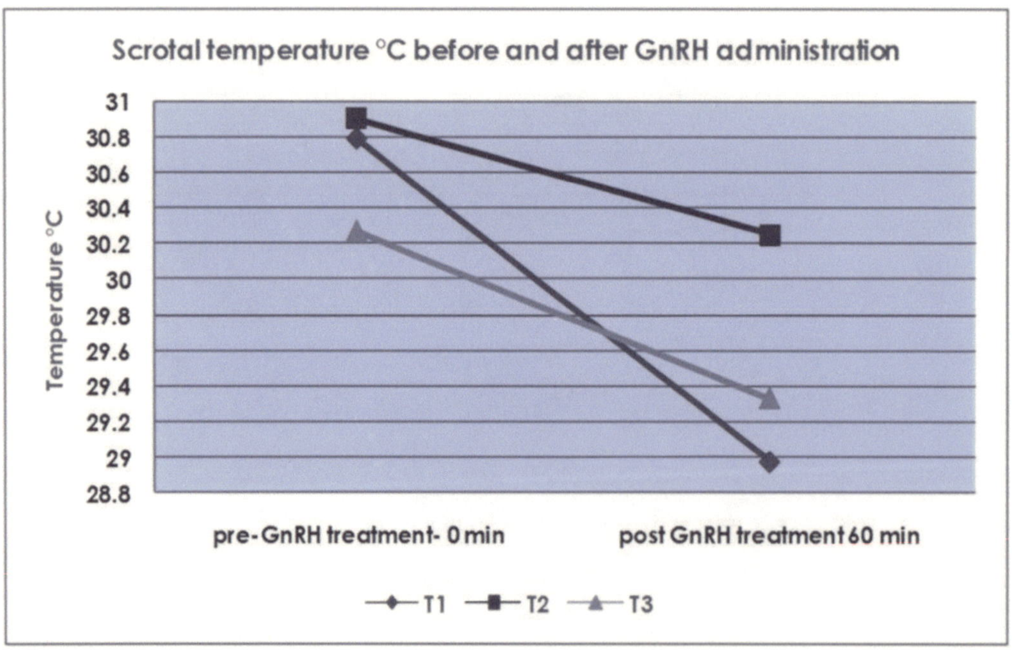

Graphic 4.

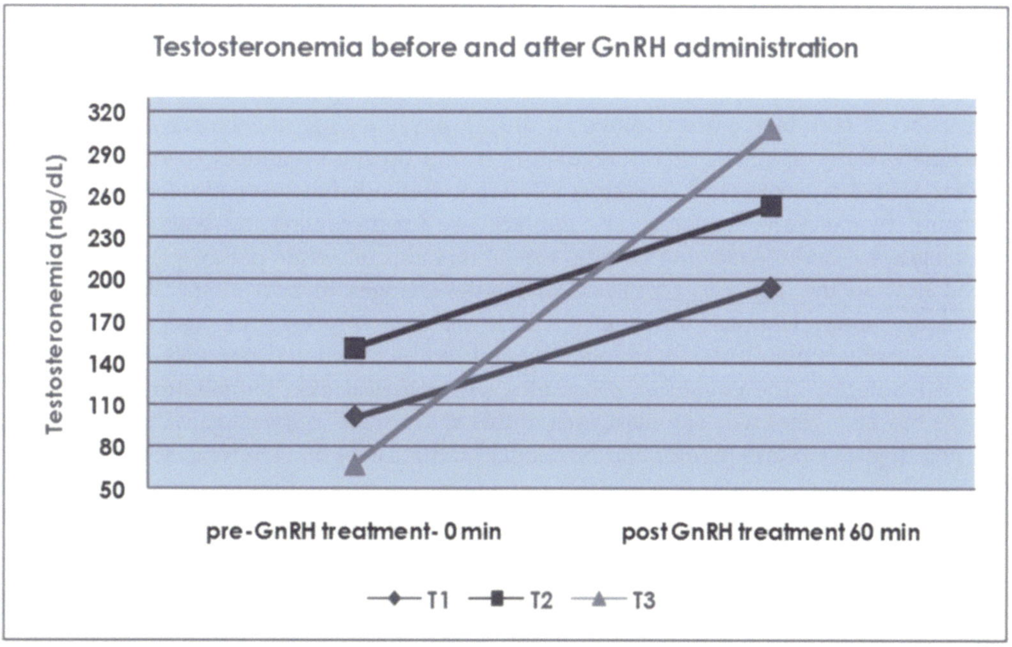

Graphic 5.

	Scrotal Temperature				Testosteronemia			
	T1	T2	T3	s.e.m.	T1	T2	T3	s.e.m
0	30,79*	30,90	30,27*		101.52	151.34	67.64*	
				0.27				30.19
60	28,98^{a**}	30,25^{b}	29,33^{ab**}		195.32^{a}	253.2^{ab}	308.2^{b**}	

Table 2. Comparisons for scrotal temperature and testosteronemia (mean ± s.em.) during the three GnRH tests.

Testosteronemia increase during the monitoring time with the higher variation after mounts.

Pearson correlation coefficients were 0.77, -0.638 and -0.378 ($P<0.05$) for testosteronemia-times, ST-times and testosteronemia-ST respectively. ST decrease starting 5 minutes after GnRH administration during 1st and 3rd test. The stability of ST after the female exposure could be related to the primary activation of the local hormonal control of the testis. It was notable individual response to the work schedule, despite the significant mean differences, probably due to hierarchical behaviour among males.

The effect of GnRH administration on scrotal surface temperature may be assessed by thermography. The correlation between temperature and testosteronemia could be indicative of the stimulating local effect of the GnRH.

3.2 Effect of food intake and first digestive phase on superficial temperature estimated by thermography in dairy cattle

Regulation of the skin circulation is also the main mechanism to control the preservation or dispersion of core temperature above all during physiological phenomena like digestion. The cutaneous sympathetic thermoregulatory neural function regulates vasomotor activity within dermal arterioles and capillaries. Different methods has been adopted to study the dynamic temperature response i.e. indirect or direct cooling of body parts (stress thermography) and re-warming time as correlation index of certain pathological conditions[2]. Blood flow in the skin can be measured by the changes in skin temperature, washout technique, laser Doppler flowmetry, ulcer healing process, or the tissue PaO2. Thermography is non-invasive, simple, safe and can be monitored remotely[21,22]. In spite of such advantages, little evidences concerning physiological skin temperature changes for bovine has been reported. The objectives of this study were to investigate the influence of the first digestive phase (neural and hormonal mechanisms) on skin temperature variation in three different classes of dry cows, divided depending on the distance from the presumed delivery data and drying days, using the analysis of thermographic sampling before and after the Total Mixed Ratio (TMR) administration.

Twenty dry Holstein cows were considered during four times of thermographic monitoring. The cows were housed in a tie-stall during all observation time. Thermographic images were taken considering a poligonal area traced from ileum wings to sacral, caudal, gluteal and perianal areas. Two thermographic sessions (5 scans every 5 min for each one), before and after total mixed ration (TMR) distribution, were performed. All images were scanned using a hand-held portable infrared camera (ThermaCam P25, FLIR Systems, Limbiate, Italy). Temperatures were recovered by processing the thermographic images in ThermaCAMResearcher Software.

Data collected were divided taking in consideration the distance from the beginning of the dry period (class 1: 0-20 days; class 2: 21-40 days; class 3: >41 days) and the distance from the predicted parturition date (class 1: 0-20 days; class 2: 21-40 days; class 3: >41 days). Data were analyzed using a two ways ANOVA (sampling period: before and after TMR distribution; classes of distance from predicted parturition date or/and beginning of the dry period) with the GLM procedure of the software SigmaStat2.03. Moreover data were analyzed using a digital infrared imaging software package to collect the temperature ranges frequencies in the selected area and to study the variations between before and after TMR administration.

Skin temperatures were dependent on the time of measurements (before and after TMR administration). Results indicate a ΔT °C of about 1.5 °C in classes of animals with more than 20 days from the predicted delivery date (Table 3 and 4).

The analysis of the temperature ranges frequencies (% of 0.25 °C ranges from 20°C to 35°C) showed a substantial frequencies variation between before and after TMR administration (graphics 6-8).

The distributions of frequencies are different in classes considered and the variations due to the TMR intake are different among the classes considered. Another interesting finding was the possibility of individual thermal mapping during different thermographic sessions.

Item	Classes of distance from the predicted parturition date					
	0-20 days	21-40 days	>40 days	*P* =		
T °C	27.55 ± 0.65	27.97 ± 0.43	28.72 ± 0.47			
ΔT °C	1.064	1.463	1.510	0.311	0.014	0.028

Table 3. Effect of distance from the predicted parturition date on skin temperature (mean ± SEM) and the difference (ΔT) between before and after TMR distribution.

Item	Classes of distance from the beginning of dry period					
	0-20 days	21-40 days	>40 days	*P* =		
T °C	28.60 ± 0.48	28.45 ± 0.50	27.38 ± 0.46			
ΔT °C	1.592	1.336	1.250	0.030	0.028	0.084

Table 4. Effect of distance from the beginning of dry period on skin temperature (mean ± SEM) and the difference (ΔT) between before and after TMR distribution.

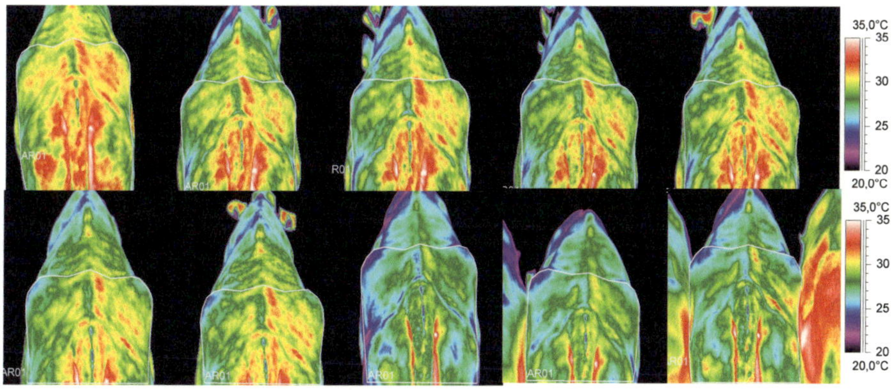

Fig. 2. Example of thermographic images of cows during the first digestive phase.

The body surface temperature is determined by the local metabolism, blood circulation underneath the skin, heat exchange between the skin and its environment. Variations in any of these parameters may induce alteration from the normal temperature or heat flux range at the skin surface. Therefore these changes may be reflecting the physiological and pathological state. In our trial, it was established that physiological changes due to the first phase of digestion may be investigated through the thermographic scanning. The choice of dry cows as animals for thermal scanning was almost imperative because it is clear the fetus influence on feed voluntary intake and the physiological changes which may be observed during the last two months of pregnancy. These changes are strictly linked to the reached maturity of fetal adrenal gland for corticosteroid production, to the hormonal variations during the last pregnancy time and, not less important, the fetal mass which limits the quantity of feed intake. Cows thermal mapping is possible even if hair length can influence

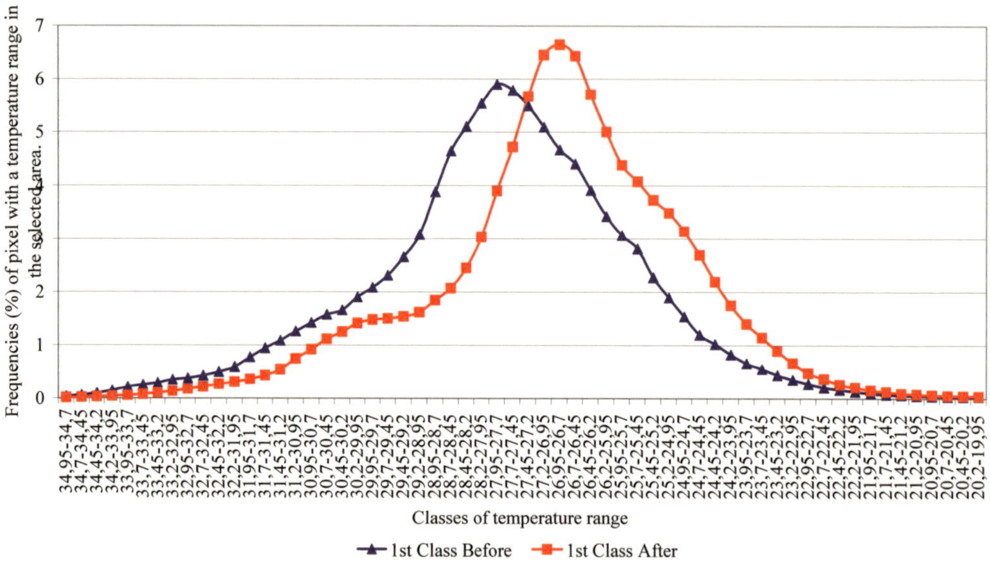

Graphic 6. Temperature ranges frequencies (%) in the 1st class (0-20 days) of distance from the presumed delivery data (red line)

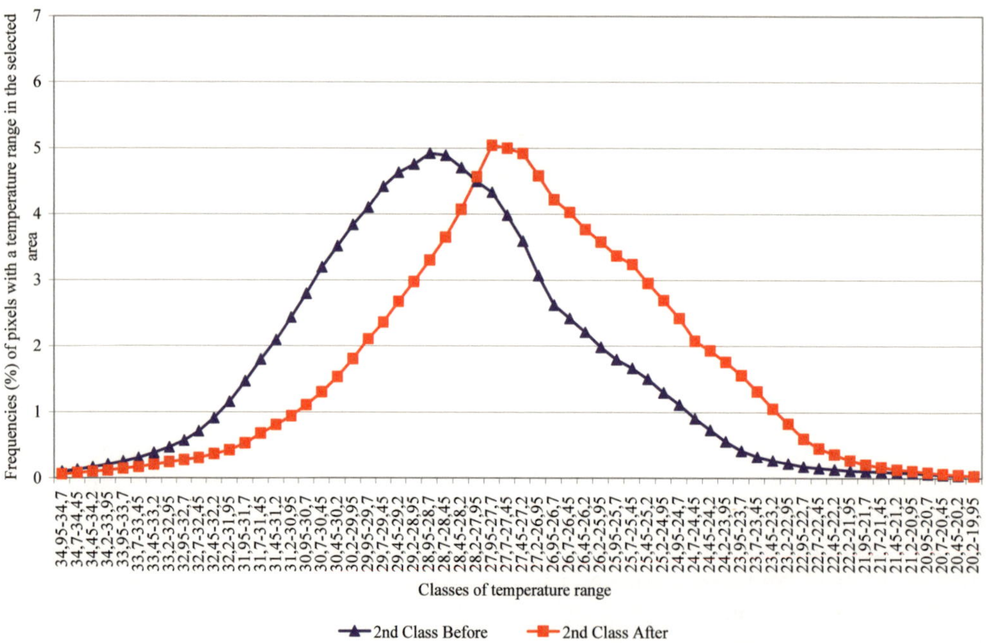

Graphic 7. Temperature ranges frequencies (%) in the 2nd class (21-40 days) of distance from the presumed delivery data

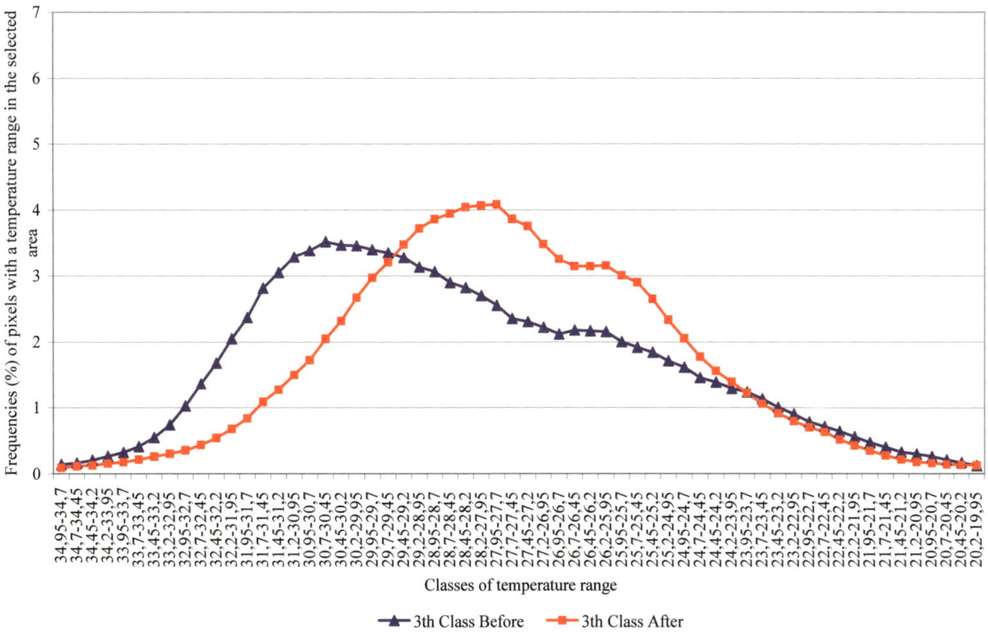

Graphic 8. Temperature ranges frequencies (%) in the 3th class (>41 days) of distance from the presumed delivery data

the skin emissivity of infrared rays and, therefore, a error margin have to take into consideration. In our trials, the cows skin emissivity was fitted to 0.95. This value is similar to previously used values for bovine's skin[2,3,4].

Thermal imaging camera appear to be reliable under field condition. In cows as other animals the body surface temperature depends by blood flow and metabolism rate of the underlying tissues. The physiological state of the underlying cells could be studied by measuring skin temperature using thermography. This study enable us to quantify, in terms of skin temperature ranges frequencies in a selected area, the redistribution of the blood during the first phase of digestive processes. In cows these variations were different during the dry period. The voluntary feed intake, controlled by numerous neural and hormonal mechanisms, changes during the last period of pregnancy because fetus body encumbrance occur. Also during the dry period there is an increment of the subcutaneous fat content which could influence the heat radiation. Hormonal variations typically evidenced to partum preparation could also changes the tissues heat transfer capability. In conclusion thermography may have potential as a study tool for specie-specific heat transfer equation. However, more data about the relationship among others physiological events affecting cows skin temperature are required.

3.3 Thermographic study of the perivulvar area in estrous and anaestrus ewes

The aim of the present study was to detect skin temperature differences of perivulvar area between ewes in estral and anestral phase. Twenty four dairy ewes - 16 in estral phase and 8 in anestral phase - were investigated. Estruses were synchronized by using intravaginal progestagen-impregnated sponges (fluorogestone acetate, FGA) for 14 days and after sponges removal ewes were treated with PMSG. Thermography sessions were carried out 50 hours after sponges removal at a distance of a meter from the vulvar area. A skin emissivity of 0.95 was assumed. The control subjects with no synchronization treatment were in a seasonal anestral period. The analysis performed on the acquired thermograms were: qualitative and quantitative analysis taking into account the mean perivulvar area temperature, quantitive analysis of temperature differences between unfleeced adjoining areas, quantitative analysis of frequencies of intervals temperature of 0,2 °C. A significant difference between the two groups in values expressed as mean temperature of perivulvar area ($P < 0.05$) was observed. The subjects in estrus and in anestrus ranging a temperature from 35,9 ° C to 37,7 °C with an average of 36,9 ± 0,5 °C and from 34,2 ° C to 36,5 °C with an average of 35,42 ± 0,63 °C respectively. The superficial temperatures detected in unfleeced areas of posterior anatomical regions may be taken into account in the study of circulatory and/or hormonal variations in ewes.The mammary skin receives only the 2% of the regional hematic flow and so this area may be used as control for other adjoining areas because the emitted heat increase is proportional to the parenchymal hematic flow. The animals in estrus show an increase in the hematic flow to the genital apparatus and hormonal changes. These variations are able to explain the increase of heat emitted by the unfleeced skin of posterior areas. Thermography is a tecnique useful to point out the different skin's ability of heat transmission from underlying areas. The subjects with estrus induced by synchronization have shown a different thermal behavior that can be detected by thermograpy sessions in adjoining areas receiving blood from common arteries (internal iliac -pudendal artery).

3.4 The use of thermography on the slaughter-line for the assessment of pork and raw ham quality

Several studies have been carried out to measure the surface temperature of pigs to evaluate the effects of environmental conditions[1,23] and to predict the pork quality of pigs immediately before they are slaughtered[24,25]. The aim of this preliminary study was to examine the possibility of applying the IRT directly on the slaughter-line for the evaluation of pork quality and ham suitability to be processed as dry-cured ham.

Thermographic images were collected on 40 carcasses of heavy pigs at 20 min after stunning. The Infrared Thermal Camera (ITC) (Flir System, Model ThermaCam P25) was placed after carcass splitting, thus left and right caudal and dorsal surface images were kept for each half carcass. The settings of the camera were as follows: emissivity of pig's skin 0.98; reflected air temperature 22°C; distance between camera and skin surface m 2.5. Temperatures were recovered by processing the thermographic images of a squared area located in the centre of the caudal side of ham using the ThermaCam Researcher Basic Software (Flir System). The pigs, consisting of commercial hybrids, were supplied from one farm located 35 km from the slaughter plant and were stunned by electronarcosis (250 V, 1.25 A) after a lairage that lasted 2 hours. At 90 min post mortem the pH (pH1) was measured on the semimembranous (SM) muscle of each left ham. After 24 hours of chilling at a temperature of 0-4°C, the measure of pH (pHu) was repeated together with the objective colour assessment (CIE Lab system, Minolta Colorimeter Cr300) after 30 min of bloom time. Moreover, a subjective evaluation of some characteristics of ham such as the veining defect (4-point scale, 1=none, 4= serious), the redness of skin defect (3-point scale, 1=none, 3=serious) and the fat cover (3-point scale, 1=insufficient, 3=excessive) was carried out on the trimmed left thighs destined to be processed as Parma dry-cured-ham[26]. The temperature range frequencies (% of 0.25°C ranges from 27°C to 30°C) and the means of the obtained distributions were calculated. In order to evaluate the possible relationship between the surface temperature of ham and the meat quality, pH and L* colour values were arranged in the following classes (pH1, < 5.80, ≥ 5.80 < 6.00, ≥ 6.00; L*, < 45, ≥ 45 < 50, ≥ 50). The means and the measures of variability of the surface temperatures and meat and ham quality traits are reported in Table 1. The distributions of the average surface temperatures on left and right hams showed a total range from 27.3°C to 29.2°C with mean values of 28.3°C and 28.1°C, respectively. Although all carcasses were processed in the same way along the slaughter chain, a variation of 1.8°C for the left hams and of 1.5°C and for the right hams were found. The widest ranges of surface temperatures using ITC were found on live pigs before stunning at the level of the back by Schaeffer et al.[25] (from 17°C to 34°C) and by Gariepy et al.[24](from 21°C to 29°C), and after sticking at the level of the ears by Warriss et al.[1] (from 27°C to 35°C). A largest range of surface temperatures, from 16°C to 21°C, was also found on pig carcasses[25]. The pigs used in the present study produced meat characterized by a large variability in terms of glycolysis speed, as demonstrated by the range of pH1 values, but also a similar extent of acidification, as confirmed by the narrow range of the ultimate pH values (Table 5). The L* colour co-ordinate values showed that final meat colour of the examined pigs was included from normal to slightly pale.

	Mean	Standard Deviation	Min.	Max.
Average left ham surface T°	28.3	0.43	27.40	29.20
Average right ham surface T°	28.1	0.36	27.30	28.80
pH_1	6.13	0.26	5.56	6.63
pH_u	5.42	0.05	5.33	5.57
L*	48.2	2.90	40.7	55.6
Veining score	2.14	0.80	1	4
Skin redness score	2.25	0.60	1	3
Fat cover score	2.11	0.57	1	3

Table 5. Means and measures of variability of the surface temperatures and meat and ham quality.

Least-squares means of surface temperature in the classes of meat and ham quality are presented in Table 6. In both hams, the differences of surface temperature among pH_1 classes were extremely small and non significant. The temperature of both hams were also very similar and not significantly different in the L* colour co-ordinate classes. These results are consistent with previous findings of Schaeffer et al.[25] showing an absence of relationship between these meat quality traits and the skin surface temperature. The veining and the red skin defect classes were not significantly related to a variation of the skin surface a temperature; although, in both hams there was a tendency for the latter defect to decrease with an increase in the surface temperature. Significant differences of temperature in both hams according to the fat cover score were found. An increase of temperature was found in hams with a decreasing of fat cover, particularly in the right hams. It is suggested that lower thermal insulation due to a thinner subcutaneous adipose tissue might be responsible to of the higher skin surface temperature. The relationship between the fat cover score of ham and the surface temperature suggests that infrared thermography could be a valuable, fast and non-invasive method to estimate its fatness. Thus, the preliminary results achieved here showed a possible application of this technique to better select the raw hams destined to the successive dry-cured processing.

	Left ham	Right ham
pH_1		
< 5.80	28.40 ± 0.09	28.12 ± 0.09
≥ 5.80 < 6.00	28.40 ± 0.19	28.26 ± 0.18
≥ 6.00	28.32 ± 0.22	28.08 ± 0.21
L*		
< 45	28.41 ± 0.18	27.99 ± 0.18
≥ 45 < 50	28.32 ± 0.10	28.19 ± 0.10
≥ 50	28.54 ± 0.15	28.12 ± 0.15
Veining score		
1	28.32 ± 0.17	28.14 ± 0.17
2	28.53 ± 0.11	28.19 ± 0.11
3+4	28.24 ± 0.13	28.05 ± 0.13
Skin redness score		
1	28.77 ± 0.27	28.57 ± 0.26
2	28.44 ± 0.10	28.19 ± 0.09
3	28.23 ± 0.13	27.96 ± 0.12
Fat cover score		
1	28.82 ± 0.19[a]	28.75 ± 0.16[A]
2	28.35 ± 0.09[b]	28.08 ± 0.08[B]
3	28.22 ± 0.15[b]	27.90 ± 0.13[B]

[a,b]= $P<0.05$.[A,B]= $P<0.01$.

Table 6. Least squares means and standard errors of surface temperature in the classes of meat and ham quality.

3.5 Use of thermography during monitoring of estrous and ovulation times in the mare

Essential prerequisites for successful artificial insemination is accurate estrus and ovulation detection, which is classically perfomed by teaser, transrectal palpation and ultrasonographic examination of reproductive tract. Alternative method could be the measurement of electrical impedance of vaginal mucus and perivulvar and vulvar temperature[27,28]. Report on using infrared thermography in detection of estrus in mare is limited. Therefore, the aim of this study was to assess perivulvar and vulvar temperature using infrared thermography as a non invasive method for the monitoring of estrous cycle in the mare.

Animals used were nine trotter mares, five with foal and four without foal. The monitoring was done in three stages : T1(follicle with Θ > 3cm), T2 (follicular growth), T3 (ovulation). During each moment, the thermography was performed first on perivulvar and vulvar regions by Thermacam P25, followed by transrectal palpation and ultrasonographic examination of reproductive tract by 8Mhz probe, and finally, the measurement of electrical impedance was done with endovaginal Draminisky probe. Ten blood samples were collected from five mares to measure serum progesteron and estrogen concentration in stages T2 (5) and T3 (5).

	T1	T2	T3
PMinT (°C)	28,6±0,80	28,84±0,32	27,72±0,54
PMaxT (°C)	34,2±0,38 ^{ab}	34,41±0,22 ^{a}	33,43±0,13 ^{b}
PMT (°C)	31,98±0,51 ^{a}	31,54±0,15 ^{ab}	30,91±0,21 ^{b}
VMinT (°C)	27,95±0,22	27,17±0,61	26,85±0,28
VMaxT (°C)	33,72±0,46 ^{ab}	34,12±0,18 ^{a}	33,14±0,14 ^{b}
VMT (°C)	31,40±0,30 ^{a}	31,31±0,19 ^{a}	30,31±0,20 ^{b}
ΔPVMT (°C)	0,58±0,23	0,23±0,16	0,60±0,19
DGF (cm)	4,31±0,28 ^{a}	4,72±0,21 ^{a}	1,10±0,13 ^{b}
EFW	2,30±0,12 ^{a}	2,45±0,11 ^{a}	1,0±0,00 ^{b}
FC	1,58±0,07 ^{a}	1,79±0,04 ^{b}	1,0±0.00 ^{c}
UE	1,50±0,29 ^{a}	1,39±0,14 ^{a}	0,92±0,05 ^{b}
EI (mOhm)	320,00±21,99	391,11±23,12	366,67±22,11
PG (ng/ml)	O,2±0,00	0,23±0,01	0,31±0,03

Different letters among groups mean a significant difference (P<0,05)
PerivulvarMinimum temperature (PMinT); Perivulvar Maximum Temperature (PMaxT); Perivulvar Mean Temperature (PMT); Vulvar Minimum Temperature (VMinT); Vulvar Maximum Temperature (VMaxT); Vulavr mean Temperature (VMT); Delta mean temperature Perivulvar-Vulvar (**ΔPVMT**); Diameter of Greater Follicle (DGF); Follicle concistency (CF) grade 1= firm, grade 2=soft, Echotexture Follicular Wall(EFW)grado 1= anecogenic, grade 2= medium ecogenicity, grade 3=ecogenic; Uterine edema (UE) grade 1=mild, 2=moderate, 3=heavy; Electrical impedance (EI); Progesterone (PG); T1(DGF>3cm); T2 (follicle growing); T3(ovulation).

Table 7. Monitoring parameters considered during estrous and ovulatory times in mare

Thermographic analysis was performed using an established range of temperature (15°-35°C°) by the software Thermacam Researcher. The data was statistically analyzed by software Sigmastat. Parameters were evaluated with R>0,4 and P<0,05.

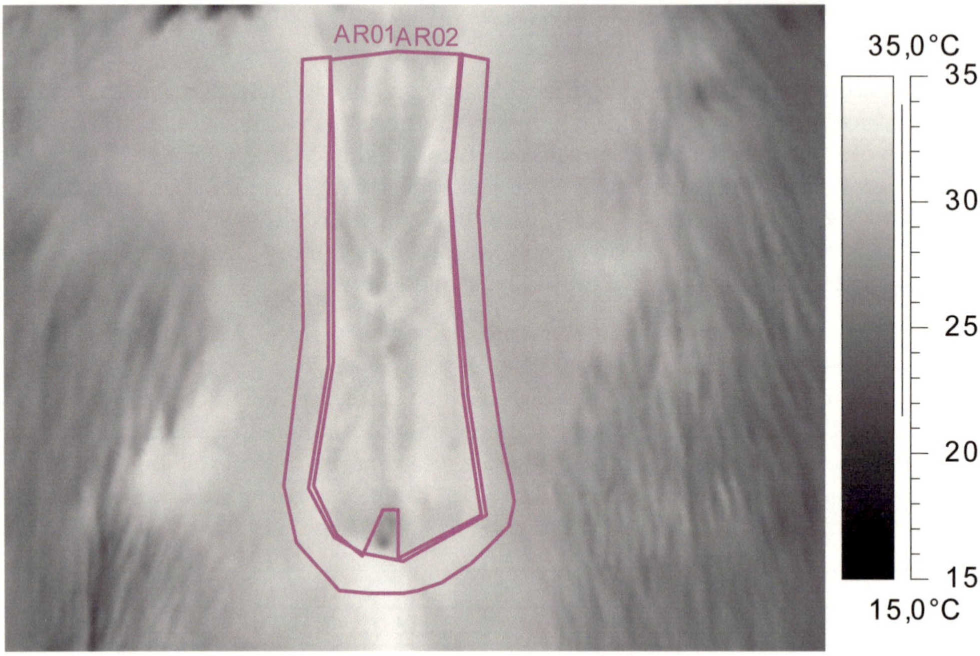

Fig. 3. Example of vulvar and perivulvar mare's thermography

The analysis showed a positive correlation within the thermographic parameters, which were Perivulvar Maximum Temperature (PMT), Perivulvar Mean Temperature (PMT), Vulvar Maximum Temperature (VMT), Vulvar Mean Temperature (VMT). There was a simultaneous increase of these parameters. These increased parameters positively correlated with Diameter of Greater Follicle (DGF) and Echotexture Follicular Wall(EFW), and negatively correlated with the presence of corpora lutea (CL). These data suggest an increase of maximum and mean perivulvular and vulvular temperature during follicular growth and a decrease of the same temperature during the establishment of CL. Probably that is because the mare under the influence of estrogen has an increase of hyperemia of the vulvar region[29]. A negative correlation was also existed between PMT and the values of Electrical Impedance (EI). So, the increase of PMT is associated with lower values of EI, which occur during the follicular growth. Moreover, the Perivulvar Minimum Temperature (PMT) was positively correlated with serum estrogen concentration and negatively correlated with serum progesteronconcentration which occur during the follicular growth and ovulation time respectively.

These preliminary results are in favour to a possible use of thermography as an auxiliary non invasive method during estrous cycle monitoring in mares.

3.6 Milking procedures and thermographic monitoring

The milking machine may affect both blood and lymphatic circulation in teat walls, because the radial and longitudinal stretching action exerted by the milking vacuum, particularly in correspondence of an inadequate massage phase. This can induce teat congestion and oedema, altering the defense mechanisms against bacterial penetration of teat duct. Milking duration (overmilking), liner and cluster characteristics, pulsation parameters and vacuum level have been recognized as factors that can affect the integrity of teat tissues[30,31].

The effect of milking procedures and liners on udder and teat skin temperature was investigated in cows through thermographic scanning, showing that thermography can be a very useful tool to evaluate, estimate and differentiate short and longer-term tissue reactions to machine milking[2,3]. In dairy sheep different methods have been adopted to measure udder blood circulation and the effect of cold exposure and lactation on the distribution of blood flow[32].

The objective of this study was to evaluate the influence of the vacuum level on udder and teat temperature changes during milking procedure, monitoring the teat recovery via digital infrared thermography.

The experiment was carried out in a 1x24 side by side milking parlour, with a low pipeline milking system, using the same pulsation parameters (120 cycles/min and 60% ratio) and a medium weight milking unit (0.49 kg) equipped by plastic shells and cylindrical rubber liners (20 mm mouthpiece bore, 101.2 mm length mounted, 2,8% extended) (AlfaLaval 961403-01). Two groups of 24 ewes were milked twice a day at 28 and 42 kPa for a period of 8 weeks, applying a standard routine without udder preparation and stripping. Individual milking-on time was on average 55.7 s when the system vacuum was set at 42 kPa and increased to 67.1 s at 28 kPa. At the 9th week six ewes of each group were monitored during two consecutive evening milkings via infrared thermography.

Thermographic images (Flir System, ThermaCam P25, sensitivity of 0.08 °C) of posterior udder area (PUA) and teats were taken pre-milking (PM), during milking (M) (only for PUA), immediately after milking (IAM) and up to 2.25 minutes after milking (AM+). Mean temperatures of the different teat areas (teat base - TB; mid teat - MT and teat tip - TT) and the distribution of frequencies of temperature ranges (0.2 °C) were recovered by processing the thermographic images in ThermaCam Researcher Basic 2.8 SR-1 Software (Flir System).

Furthermore, digital pictures of each teat have been taken to examine teat apex condition and, after a classification in three classes of phenotype (Class 1, 2, 3), find any connection to the morphology and teat skin temperature (figure 5). The classification was based on teat cistern height and on sphincter protrusion, both linked to the intra-mammary pression before milking, to the teat end wall thickness and to papillary canal length.

Some IR images of the udder posterior area taken at low and high vacuum level are shown in figure 4, where the different thermal patterns of the teats before and after milking can be easily individuated.

Thermographic analysis of all different teat areas (teat base, mid teat and teat tip) shown that skin temperatures before milking were characterized by decreasing values between TB and TT. During milking the skin temperature had an average drop of about 2,2 and 1,9 °C, respectively at low and high vacuum level. After milking, temperature differences among all teat locations were more evident and specifically starting from IAM for high vacuum level group, while from AM+15s for the low ones (table 8).

The temporal variation of temperatures for each teat location is reported in graphics 9 and 10, for low and high vacuum level respectively. It was evident that low vacuum level

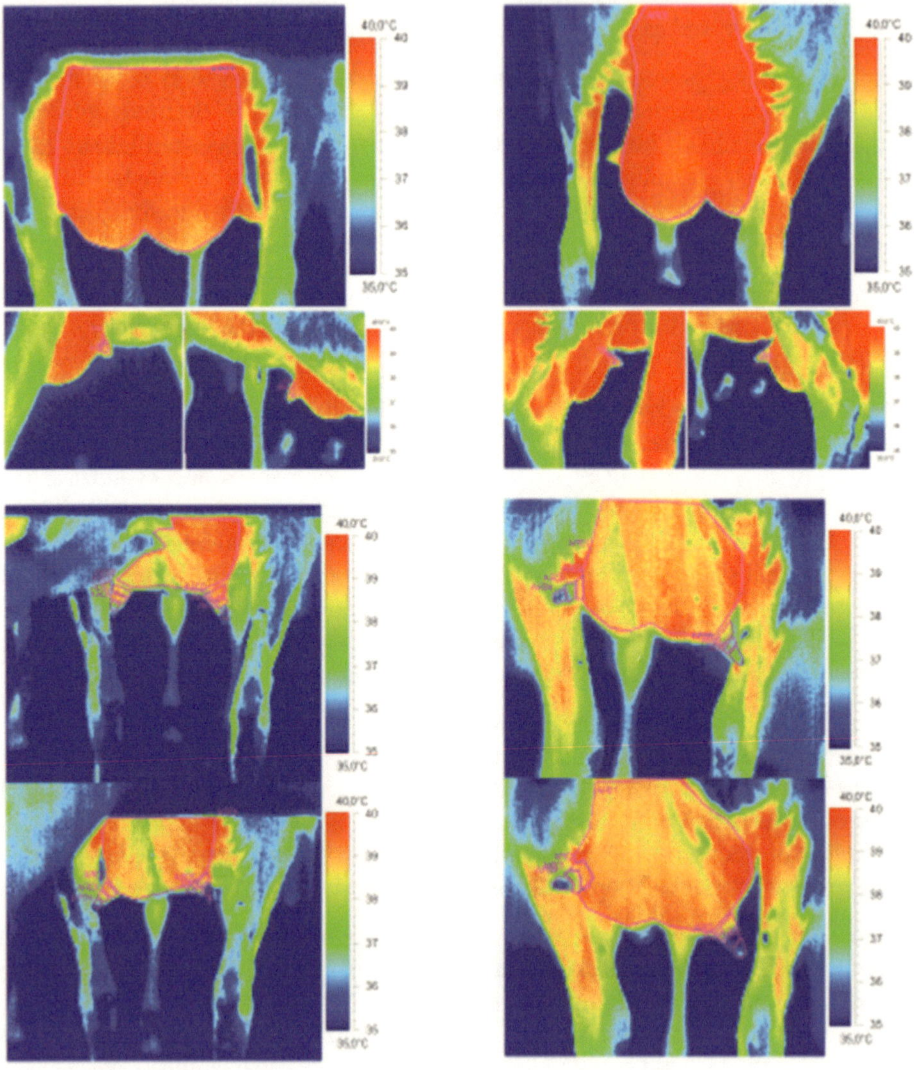

(a) (b)

Fig. 4. Thermograms of the udder and teats milked with high (a) and low (b) vacuum level

maintains a higher difference among the teat locations. Graphic 11 reports the distribution of the frequencies of temperature ranges (0.2°C) for each teat location and both vacuum levels. The temperature recovery time was shorter (1 min vs 2 min)) at the teat base for the group milked at 28 kPa, and this can be attributable to a faster return to a normal blood flow, as a result of a lighter mechanical action exerted by the machine during milking at a low vacuum.

The phenotypic teat classification, carried out independently from vacuum levels, gives more information about the influence of the teat conformation on the mechanical milking aptitude (figure 5). Class 1 was the group of animal which have had the lower temperature gap between PM and IAM to AM+60sec while the class 2 have had the slower recovery time.

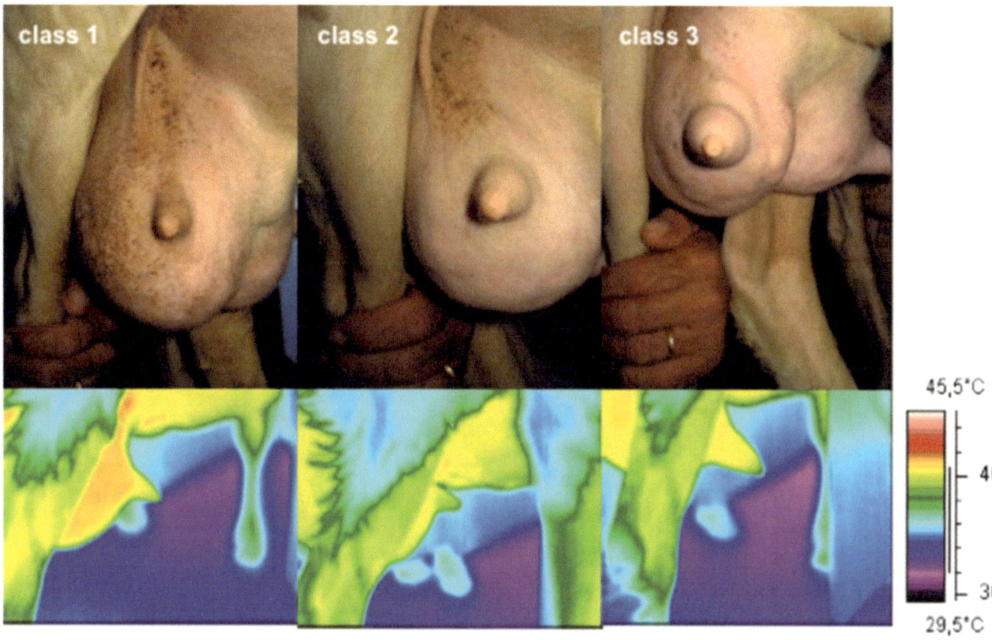

Fig. 5. Classification of the teat conformation

	Low Vacuum	High Vacuum	SEM	Contrast					
				Low			High		
				A	B	C	A	B	C
PM	39.59	39.76	0.23			*			*
IAM	37.36	37.89	0.23	*		*	*	*	*
AM+15 s	37.30	37.61	0.23	*	*	*	*	*	*
AM+30 s	37.34	37.35	0.23	*	*	*	*	*	*
AM+45 s	37.49	37.24	0.23	*	*	*	*	*	*
AM+60 s	37.52	37.61	0.23	*	*	*	*		*
AM+75 s	37.70	37.81	0.23	*	*	*	*		*
AM+90 s	37.86	37.94	0.23	*	*	*			*
AM+105 s	38.00	38.20	0.23	*	*	*			*
AM+120 s	38.06	38.36	0.23	*	*	*			*
AM+135 s	38.12	38.48	0.23	*	*	*			*

Contrast: A = TB vs MT; B = MT vs TT; C = TB vs TT *=$P<0.05$

Table 8. Least Squares Means of temperatures of teats milked with low and high vacuum level during milking procedures (pre-milking-PM; immediately after milking-IAM; after milking AM+) and contrast among teat locations.

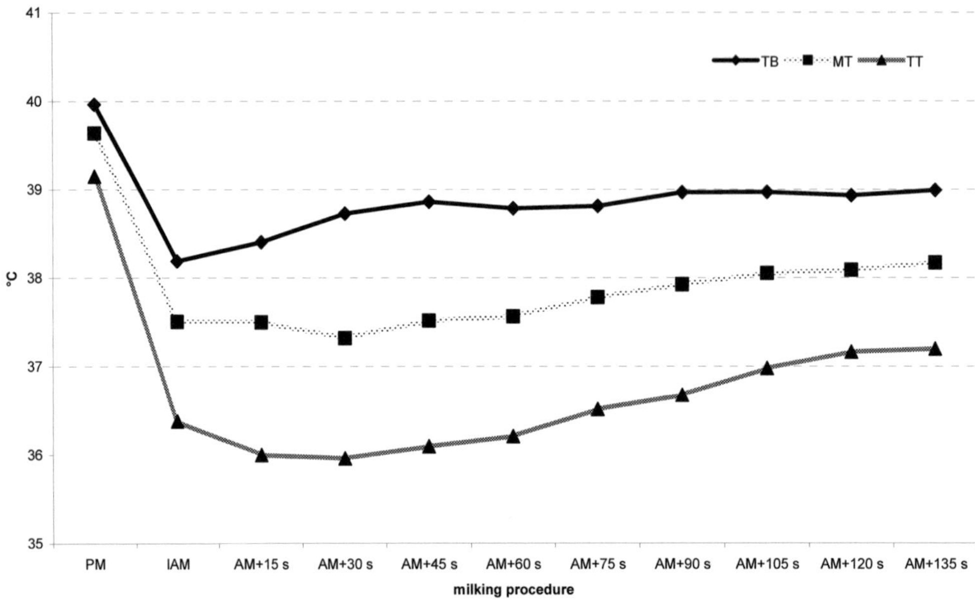

Graphic 9. Temperature variation of teat locations (Teat base-TB, Middle teat-MT, Tip teat-TT) during milking procedure at low vacuum level

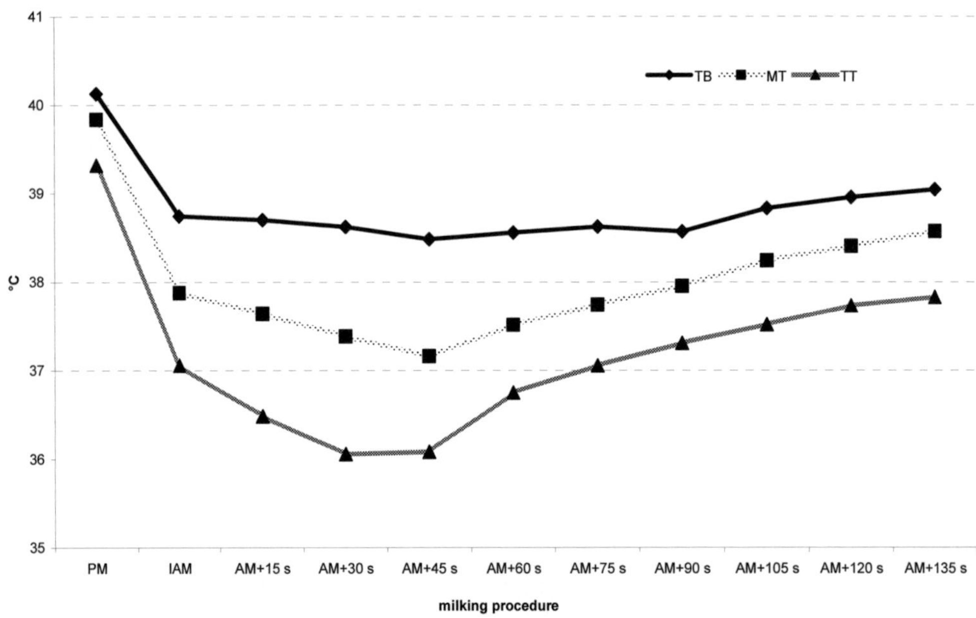

Graphic 10. Temperature variation of teat locations (Teat base-TB, Middle teat-MT, Tip teat-TT) during milking procedure at high vacuum level.

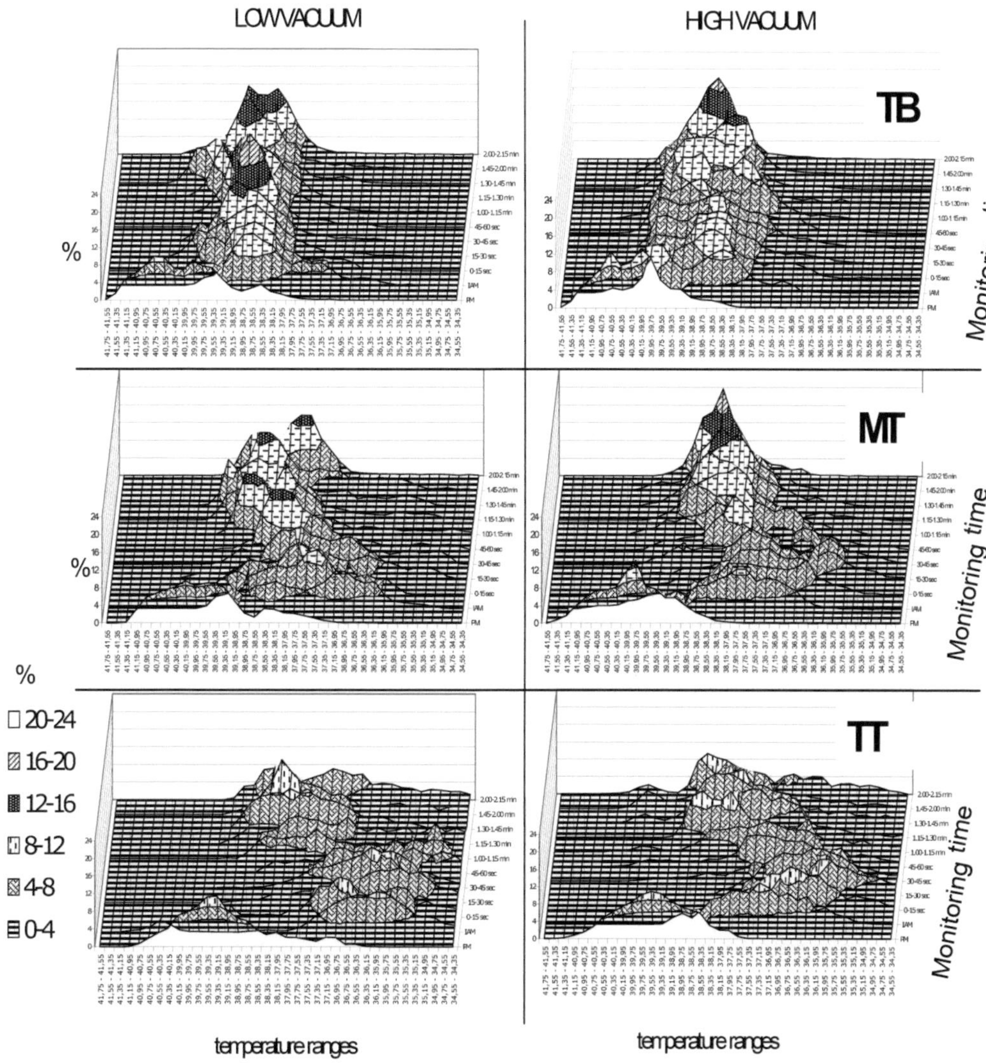

Graphic 11. Distribution of frequencies of the temperature ranges (0.2°C) for the teat locations (teat base –TB; mid teat – MT and tip teat – TT) at low and high vacuum levelsduring the milking monitoring time.

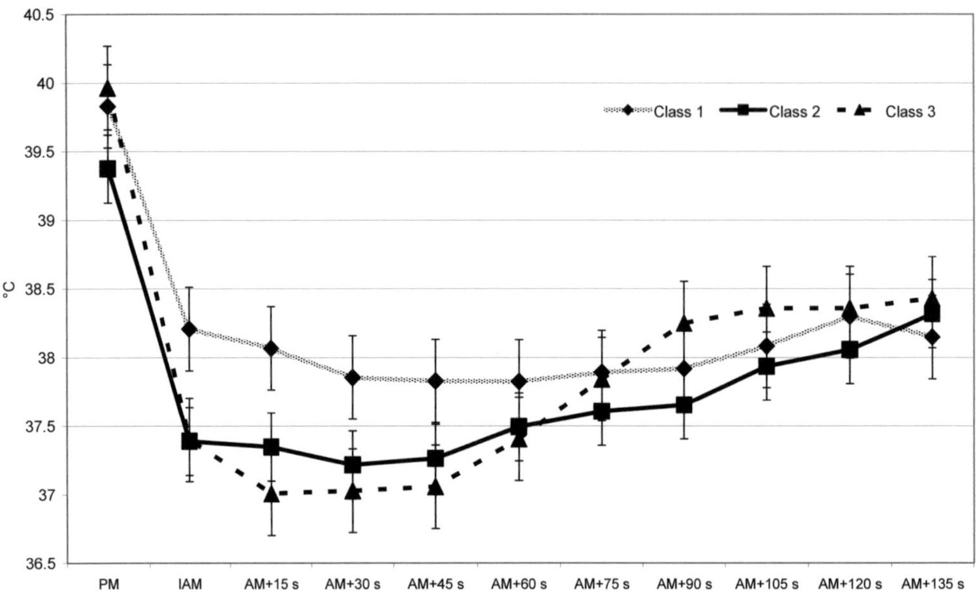

Graphic 12. Temperature variation of three phenotypic teat classes during milking monitoring time.

Mechanical milking affects teat temperature of dairy ewes in a different way if compared to the effect on cows, where milking caused a marked increase in teat temperature after an initial drop due to preparation massage (3, 4). In dairy ewes there is no manual udder stimulation before milking and the decrease in teat temperature during milking can be comparable to the effect of the udder massage.

A low vacuum level seems to be more physiological than a high level because the faster recovery time and the more stable temperatures at both TB and MT after milking. Further surveys could provide more detailed information about the role played by machine milking parameters and teat morphology on circulation impairments.

4. References

[1] Warriss P.D., Pope S.J., Brown S.N., Wilkins L.J., Knowles T.G.(2006). Estimating the body temperature of groups of pigs by thermal imaging. Vet. Rec.158:331-334.

[2] Paulrud C.O., Rasmussen M.D. (2004). Teat thermography, hot or not. NMC Annual Meeting Proceeding. 159-168.

[3] Paulrud C.O., Clausen S., Andersen P.E., Rasmussen M.D. (2005). Infrared thermography and ultrasonography to indirectly monitor the influence of linear type and overmilking on teat tissue recovery. Acta Vet. Scand. 46:137-147.

[4] Whay H.R., Bell M.J., Main D.C.J. (2004). Validation of limb identification though thermal imaging. Proc. 13th Intern. Symposium and 5th Conference on lameness in Ruminants.

[5] Nikkhah A., Plaizier J.C., Einarson M.S., Berry R.J., Scott S.L., Kennedy A.D.(2005). Infrared thermography and visual examination of hooves of dairy cows in two stages of lactation. J. Dairy Sci. 88:2749-2753.
[6] Eddy A.L., Van Hoogmoed L.M., Snyder J.R. (2001). Review: The role of thermography in the management of equine lameness. Vet. J. 162:172-181
[7] Spire M.F., Drouillard J.S., Galland J.C., Sargeant J.M. (1999). Use of infrared thermography to detect inflammation caused by contaminated growth promotant ear implants in cattle. J. Am. Vet. Med. Assoc. 215(9):1329-4.
[8] Turner T.A. Diagnostic thermography. (2001). Vet. Clin. North Am. Equine Pract. 17(1):95-113.
[9] Berry R.J., Kennedy A.D., Scott S.L., Kyle B.L., Schaefer A.L. (2003). Daily variation in the udder surface temperature of dairy cows measured by infrared thermography: potential for mastitis detection. Can. J. Anim. Sci. 83:687-693.
[10] Schwartzkopf-Genswein K.S., Stookey. (1997). The use of infrared thermography to assess inflammation associated with hot-iron and freeze branding in cattle. Can. J. Anim. Sci. 77:577-583
[11] Schaefer A.L., Cook N.J., Tessaro S.V., Deregt D., Desroches G., Dubeski P.L., Tong A.K.W., Godson D.L. (2003). Early detection and prediction of infection using infrared thermography. Can. J. Anim. Sci. 84:73-80.
[12] Nocek J.E. Bovine acidosis: implication on laminitis. (1997). J. Dairy Sci. 80:1005-1028
[13] Nanni Costa L, Stelletta C, Cannizzo C, Gianesella M, Pietro Lo Fiego D, Morgante M (2007). The use of thermography on the slaughter-line for the assessment of pork and raw ham quality. Italian Journal ofAnimal Science, vol. 6, p. 704-706, ISSN: 1594-4077
[14] Stelletta C, Murgia L, Caria M, Gianesella M, Pazzona A, Morgante M (2007). Thermographic study of the ovinemammary gland during different working vacuum levels. Italian Journal ofAnimal Science, vol. 6, ISSN: 1594-4077
[15] Calabria A, Corcillo G, Valentini S, Stelletta C (2010). Utilizzo della termografiadurante la determinazionedell'estro e del momento di ovulazionenellacavalla. In: Atti VIII CongressoNazionale S.I.R.A. Ozzanodell'Emilia, 17-18 Giugno 2010
[16] Stelletta C, Stefani A, Marion I, Bellicanta Y, Cannizzo C, Romagnoli S. (2009). Effect of GnRH test on scrotal surface temperature in Alpaca. In: Proceedingof 2nd Conference of ISOCARD. DJERBA, 12th-14th march 2009
[17] Murgia L, Stelletta C, Caria M, Gianesella M, Gatto M, Pazzona A, Morgante M. (2008). Using infra red thermography to monitor the effect of different milking vacuum levels on teat tissue in dairyewes. In: International Conference on Agricultural Engineering. Creta, 23-25 June
[18] Stelletta C, Speroni M, Gianesella M, Morgante M (2006). Thermography applied on pregnant dairy cows during dry period as physiological status indicator. In: XXIV World BuiatricsCongress. Nice, 15/19 October
[19] Stelletta C, Stradaioli G, Gianesella M, Mayorga Munoz I.M, Morgante (2006). Studio termografico della areaperivulvare in pecore in fase estraleed in fase anaestrale. In: XVII CongressoSipaoc. Lamezia Terme (Cz), 25/28 Ottobre
[20] Morgante M, Stelletta C, Gianesella M, D'alterio G, Stradaioli G. (2006). Use of thermography for evaluating ovine health and welfare status: preliminary investigations. In: XXIV World BuiatricsCongresS. NICE, 15/19 OCTOBER

[21] Skagen, K., Haxholdt O., Henriksen O., Dyrberg V.(1982). Effect of spinal sympathetic blockade upon postural changes of blood flow in human peripheral tissues. Acta Physiol. Scand. 114:165-170.

[22] Holloway G.A., Watkins D.W.(1977). Laser Doppler measurement of cutaneous blood flow. J. Invest. Dermatol. 69:306-309.

[23] Loughmiller J.A., Spire M.F., Dritz S.S., Fenwick B.W., Hosni M.H., Hogge S.B.(2001). Relationship between mean body surface temperature measured by use of infrared thermography and ambient temperature in clinically normal pigs and pigs inoculated with *Actinobacilluspleuropneumoniae*. Am. J. Vet. Res.62:676-681.

[24] Gariepy C., Amiot J.,Nadai S.(1989). Ante-mortem detection of PSE and DFD by infrared thermography of pigs before stunning. Meat Sci.25:3-41.

[25] Schaeffer A.L., Jones S.D.M., Murray A.C., Sather A.P., Tong A.K.W.(1989). Infrared thermography of pigs with known genotypes for stress susceptibility in relation to pork quality. Can. J. Anim. Sci. 69:491-495.

[26] Tassone F., Ielo M.C., Bertolini D., Lo Fiego D.P., Nanni Costa L., Russo V.(2006). Caratterizzazione morfologica di prosciuttifreschicolpiti dal difetto di venatura in treimpianti di macellazione. Atti Soc. Ital. Sci. Vet. 60:495-496.

[27] Řezáč P.(2008). Potential applications of electrical impedance techniques in female mammalian reproduction.*Theriogenology* 70 1–14.

[28] Hurnik J.F., Webster A.B., DeBoar S.(1985). An investigation of skin temperature differentials in relation to estrus in dairy cattle using a thermal infrared scanning technique,*J Anim Sci*. 61:1095-1102.

[29] Samper J.C.(1997). Ultrasonographic appearance and the pattern of uterine edema to time ovulation in mares*Proceedings of the Annual Convention of the AAEP*.

[30] Hamann J.(1997). Machine induced teat tissue changes and new infection risk. Int. Conf. on Machine Milking and Mastitis, Cork, 78-98.

[31] Mein G.A.(1992). Action of the cluster during milking in Machine Milking and Lactation, Bramley A.J et al. Ed.; Insight Books, Vermont, 97-140

[32] Thompson G.E. (1980). The distribution of blood flow in the udder of the sheep and changes brought about by cold exposure and lactation. J. Phisiol. 302:379-386.

7

Infrared Thermography in Sports Activity

Ahlem Arfaoui[1], Guillaume Polidori[1], Redha Taiar[2] and Catalin Popa[1]
[1]*Université de Reims Champagne-Ardenne,*
GRESPI/Thermomécanique (EA4301)
Moulin de la Housse, Reims Cedex 2
[2]*UFR STAPS, Moulin de la Housse, Reims Cedex 2*
France

1. Introduction

The origin of infrared thermography comes in 1800 when William Herschel discovered thermal radiation, the invisible light later called infrared, but only in the mid-sixties infrared thermography became a technique of temperature cartography. He proved that this radiation, called infrared, followed the same law as visible light. Later, this phenomenon was connected with the laws of Planck and Stefan. The first detectors for this type of radiation, based on the principle of the thermocouple and thermopile called, were developed around 1830. In 1970, the first cameras appeared for commercial. The first models were made up of a technology-based pyroelectric tube with an optical IR instead of the classical elements. Today, these concepts have been improved with new technologies in electronics and computing. Infrared acquisition systems can arrive at very high frame rates. The major argument is whether infrared thermography can determine thermal variations to enable sufficient quantitative analyses. The creation of computerized systems using complex statistical data analysis, which ensure high quality results, and the enhancement of thermal sensitivity have increased the development of technology of infrared thermography.

For years, infrared thermography has become a powerful investigation tool to inspect in many applications, from mechanical, electrical, military, to building and medical applications. Due to its non-intrusive feature (Kaminski et al 1999; Hoover et al, 2004; Wu et al, 2009; Hildebrandt et al, 2010), infrared thermography (IRT) can be defined as the science of analysis of data from non contact thermal imaging devices. Thermal imaging cameras detect radiation in the infrared range of the electromagnetic spectrum and produce images of that radiation, called thermograms. This method provides real-time, instantaneous visual images with measurements of surface temperatures over a greater distance.

Few studies using infrared thermography have been devoted to sports performance diagnostic and to sports pathology diagnostic .It is well known that sports activity induces a complex thermoregulation process where part of heat is given off by the skin of athletes. As not all the heat produced can be entirely given off, there follows a muscular heating resulting in an increase in the cutaneous temperature. In particular, the IRT method will enable, in the long term, to quantify the heat loss according to the swimming style, and to consider the muscular and energy outputs during the stroke.

A close examination of the literature shows that no study has been devoted to these problems. Though, for example, Brandt and Pichowski (1995) determined the temperature of a swimmer to be 33 °C after exercise, it was a local measurement only, therefore very partial, obtained by means of a thermocouple placed at the deltoid. In the same way, Huttunen et al. (2000) studied variations in the internal temperature of long distance swimmers in cold water. In no way did they study the cutaneous temperature. Whereas the technique of infrared thermography is usually used under thermal conditions of living (Jansky et al, 2003), no mention of its application to swimming or cycling can be found in the literature. The temperature of skin is a significant parameter, which conditions the evolution of physiological parameters such as, for example, the production of lactate or the heart rate (Mougios et al, 1993), which have a direct influence on the process of body thermoregulation.

Understanding spatial and temporal gradients in latent heat loss of the human body allows optimization for thermal comfort. Recently, De Bruyne et al. (2010) carried out a search thataims at quantifying transient spatial gradients in sweat production on a human head while cycling. The results of their research aims to enhance physiological insight of the sweating process and it can also help to develop sweating thermal manikins (Hsu et al, 2000; Davis et al 2001; Bruhwiler et al 2003; Bruhwiler et al, 2004) that behave more realistically to thermal changes (Bruhwiler et al, 2006; Buyan et al, 2006; Bogerd et al, 2008; Brühwiler et al, 2009). Latent heat loss of the human head has been quantified in many experiments. Few experiments tested the hypothesis that gradients in latent heat loss on a human head can be observed. However, in any case they have studied the distribution of skin temperature.

Infrared thermography allows for this distribution. But In the field of cycling, no application is mentioned in the literature.

The first medical application of infrared thermography for skin temperature measurement was in 1960. In 1980, the early detection of diseases was quickly developed. In the field of pathology diagnostic, applied to sports activity, the application of infrared thermography (IRT) has a long history (Jiang et al, 2005), mainly musculoskeletal trauma, pathologic processes such as pain in the lumbosacral region, intervertebral disc prolapse, spinal cord lesion, traumatic lesions, fractures, cardiovascular…

ALBERT et al. (1964) were the first to assess pain by infrared thermography. This technique is a diagnostic method providing information on the normal and abnormal sensory and nervous systems, trauma, or inflammation locally and globally. Infrared thermography shows physiological changes rather than anatomical changes and could be a new diagnostic tool to detect the pathology of the knee (Selfe et al, 2010).

However, interest in IRT is up today because of improved devices and methods of calculation. The special characteristic of IRT studies is that we can get additional information about the skin's thermal aspect and about the complex thermoregulatory process. IRT gives a possibility to evaluate the effect of the sporting activity and to detect possible trauma or dysfunctions, which cannot be shown by present conventional methods.

It can measure skin temperature over inflamed joints. This developing technology is used to detect thermal abnormalities characterized by a temperature increase or decrease found at the skin surface. The technique involves the detection of infrared radiation that can be directly correlated with the temperature distribution of a body region (Melnizky et al 1997).

The question is whether infrared thermography can accurately determine thermal variations to enable sufficient quantitative analyses. The purpose of this study is, on the one hand, to assess the usefulness of infrared thermography in sport activity and in sport medicine and, on the other hand, to show up to what point the cutaneous temperatures are influenced by the nature of activity.

2. Principe of IRT

Based on the measurement of infrared radiation of the body, infrared thermography is the most accessible technique to obtain images of the temperature profiling of a surface or a point. The principle of this technique is that every corps emits an amount of IR energy and the intensity of this IR radiation is a function of temperature.

Infrared thermography transforms the thermal energy, radiated from body in the infrared band of the electromagnetic spectrum (roughly 0.8 μm - 1000 μm in figure 1) into a visible image of that radiation, called a thermogram.

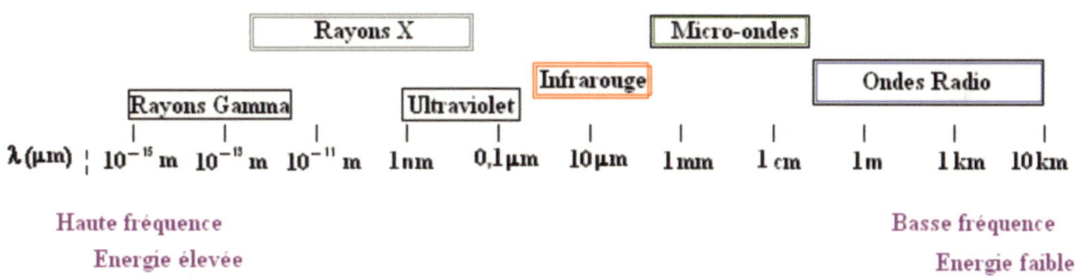

Fig. 1. Electromagnetic spectrum

The measuring range is restricted to infrared radiation classified into three categories:

- Near infrared of 0.75 μm – 1 μm
- Middle infrared of 1.5 μm - 4 μm
- Far infrared of 4 μm - 1000 μm

2.1 Fundamentals of thermal radiation

The radiation measured does not depend only on the body surface temperature but is also a function of the emissivity under measurement and on the environment. Emissivity is the relative ability of its surface to emit energy by radiation.

The total energy radiated per unit surface area of a black body per unit time is directly proportional to the fourth power of the black body's absolute temperature T.

2.1.1 Stephan-Boltzmann law

The equation for the radiation of an object is governed by Stefan–Boltzmann law (Schmidt et al, 1993; Gaussorgues, 1999):

$$R = \varepsilon.\sigma.T^4 \tag{1}$$

Where

R: radiation (W/m2),
T: temperature (K),
σ: Stefan–Boltzmann constant = 5.67 10^{-8} W/m^2 K^4
ε: emissivity

2.1.2 Black body

The radiation emitted by a body depends on its nature. The thermal emission is referred to the notion of the black body, defined as being able to completely absorb all incident radiation, regardless of its wavelength.

2.1.3 Planck's law

The characteristics of infrared radiation emitted by an object in terms of spectral radiant emittance are described by Planck's law (Planck, 1901):

$$W(\lambda,T) = \frac{\overline{2\pi hc^2}}{\lambda^4}\left[\exp\left(\frac{hc}{\lambda kT}\right)-1\right]^{-1} \mathrm{Wcm^{-2}\mu m^{-1}} \qquad (2)$$

Where :

H = 6.6256 .10^{-34} Js : Planck's constant
K = 1.38054 .10^{-23} WsK^{-1} : Boltzmann's constant
C = 2.9979 .10^{8} ms^{-1} : velocity of light in vacuum
μ: wavelength in μm
T: temperature in K

Human skin emits infrared radiation mainly in the wavelength range of 2–20 μm with an average peak of 9–10 μm (Steketee, 1973). 90% of the emitted infrared radiation in humans is between 8 and 15 μm (the emitted infrared in this case is based on Plank's Law roughly). Human skin is a black body radiator with an emissivity factor of 0.98 (Steketee, 1973) and is therefore a perfect emitter of infrared radiation at room temperature

2.2 Passive and active thermography

The quality of thermal image depends on the variation in surface temperatures. Thermal images can be obtained under ambient conditions. Two types of schemes are deployed to make measurements by infrared thermography.

Passive thermography is essentially a qualitative or quantitative approach where the thermal model is available. It consists in testing a body or element whose surface temperatures are naturally different (often higher) than ambient.

However, in the case of the active thermography, an external stimulus is necessary to induce relevant thermal contrasts which are not available otherwise. Create a thermal gradient allows to highlight the presence of internal defects or simply define the thermophysical properties of material.

In this chapter, the accent is on the passive scheme.

3. Technical description of the infrared cameras

The Infrared cameras used in this investigation operate on the principle of high spatial resolution and highest sensitivity and accuracy. Tests were performed in by using two IR cameras: **FLIR-SC1000, CEDIP TITANIUM HD560M** coupled with thermographic software: **ThermaCAM Researcher** and **ALTAIR 5.50.**

3.1 FLIR-SC1000

The IR camera FLIR SC1000 has a range of temperature measurement between -40 ° C and 1500 ° C with an accuracy of ± 2% of measuring range (www.flirthermography.com).

Its 256x256 pixels format InSb focal plane array delivers respectively an outstanding 200 Hz frame rate while keeping extraordinary linearity and sensitivity figures, with high quantum efficiency (> 70%). The SC1000 has a sensitive to wavelengths from 3.4 to 5 µm ± 0.25 µm. It is equipped with a system of instant integration variable from 1 µs to 10ms and speed frame rate of up to 50 images / s in full speed.

The acquisition system is designed for ease-of-use, this real-time data; acquisition program communicates with and acquires data from numerous off-the-shelf and custom radiometric instruments. It transforms an image captured in the infrared according to the brightness of the object observed in a visible image. The heat maps are stored and processed in real time using the software "ThermaCAM Researcher" that allows:

- To deal with static images,
- To deal with video and infrared data in real time
- To process and analyze digital data at very high speed infrared.

3.2 CEDIP TITANIUM HD 560M

The IR camera CEDIP TITANIUM HD560M has a sensitive to wavelengths from 3.5 to 5 µm ± 0.25 µm (www.flirthermography.com). The focus of the camera is between 10 and 100 mm ± 0.5 mm, and the resolution of the camera is 640 × 512 features high quantum efficiency (> 70%). Thermal sensitivity was 0.04 °C per grayscale level over the physiological temperature range. Camera is equipped with a system integration instantaneous variable of 1µs to 10 ms and speed frame rate of up to 100 frames / s in full window.

The thermal image obtained was sent instantly to a portable computer connected to the camera, and it was immediately processed in real time using the software "ALTAIR 5.50" that allows:

- To deal with static images,
- Treat video and infrared data in real time,

3.3 Calibration and optimization of thermal resolution

Calibration of the infrared system by simulating real operating conditions must be performed, to take into account many variables.

The energy really emitted by the surface of body and detected by the infrared camera depends on the emissivity of the surface under measurement, on properties of body and on the environment. It is important not to take thermal cartographies in a too bright or when the body is exposed to other radiations.

For this, the choice of parameters of a system of infrared thermography as an instrument for measuring temperature must meet several compromises:

- Emissivity in the infrared and the spectral bands corresponding,
- Nature of the body,
- Ambient temperature,
- Temperature range in witch progress the body
- Objects parasites on the environment that can shine directly on the detector, or indirectly by reflection,
- Minimum and maximum distance linked to environmental constraints…

Thus, optimizing the thermal resolution is more than necessary. We must proceed step by step to get an optimal solution of the sizing system, by:

- Evaluating the spectral emissivity of the body,
- Evaluating the size of the body,
- Choosing the surface pupil and lens to obtain spatial resolution desired.
- Evaluating temperature range in witch progress the body…

4. Infrared thermography in sport

4.1 The influence of swimming type on the skin-temperature maps of a competitive swimmer from infrared thermography

This study aims, on the one hand, to highlight the feasibility and applicability of infrared thermography in swimming for the purpose of quantifying the influence of the swimming style on the cartographies of cutaneous temperatures of a swimmer.

Whereas the infrared thermography is used under thermal conditions of living (Jansky et al, 2003), no mention of its application to swimming is found in the literature. The study was carried out in a swimming pool. The temperature of water in the pool is a significant parameter, which conditions the evolution of physiological parameters such as, for example, the production of lactate or the heart rate (Mougios et al, 1993) and which has a direct influence on the process of body thermoregulation. Swimming induces a complex thermoregulation process where part of heat is given off by the skin of a swimmer. As not all the heat produced can be entirely given off, there follows a muscular heating resulting in an increase in the cutaneous temperature. An analysis carried out by Robinson and Somers (1971) showed that in swimming, the optimal temperature of water for 20 minutes of freestyle must range between 21 and 33 °C, with an optimal breakeven point at 29 °C. This value is slightly higher than the one we measured at the time of the study in the swimming pool. The investigation being a preliminary study, only one male swimmer took part in it.

4.1.1 Protocol and method

The experimentation took place in a covered swimming pool of 25 m in length. The temperature of the water was 27 °C and that of the ambient air was 24 °C. The experimental protocol, summarized in Table 1, is defined as follows:

- At the beginning, the swimmer is immersed up to the neck in the static position for 10 minutes.
- At the end of this period, the swimmer leaves water and then is rapidly dried (water is opaque to infrared radiation).
- The following task was the first recordings of the body surface thermal maps, that constitute the thermal reference level of the swimmer at rest.
- Next, the swimmer executes his first 100 m butterfly, leaves water, is dried and then is subjected to a second set of thermal acquisitions that will provide the cartography of the body surface temperatures after the exercise.
- After that, the swimmer is immersed in water again. The duration between two swimming series is sufficiently long (10 minutes) to allow a return to the thermal balance of the swimmer.

This cycle is reproduced for the three other strokes. It is to be specified that the swimmer was subjected to a test during the period of recovery according to his training program, which explains the average results obtained during the test of a 4 × 100 m medley with a departure in water.

	Thermal balance	100 m butterfly	Thermal balance	100 m backstroke	Thermal balance	100 m breast stroke	Thermal balance	100 m free style	End test
Duration of sequences	10'	1'03	10'	1'08	10'	1'22	10'	1'03	–
Test chronology →									
Data acquisition	x (reference)		x		x		x		x

Table 1. Infrared thermography data acquisition

One subject participated in our experimentation. He is a swimmer of national level specialist in the 400 m medley who is training on average 10 to 12 hours per week. The principal anthropometric characteristics of the subject are summarized in the following table (Table 2):

	Age	Height (m)	Mass (kg)	Body fat (%)
Swimmer (man)	19	1.78	67	12.4

Table 2. Morphometric data for the swimmer

In order to better assess the influence of the swimming style on the muscular heating, the cutaneous surface was divided into closed polygonal surfaces A, B, ..., J, according to the distribution represented in Figure 2. The limb extremities and the joint regions, which represent poorly the contribution of the body thermoregulation process, were excluded from this geometrical body cutting. The body cutting used in our study allows for the elementary geometry approach adopted by YANAI (2001) for the swimmer representation to numerical simulation purposes.

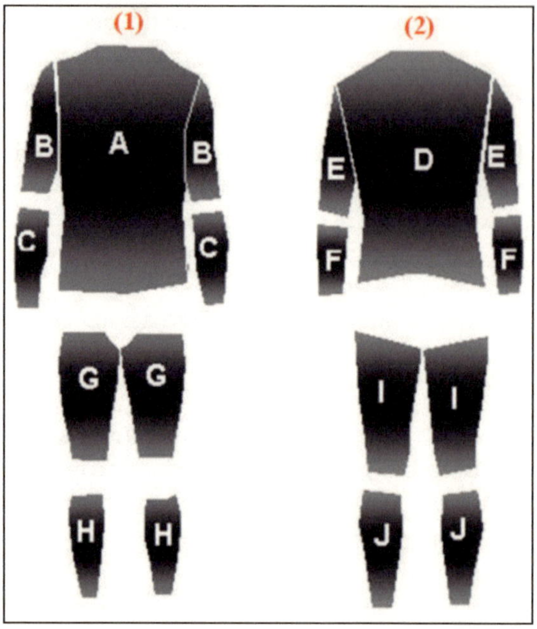

Fig. 2. Body zone definitions: frontal view (1) and dorsal view (2)

4.1.2 Reference temperature

Figure 3 presents infrared cartographies set at rest for the upper and lower limbs of the swimmer, after the swimmer had spent 10 minutes in the pool water at 27 ° C. From these

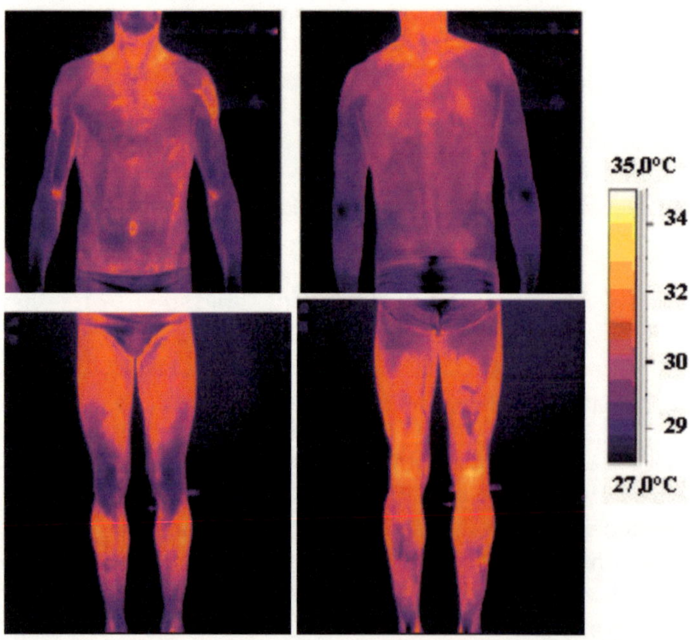

Fig. 3. Infrared cartographies of the upper and the lower limbs before effort

cartographies and using the software ThermaCAM Researcher, temperatures at rest were determined for areas A, B, C, D, E, F, G and H defined in Figure 2.

To quantify this distribution, let us introduce the histogram of the cutaneous temperature of the swimmer at rest shown in Figure 3. These temperatures constitute the case of reference of the study. There appear disparities in the distribution of the average temperatures of the zones; the highest temperatures correspond to the zones closest to the vital organs of the swimmer (abdomen, chest, back), whereas the lowest are those of the forearms of the swimmer. The maximum variation observed is about 1.7 °C.

4.1.3 Influence of the swimming style: Temperature after effort

Figures 4 and 5 show the various infrared cartographies established for the four styles and for the upper and lower limbs of the swimmer, after the swimmer had spent 10 minutes in the pool water at 27 ° C. Notable differences appear in the images, which allows us to predict a considerable influence of the swimming style on the distributions of the cutaneous temperature.

The average temperature difference (ΔT), defined as the difference between the temperature measured after effort and that measured at rest allows quantifying the influence of swimming style. The corresponding histograms are shown in Figure 6.

The histograms indicate that a significant increase in the cutaneous temperature is possible in accordance with the swimming style and the body zone considered. Indeed, it appears that the highest temperature is reached in the upper part of the body corresponding to zones A, B, C, D, E, F for the backstroke with $2.50 \pm 0.10 \leq \Delta T \leq 4.55 \pm 0.10$, whereas on the level of

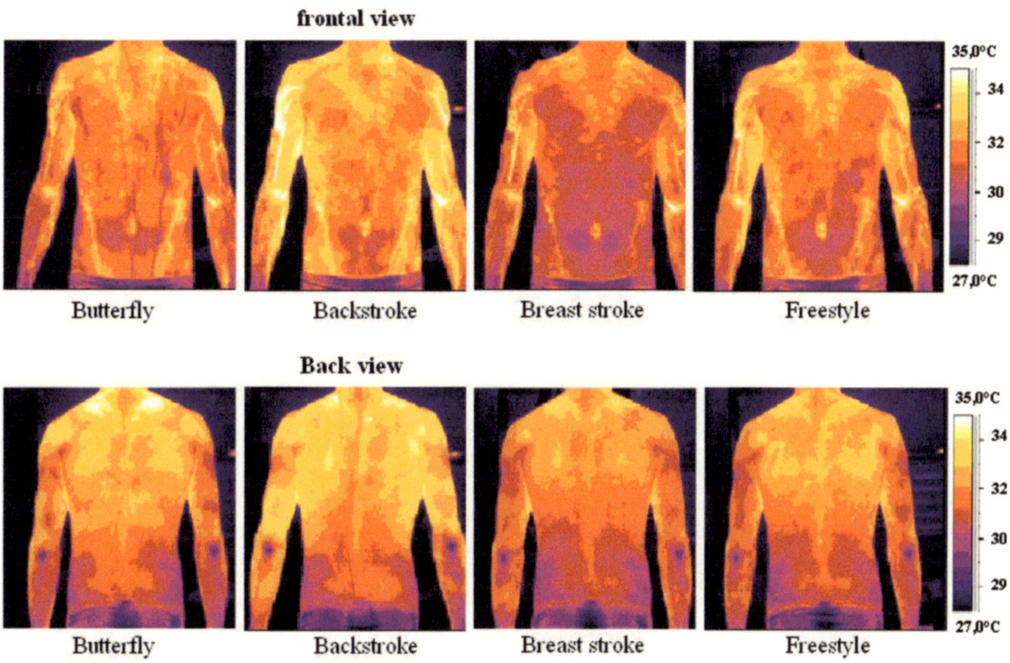

Fig. 4. Infrared cartographies of the trunk and the upper limbs

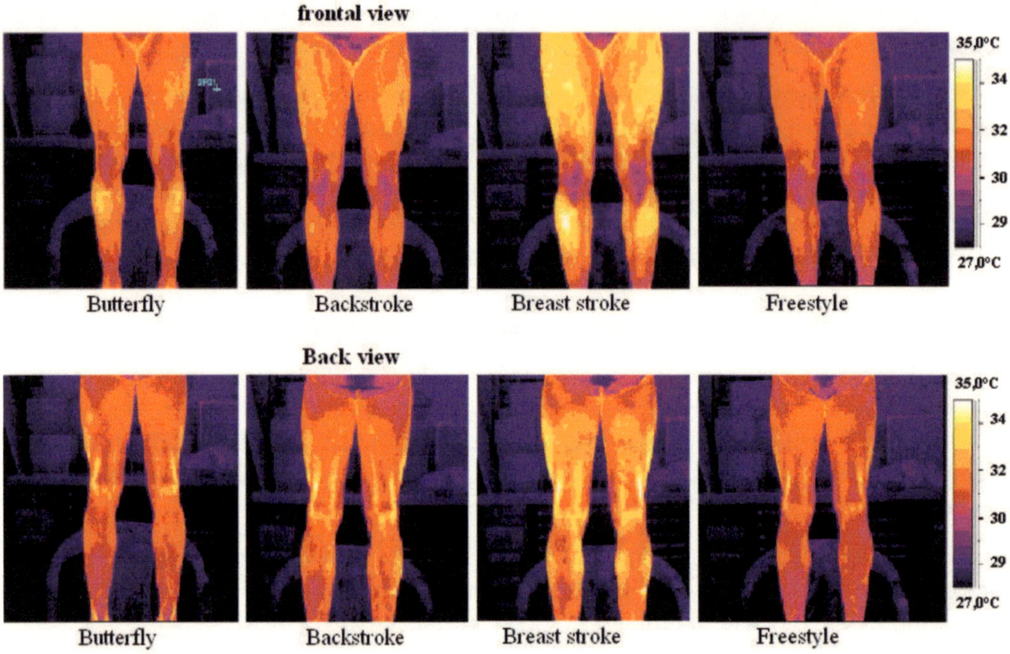

Fig. 5. Infrared cartographies of the lower limbs

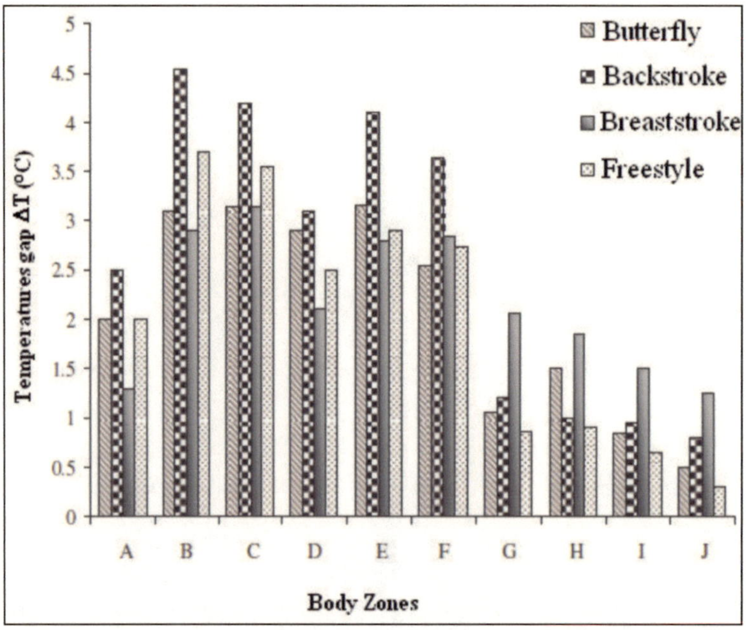

Fig. 6. Influence of the swimming style on the temperature gaps measured after effort and at rest

zones G, H, I, J, corresponding to the lower limbs, it is the breast stroke that generates the greatest increments in the cutaneous temperature, with $1.25 \pm 0.10 \leq \Delta T \leq 2.05 \pm 0.10$. On the contrary, freestyle induces the weakest variations in temperature on the lower limbs, with $0.30 \pm 0.10 \leq \Delta T \leq 0.90 \pm 0.10$.

Obviously, these results appear to be adequate with the intensity of the muscular activity related to the type of stroke. These are summarized in Table 3.

	Zone A	Zone B	Zone C	Zone D	Zone E	Zone F	Zone G	Zone H	Zone I	Zone J
ΔT max	Back-stroke	Breast stroke	Back-stroke	Back-stroke	Back-stroke	Back-stroke	Breast stroke	Breast stroke	Breast stroke	Breast stroke
ΔT min	Breast-stroke	Breast stroke	Breast stroke butterfly	Breast stroke	Breast stroke	Butterfly	Freestyle	Freestyle	Freestyle	Freestyle

Table 3. Distribution of temperature difference (ΔT) minimum and maximum in accordance with the swimming style and the body zone considered.

4.1.4 Global cutaneous temperature

After calculating all the average temperatures for all body areas previously defined for each swimming style, we were interested in defining a global average cutaneous temperature $\overline{T}_{overall}$ calculated over of the zones as a whole. It is given by the relation:

$$\overline{T}_{overall} = \frac{\sum_{i=A}^{j} T_i S_i}{\sum_{i=A}^{j} S_i} \qquad (3)$$

Where Si is the number of pixels defining each polygonal zone, and Ti is the average temperature for each zone.

Table 4 contains the global temperatures obtained for the four styles. It is found that the highest skin temperature average determined in all selected zones corresponds to the case of swimming "backstroke". From an energy point of view, this swimming seems to be the most demanding, in our study. The lowest temperature corresponding to the case of swimming "Breaststroke" seems that of less expensive overall energy. The temperature difference induced by the practice of these two swimming is, in this case and according to the protocol established by 0.78 ° C ± 0.10.

	Butterfly	**Backstroke**	**Breast stroke**	**Freestyle**
$\overline{T}_{overall}$	31.73 ± 0.10	32.14 ± 0.10	31.43 ± 0.10	31.58 ± 0.10

Table 4. Overall cutaneous temperature values

The difference in global average temperature, calculated after every effort to swimming style is shown in figure 7.

We note that the temperature averaged over the entire surface cutaneous, compared to the position of rest, increased by 2.16 ° C for the butterfly swimming, 2.56 ° C for backstroke, 1.78 ° C for breaststroke and 2.00 ° C for the freestyle, after the test performed.

Holmer (1974) carried out experiments in the experimental pool of Stockholm. At a given speed, significant differences in energy expenditive were observed between the four swimming styles. Techniques with alternate locomotive cycles (crawl and backstroke) were more efficient than techniques with simultaneous cycles (butterfly and breaststroke).

Thereafter, these results were confirmed by Lavoie and Montpetit (1986). From the results of Table 4, one can note that the highest global average temperature corresponds to the backstroke case. This stroke seems to show, as for this study, the greatest expenditure of energy. The lowest temperature corresponds to the case of the breast stroke whose overall expenditure of energy seems to be the least. In the present case and according to the protocol drawn up, the temperature difference induced by the practice of these two strokes is 0.78 ± 0.10 °C. One should point out that by no means can our results be compared with those of Holmer (1974) and Lavoie & Monpetit (1986), which were established statistically on the basis of different experimental protocols.

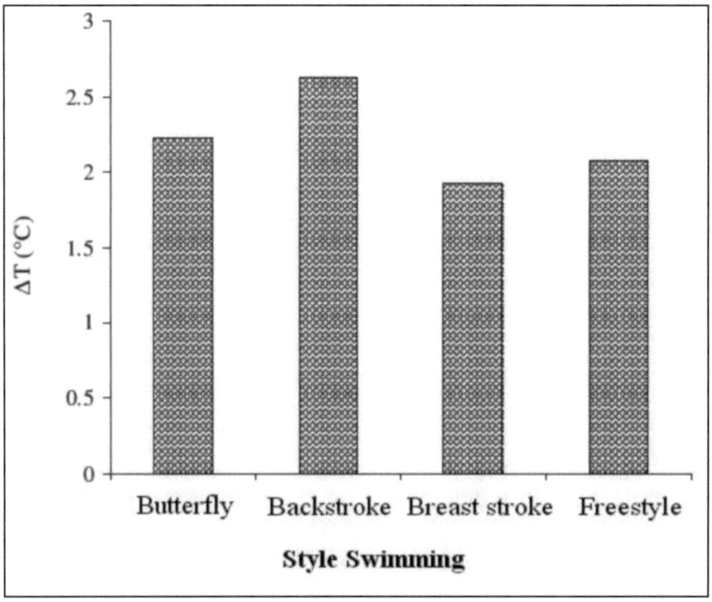

Fig. 7. Increase in mean body temperature after effort for the four swimming styles

A preliminary experimental study was undertaken, on the one hand, for studying the feasibility of using IR in the discipline of swimming and, on the other hand, for quantifying the influence of the type of stroke (within the framework of a well defined protocol) on the distributions of cutaneous temperatures. To the best of our knowledge, it is for the first time that such a study has been undertaken. In particular, this study shows significant variations in the cutaneous temperature according to the swimming styles. From the examination of infrared cartographies, one can note that the temperature, averaged over the whole body surface, is respectively increased by 2.16 °C for the butterfly, 2.56 °C for the backstroke, 1.78°C for the breast stroke and 2.00°C for the freestyle, after the practiced test.

One should recall that these conclusions cannot be considered as universal as far as only one subject, a male swimmer of national level, took part in this study. Nevertheless, the conclusions make us think of considering a statistical study that would also account for the initial temperature of the water in the swimming pool

4.2 On the use of thermography IR in cycling activities

Another possible use of thermography IR concerns the relationship analysis of both thermal and mechanical behaviors in the cycling technique of senior cyclists for example. The analysis of the driving force coupled with the acquisition of the skin surface temperature field of active legs can lead:

1. to highlight a systematic significant deficit of the force developed by one of the lower limbs,
2. to apprehend the mechanisms of muscular thermoregulation.

To fight against the different resistances to progress, the cyclist generates forces to the pedals during the pedaling motion. Pedaling results from actions of antagonistic muscle groups to mobilize the joints of the hip, knee and ankle, so as to achieve a full rotation of the crank around the bracket (Moussay et al, 2003). The use of footrests (pedal straps) or clipless pedals, however, allows the cyclist to generate a propulsive work even during the rise of the pedal. The use of the link shoe/pedal requires to describe the four phases in cycling.

The first phase of pedaling corresponds to a crank angle varying from 315 to 45°, generating what we can call a "dead pedaling zone". This is a transitional phase during which almost passive knee joints and hip are flexed.

The second phase of pedaling corresponds to a crank angle from 45 to 135°. This phase is the thrust phase during which the main driving pedaling efficiency is more important. During this phase, there is the extension of the lower limbs

The third phase is for a crank angle in the range 135-225°. This sector corresponds to the "bottom dead center". This is the transition between the pushing phase and the phase of the draw.

The fourth phase is a crank angle from 225 to 315°. In the area corresponding to the rise of the pedal, several scenarios are possible: 1) No action is taken on the pedal; the leg creates a resisting torque, 2) A driving force is created at the pedal to compensate some of all of the weight of the rising leg, 3) A driving force greater than the weight of the lower limb is created.

During Phases 1 and 3 it is anatomically difficult to have efficient propulsion.

The difference in performance between the 40 km cycling specialists against the clock is not entirely dependent on some physiological variables (Coyle et al, 1991). A study in which were measured: 1) The maximum oxygen consumption, 2) The lactic anaerobic threshold, 3) The use of muscle glycogen, 4) The type of muscle, 5) Enzyme activity, has allowed drawing the hypothesis that the cycling performance could be partially related to biomechanical factors related to individual pedaling technique (Coyle et al, 1991). Moreover, the experienced cyclists consume less oxygen per unit of power output, than cyclists of lower level (Coyle et al, 1992). It seems that these differences in oxygen consumption are not entirely due to physiological factors (muscle type), but also in biomechanical parameters (Coyle et al, 1991; Coyle et al, 1992; Kautz & Hull, 1995). It seems obvious that the physical potential of the athlete is a major parameter in the performance in cycling. However, it seems important to study how the energy generated by muscular contraction is converted into a propulsive energy on the pedal. The application of forces on the pedal is the last link in the conversion of metabolic energy into mechanical energy to drive the pedal.

During a competition on a very long time (eg Tour de France), a minimal improvement in motor efficiency of the athlete can make a difference in the way of performance. Indeed, a change in the pattern of pedaling can: 1) Change the distribution of work product, 2) Potentially reduce fatigue, 3) Increase the performance (Coyle et al, 1991).

A production of equivalent force on each pedal is also a factor in improving the efficiency of pedaling. The majority of elite cyclists has a "round and smooth" pedaling even though some may have an asymmetry greater than 10%. A population of cyclists are most vulnerable to these risks is the master population (50-60 years). In addition, this group of cyclists concerns the majority of practitioners. The purpose of this study is to study the biomechanics of cycling in masters cyclists during incremental test. At the same time acquiring a mapping of skin temperatures of active members will be conducted to better understand the mechanisms of thermoregulation muscle.

4.2.1 Materials and methods

Eleven masters cyclists (Table 5) performing tests of long distances (200 km) voluntarily participated in this study. Subjects were informed in detail of the study protocol, signed a letter of consent and could stop at any time if the wanted to abort the protocol.

Mechanical power and the pattern of pedaling (engine torque depending on the angle of the crank) were measured (200 Hz) using the system SRM Training System (scientific model,

Subjects		Age (years)	Distance traveled per year (km)	Height (cm)	Mass (kg)	Body fat (%)
Subject 1	B. P.	56	7000	172	79.1	18.7
Subject 2	C. J.-F.	47	4500	181	72.5	14.7
Subject 3	V. A.	53	5000	168	68.7	19.1
Subject 4	B. J.-M.	58	6000	176	66.4	13.9
Subject 5	B. J.	59	5500	170	72.6	21.3
Subject 6	M. B.	56	10000	178	74.7	18.9
Subject 7	B. J.-P.	51	7000	179	75.3	18.9
Subject 8	P. J.	57	14000	170	69.1	18.2
Subject 9	D. J.-P.	52	6000	169	68.2	18.6
Subject 10	R. R.	53	10000	180	70.6	15.3
Subject 11	T. J.-P.	47	8000	183	72.8	19
Average		53.5 ± 4.1	7545 ± 2815	175 ± 5	71.8 ± 3.7	17.9 ± 2.2

Table 5. Physiological parameters of the subjects

precision 0.5%, Germany). The validity of the SRM has been previously shown by Jones et al. (1998). The SRM system is a pedal equipped with 20 strain gauges that transmit data by induction to a control box located on the handlebars. Before each measurement, the SRM and the analysis software were calibrated according to manufacturer's recommendations. The SRM was mounted on a bicycle race (10.2 kg) equipped with clipless pedals. Before the start of each test, each cyclist has adjusted the bike to the position he usually uses. The tires were inflated to a pressure of 700 kPa.

Heart rate beat by beat (RR interval) was recorded during all experimental sessions using the Polar S810 heart rate monitor (Finland).

4.2.2 Protocol

The heart rate of subjects was recorded at rest for 5 minutes in a sitting position. Once the rider positioned on the bicycle, the skin temperature of the skin of the gastrocnemius muscle was measured. This temperature will be the reference temperature.

During exercise, cyclists had to perform an incremental progressive exercise on the bicycle in 18 minutes. The exercise was scheduled for external mechanical power of 100 W for 10 min, then the intensity increased in increments of 50 W for 3 min up to 200 W, then the last step was performed at 250 W for 2 min (Figure 8).

Once the exercise is performed, heart rate (standing) and skin temperature of the gastrocnemius muscles were measured over a period of 10 min.

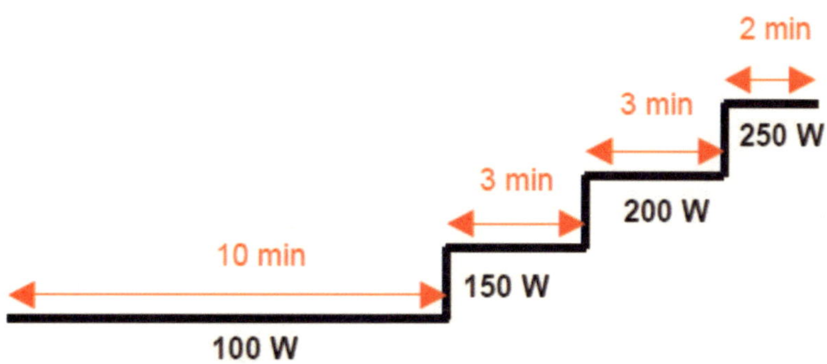

Fig. 8. Incremental bicycle exercise

4.2.3 Results

The objective of this preliminary study was to analyze in masters cyclists pedaling mechanics and to measure the skin temperature of the gastrocnemius muscle during incremental exercise.

Our results indicate (Table 6) that at a certain level of power (150 W) a significant difference ($P < 0.05$) between the peak maximum of the right and left engine torques (+ 10% on the right limb).

Figure 9 shows an example of pedaling pattern on one of the subjects at the level at 150 W showing how the pedaling pattern becomes asymmetric. Using a mathematical model

Cycling Power (W)	Torque Max Peak Left (N.m)	Max Peak Right (N.m)	Min Peak Left (N.m)	Min Peak Right (N.m)
100	19.2 ± 3.7	21.8 ± 2.5	4.3 ± 1.5	4.6 ± 1.9
150	27.5 ± 2.9	* 34.8 ± 4.2	7.5 ± 1.2	7.7 ± 2.3
200	34.4 ± 4.7	* 42.9 ± 3.5	10.4 ± 3.5	11.0 ± 3.3
250	43.0 ± 2.6	* 52.5 ± 6.6	12.9 ± 3.4	14.2 ± 4.7

Table 6. Min and max values of the torque in left and right lower limbs for different power levels

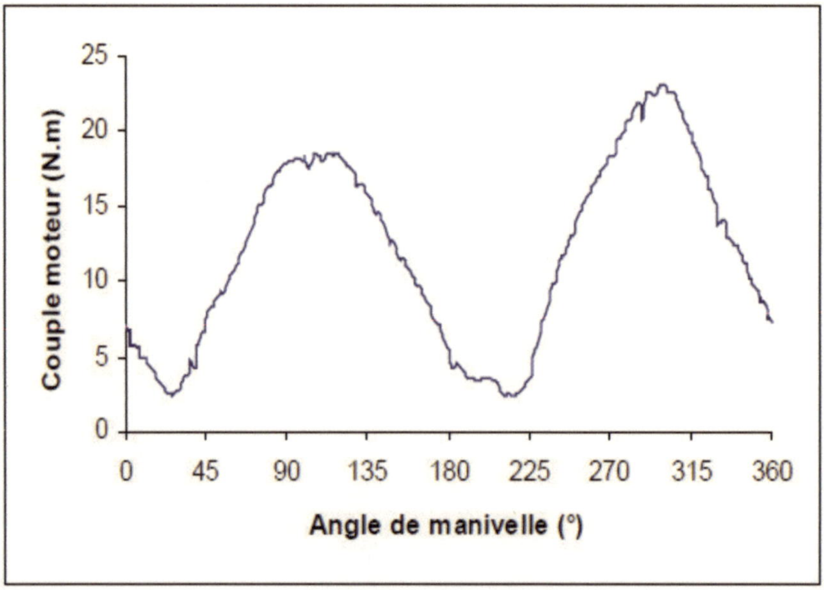

Fig. 9. Torque versus the crank angle for subject 1 at 150 W.

(Grappe et al, 1999), it is possible to estimate that a speed of 25 km/h on the flat (100 W) induces a symmetric pedaling while at 30 km/h (150 W) it becomes asymmetric. Cyclist masters of this study are specialized in long distance races (200 km or more) at a speed about 25 km/h. These results therefore suggest that cyclists have optimized their pedaling mechanics (kinetics and kinematics) only at the intensity they use most frequently. It is very likely that when the road profile is rugged, the power these riders develop is more than 100 W because this is a very low speed uphill (9 km/h on a 5% slope). This suggests that in each ascent the pedaling becomes asymmetric. These asymmetries may have several origins: 1). A marked muscle atrophy on one of the limbs, 2) A lesion of the motor command, or 3) An inappropriate gesture technique.

An imbalance in a pedal cycle repeated several thousand times may have negative effects especially in the cartilage of the knee and hip, and of course on performance. It is possible to improve the efficiency of the pedaling features, for example by performing exercises with feedback on the torque (Henke, 1998). In this way the rider can adjust itself the asymmetry.

Our study also highlight that the thermoregulatory mechanisms used by cyclists masters were variable among individuals. Even if a general trend, corresponding to the average curve of temperature, emerges, one can note the great variability of responses of the human body, and thus its adaptation from one subject to another (Figure 10).

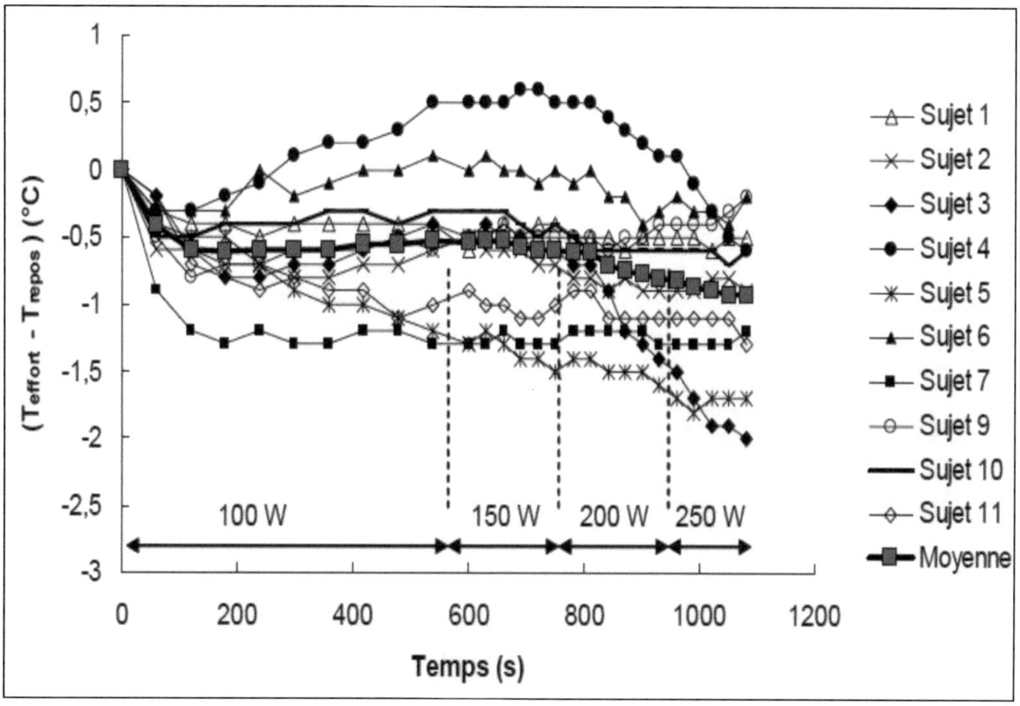

Fig. 10. Skin temperature evolution of the gastrocnemius muscles

For the majority of subjects (except subjects 4 and 6), calf skin temperature decreases with exercise intensity. The average temperature curve thus shows the levels for each power. This decrease in temperature can be related to mechanisms of heat removal employed by the human body. More the energy to evacuate is, the greater the temperature gradient across the fat must be high to allow the transfer by conduction. Then, this flow of heat must be transferred from the skin surface to the environment. During pedaling, and for low external mechanical powers, the thermal transfers are mainly convective, even if a share exchange radiation should not be overlooked. A contrario, if the exercise intensity increases, a sweating mechanism appears, allowing a much larger exchange by evaporation into the environment. The temperature difference between the surface of the skin and the air does not need to be as important as when only the convective heat exchange occurs. The cooling of the calf during the effort is clearly visible on maps of temperature in Figure 11. We also note that the cooling is not uniform over the entire surface of the skin. It will also differ from one individual to another.

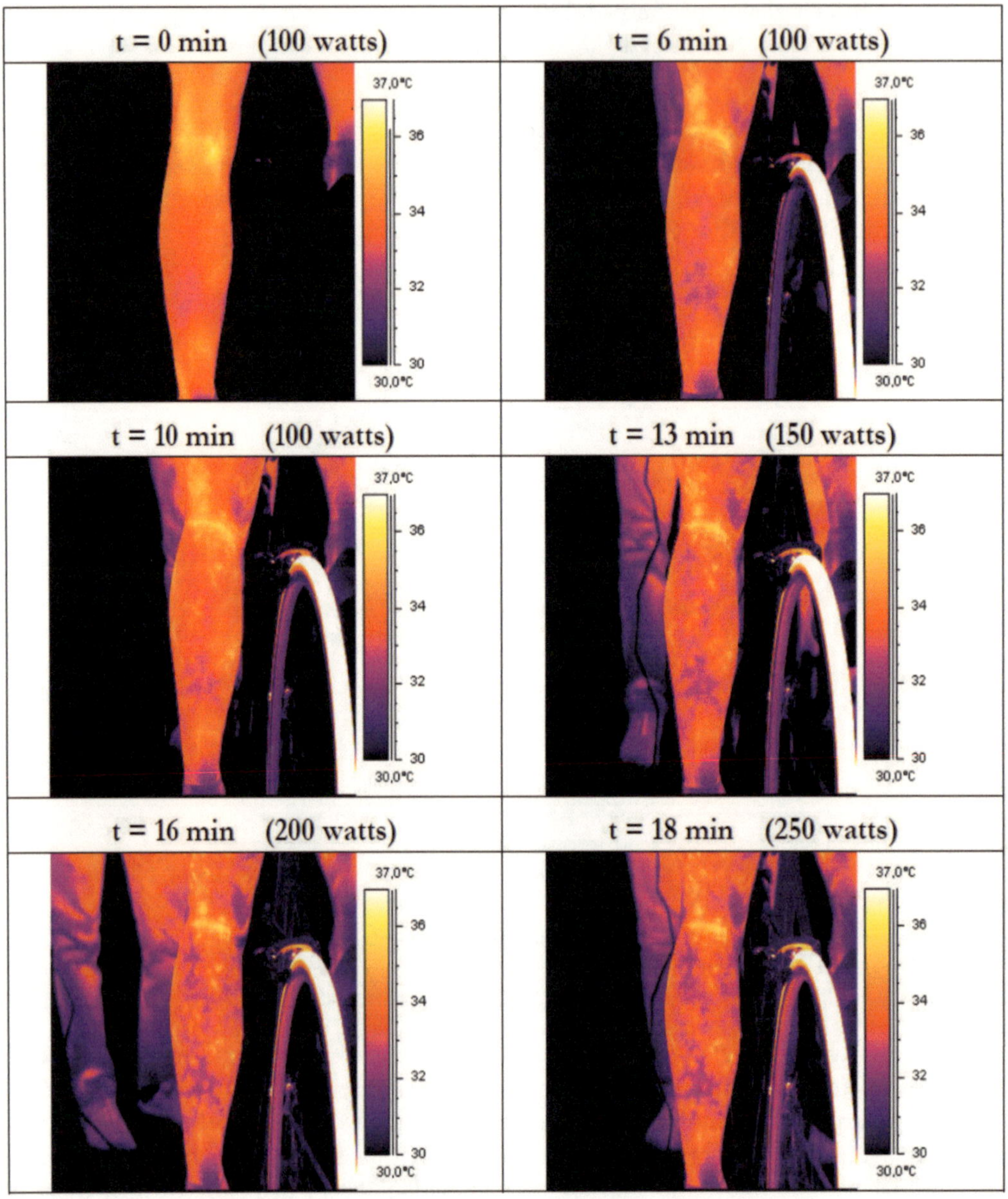

Fig. 11. Temperature maps of the calf during the protocol for subject 10

A thermomechanical analysis of Masters Cyclists pedaling was conducted in the laboratory. This preliminary study indicates that cyclists have an asymmetrical cycling from a relatively low level of power (100 W). Regarding thermoregulation during exercise, a gradual cooling of the calf as a function of external mechanical power is observed. This cooling is accompanied by strong inhomogeneities in the distribution of surface temperatures. These complex mechanisms of thermoregulation should be analyzed in future studies especially with regard to changes in heart rate and thickness of adipose tissue.

5. Infrared thermography and pathology diagnostic applied to sport activity

In this part, we aimed to study the feasibility of application of infrared thermography to detect osteoarthritis of the knee related to sport and to compare the distribution of skin temperature between participants with osteoarthritis and those without pathology. High performance training pushes the locomotor system to the edge of its anatomical and physiological limits. The knee is the most frequently affected joint in sports. Knee injuries are common in sports that involve jumping and abrupt direction changes such as football (Davidson & Laliotis, 1994). The need for further research in the field of injury prevention and management is crucial to counteract severe injuries.

Through its involvement in the activities of most humans, the knee is likely to be the seat to many diseases and injuries like osteoarthritis which is the participant of several clinical and scientific researches. Osteoarthritis is a degeneration of the cartilage without infection or special inflammation. This multi-factorial disease led to a more or less rapid destruction of cartilage that coats the ends of bones. Anatomically, this destruction is accompanied by a proliferation of bone under the cartilage. During cartilage destruction, small pieces of cartilage may break off and "float" in the articular pocket: then they trigger inflammatory attacks that result in mechanical hyper-secretion of fluid and swelling of the joint.

Horvath and Hollander (1949) measured the intra-articular temperature in patients with rheumatoid arthritis and noted that it could be used as a guide to the acuteness of inflammation. Bacon et al (1976) showed that measurement of mean skin temperature could be used as a measure of disease activity.

Infrared thermography is a diagnostic method providing information on the normal and abnormal sensory and nervous systems, trauma, or inflammation locally and globally. Infrared thermography shows physiological changes rather than anatomical changes and could be a new diagnostic tool to detect the pathology of the knee. The objective assessment of disease activity in OA is difficult. Many parameters are based on patients' symptoms, which may not give an accurate indication of the progress of the disease, and laboratory evaluation can be unhelpful.

The thermographic images have previously been used to examine anterior knee (Devereaux et al, 1986; Mangine et al 1987; Ben-Eliyahu, 1992). Ben-Eliyahu et al. (1992) investigated the clinical utility of infrared thermography in the detection of sympathetic dysautonomia in patients with patellofemoral pain syndrome. They have shown that the incidence of patellar thermal asymmetry was found to be statistically significant when tested by chi-2 analysis. More recently, the researches of Selfe et al. (2010) were aimed to investigate if palpation of the knee could classify patients into those with and those without cold knees and if this classification could be objectively validated using thermal imaging. They were unable to deduce a different response in skin temperature with cold stimuli between females whatever the initial temperature of knees is, namely cold and not. Self et al. (2010) concluded that further research was needed to assess the validity and reliability of the methods used to identify this subgroup of patients, to confirm the clinical profile.

5.1 Protocol and method

Ten participants with unilateral knee osteoarthritis (Men, between the ages of 17-26 years) and twelve reference participants without OA (Men, age range, 18-30 years) participated in this study (Table 7). For each participant, we reported the age, weight, height, body mass index (BMI), the dominant side, the existence of pain or other symptoms of functional knee.

Group	GA (Participants with knee osteoarthritis)	GB (Healthy participants)
Sex	Men	Men
Age (y) (Mean ± DS)	21,50 ± 2,51	23,08 ± 3,6
Weight (kg) (Mean ± DS)	73,46 ± 5,61	72,17 ± 3,67
Height (m) (Mean ± DS)	1,75 ± 0,06	1,76 ± 0,04
BMI (kg/m²) (Mean ± DS)	11,76 ± 2,27	12 ± 2,8
Assessment mean of pain (scale 0-4)	2,5±1,5	0

Table 7. Characteristics of the participants by subgroups

Patients were excluded if they had other pain. To exclude other causes of knee pain, all patients underwent a thorough history taking, a physical evaluation, as well as standard and dynamic series of plain radiographs of femur, patella, and tibia. No medication or other additional conservative treatments were given after enrollment.

The assessment of pain was based on the simple verbal scale on a scale of 0 to 4 (0=none; 1=low; 2=moderate; 3=severe; 4= Maximum).

All tests were conducted at the University of Science and Technology of Physical Activities and Sports in Reims (France). The participants of two groups (with and without osteoarthritis of the knee) ran on a treadmill (slope 0%) during 5 minutes, at a fixed speed of 8 km/h.

The patients were asked to avoid smoking, alcohol, coffee, and exercise for at least 5 hours before testing. We checked each patient's body temperature to ensure that there was no one with extreme body temperature (below 36.4 °C or above 37.2 °C). Air temperature and relative humidity were recorded at the start of measurement periods. IRT has been carried out; using the IR camera CEDIP TITANIUM HD560M, in a room where the temperature was maintained at 18 °C ± 0.5 °C and the relative humidity was 60%. It is important to ensure that the patient was relaxed before imaging so that his emotional state will not influence the measurements.

Before the testing, the patient must wear a short to allow the taking of thermograms of the knees. 30 minutes were needed to balance the patient's body temperature with the environment before resuming testing.

The participants remained motionless and the recording was made in the anatomical position: the body upright, feet in the longitudinal axis of the leg, forearm supination and the palms facing forward. IR thermograms of right and left knees were taken before and after the race at a distance of 1.50 m:

- Before the effort: IR images were taken
- After effort, recordings were taken during 5 minutes.

According to Maly et al. (2002) in normal conditions, 71-91% of body weight is transmitted to the tibia-femoral junction and can reach 100% in the presence of osteoarthritis. Average temperatures were recorded in two areas of each knee (points: 1 to 4 for right knees, and points: 5 to 8 for left knees) as shown in Figure 12. For the healthy participants, the temperature was averaged from these eight points, while only four points were considered for a participant having pathology of the knee.

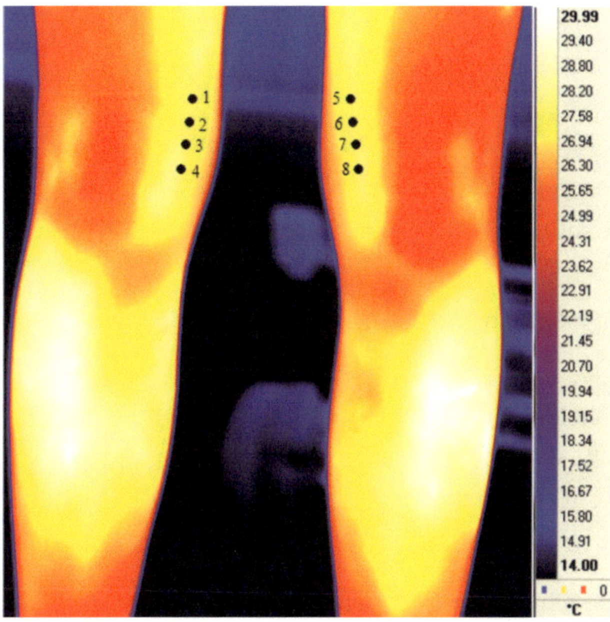

Fig. 12. Infrared thermography of right and left knees with the points selected for measuring the local temperature

5.2 Cartographies analysis

To ensure statistical conclusions, the temperature deduced from thermographic images is averaged over the study area and on all the participants of each group. It is recalled that 10 participants had knee osteoarthritis diseases while 12 participants were healthy. It is clearly shown in Fig. 13 that infrared thermography technique qualitatively enables highly visual estimate of such pathologies. For example in the case of a participant having osteoarthritis pathology in the right knee, the IR thermography in Fig. 13 reveals relevant disease by highlighting asymmetrical behavior in thermal color maps of both knees. It clearly appears by comparing a participant at rest with the same participant undergoing sporting activity that the more the knee is loaded, the more the warm thermal zone is extended. This is

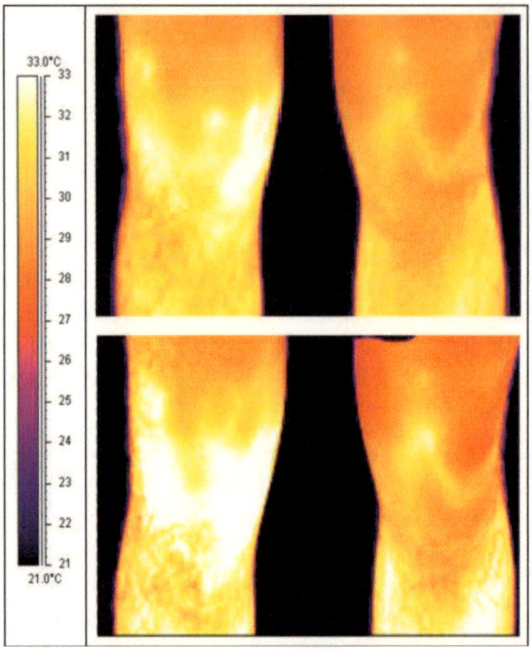

Fig. 13. Infrared thermography of right and left knees before and after race for a participant with right knee OA

probably due to the prompting of the knee, repetitive strain injury during the race increases inflammation in the knee and consequently the temperature of the skin. These findings have been observed in all participants of the GA group, while no temperature gradients appear between right and left knee thermal maps in healthy participants.

Qualitatively, and only from imagery examination, it seems that IR thermography as a simple and non-intrusive experimental device can be easily used as a powerful tool for rapid diagnosis of osteoarthritis of the knee.

5.3 Temperature distributions versus time

In Fig. 14 is represented the arithmetic average of the measured skin temperature in the knee area(s) versus time before and after the running protocol in place for both participant groups. At rest, significant differences occur depending on whether the knee is healthy or not. Thus we get an average measure of 25.83 °C in case of healthy knees while the presence of OA results in a 28.75 °C average temperature. This difference of almost 2 °C far exceeds the accuracy of the IR process, making it efficient for a reliable visual diagnosis of this disease.

After the race and whatever the participant group is, one observes a gap in the average temperature, namely about 1.15 °C for group with OA and 2.31 °C for healthy participants. This result is not paradoxical. It reflects only the fact that initially, at rest, the knee surface temperature was already high for this group of participants. During the acquisition time limited to five minutes, a similar behavior is observed in the temporal temperature evolution for both groups.

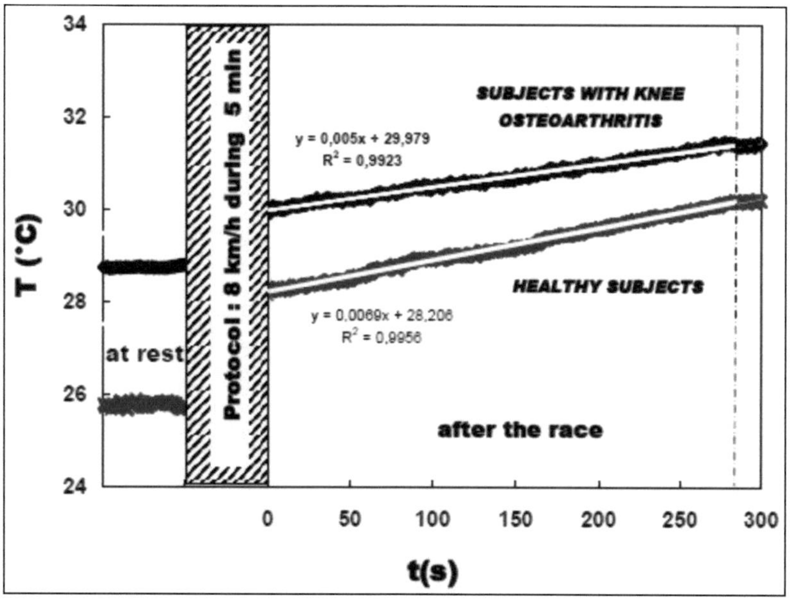

Fig. 14. Average temperature evolution versus time

One observes a perfect linear evolution (with a determination coefficient very close to the unit) of the temperature versus time during less than 5 minutes, followed by a tendency to an asymptotical behavior traducing the beginning of a relaxation thermal process. The slope of the linear regression is roughly the same order of magnitude whatever the group is. This means that there is no over-inflammation of the synovial fluid in OA after the race.

5.4 Relationship between pain intensity and the knee skin temperatures

It seems reasonable and interesting to answer the question of pain experienced by participants with OA, even if this question is based on a subjective analysis. Is there a relationship between the pain intensity and the knee temperature, itself being a feature of inflammation? Obviously, this study did not aim to establish a universal law but to draw a trend of feeling pain in participants with OA. In Fig. 15 is reported the pain intensity given by the participants versus the corresponding average temperature during the protocol. As a reference, is also reported the pain sensation in the case where participants are in rest. Although approximations are rough, these developments clearly show that the temperature can be regarded as a key parameter for evaluating pain. This reinforces the idea that the capture of temperature maps by infrared thermography as a diagnostic tool is certainly an interesting way to develop.

We measured surface temperature in normal and OA joints of the knee for 22 participants. IRT was performed in a room at constant temperatureduring experiments and patients were asked to follow strict instructions before examination. Many studies (Darton & Black 1990; Huygen et al, 2004; Varju et al, 2004; Zaproudina et al, 2009) have shown very good reproducibility of IRT in a temperature-controlled environment. The IRT was reliable to detect osteoarthritis of the knee by the way of distributions of skin temperature. In addition,

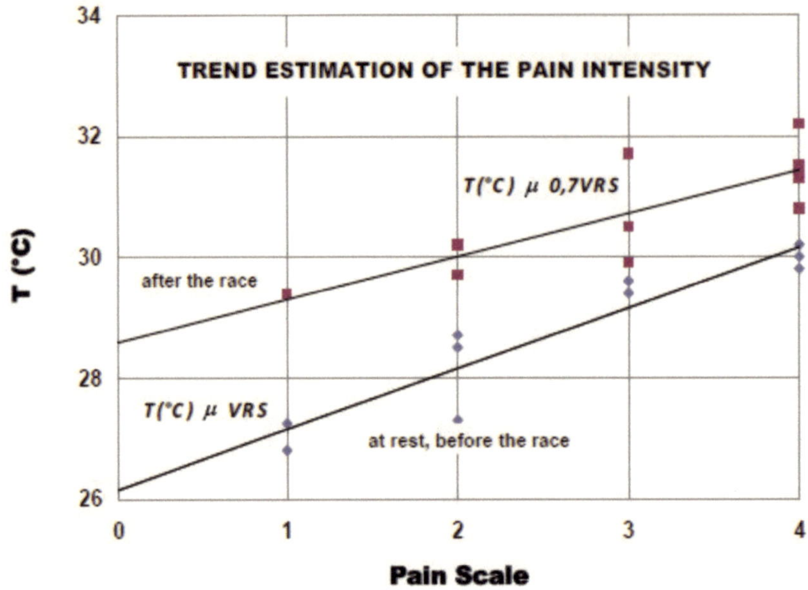

Fig. 15. Correlation between knee skin temperature and estimation of pain intensity

a correlation has been observed between knee skin temperature and the pain intensity due to OA.

Our study has demonstrated the stability of thermographic measurements over a very short period. It will be interesting to predict the disease progression of osteoarthritis and eventually determine a correlation between thermographic and radiographic images.

This study has shown that IRT appears to be a reliable diagnostic tool to detect quantifiable patterns of skin temperatures in participants with OA. It has been demonstrated that the temperature variation can be correlated with changes in pain intensity for the group GA that has osteoarthritis. We think that this non-intrusive technique enables to detect the early clinical manifestations of knee OA.

6. Conclusions

Due to its non-intrusive feature, infrared thermography (IRT) is a powerful investigation tool to be applied as well in sports performance diagnostic (due to the relationship between muscular energy and thermoregulation process) as in sports pathology diagnostic.

It is well known that sports activity induces a complex thermoregulation process where part of heat is given off by the skin of athletes. As not all the heat produced can be entirely given off, this follows a muscular heating resulting in an increase in the cutaneous temperature. For example, in sports activity, we have presented the usability of infrared thermography in swimming for the purpose of quantifying the influence of the swimming style on the cartographies of cutaneous temperatures of a swimmer similar analysis has concerned cycling activity. In particular, the IRT method will enable, in the long term, to quantify the heat loss according to the swimming style, and to consider the muscular and energy outputs during the stroke.

In the field of pathology diagnostic, applied to sports activity, the application of infrared thermography (IRT) has a long history in musculoskeletal trauma. Infrared thermography is a diagnostic method providing information on the normal and abnormal sensory and nervous systems, trauma, or local and global inflammation. Infrared thermography shows physiological changes rather than anatomical changes and could be a new diagnostic tool to detect the pathology of the knee. For example, IRT appears to be a reliable diagnostic tool to detect quantifiable patterns of skin temperatures in participants with knee osteoarthritis.

7. References

Albert, S.M., Glickman, M., Kallish, M. (1964). Thermography inorthopedics. *Ann N Y Acad Sci,* Vol.70. pp.121-157.

Bacon, P.A, Collins, A.J, Ring, F.J, Cosh, J. (1964). Thermography in the assessment of inflammatory arthritis. *Clin Rheum,* Vol.2. pp.51-65.

Ben-Eliyahu, D.J. (1992). Infrared Thermography and the Sports Injury Practice. *Dynamic Chiropractic,* Vol.10(7).

Ben-Eliyahu, D.J. (1992). Infrared thermographic imaging in the detection of sympathetic dysfunction in patients with patellofemoral pain syndrome. *J Manipulative Physiol Ther.,* Vol.15(3). pp.164-70.

Bogerd, C.P., Bruhwiler, P.A. (2008). The role of head tilt, hair and wind speed on forced convective heat loss through full-face motorcycle helmets: a thermal manikin study. *International Journal of Industrial Ergonomics,* Vol.38. pp.346-353.

Brandt, R.A., Pichowski, M.A. (1995). Conservation of energy in competitive swimming. *Journal of Biomechanics,* Vol.28. pp.925–933.

Bruhwiler, P.A., (2003). Heated, perspiring manikin headform for the measurement of headgear ventilation characteristics. *Measurement Science and Technology,* Vol.14 (2). pp.217-227.

Bruhwiler, P.A., Ducas, C., Huber, R., Bishop, P.A. (2004). Bicycle helmet ventilation and comfort angle dependence. *European Journal of Applied Physiology,* Vol.92 (6). pp.698-701.

Bruhwiler, P.A., Buyan, M., Huber, R., Bogerd, P., Szitman, J., Graf, S.F., Rosgen, T. (2006). Heat transfer variations of bicycle helmets. *Journal of Sports Sciences,* Vol.24 (9). pp.999-1011.

Bruhwiler, P.A. (2009). Role of the visor in forced convective heat loss with bicycle helmets. *International Journal of Industrial Ergonomics,*Vol.39 (1). pp.255-259.

Buyan, M., Brühwiler, P.A., Azens, A., Gustavsson, G., Karmhag, R., Granqvist, C.G. (2006). Facial warming and tinted helmet visors. *International Journal of Industrial Ergonomics,* Vol.36. pp.11-16.

Coyle, E., Feltner, M., Kautz, S., Hamilton, M., Montain, S., Baylor, A., Abraham, L., Petrek, G. (1991). Physiological and biomechanical factors associated with elite endurance cycling performance. *Med. Sci. Sports Exerc.* Vol.23. pp.93-107.

Coyle, E., Sidossis, L., Horowitz, J., and Beltz, J(1992). Cycling efficiency is related to the percentage of type 1 muscle fibers. *Med. Sci. Sports Exerc.* Vol.24. pp.782-788.

Darton, K, Black, CM. (1990). The use of infra-red thermography in a rheumatology unit. *Br J Rheumatol*, Vol.29. pp.291-293.

Davis, G.A., Edmisten, E.D., Thomas, R.E., Rummer, R.B., Pascoe, D.D., 2001. Effects of ventilated safety helmets in a hot environment. *International Journal of Industrial Ergonomics*, Vol.27. pp.321-329.

Davidson, T.M.; Laliotis, A.T. (1996). Alpine skiing injuries. *A nine year study. West. J. Med.*, Vol.164. pp.310–314.

De Bruyne, G., Aertsa, J.M., Slotenb, J.V., Goffinc, J., Verpoestd, I., Berckmans, D. (2010). Transient sweat response of the human head during cycling. *International Journal of Industrial Ergonomics*, Vol.40. pp.406-413.

Devereaux, M.D, Parr, G.R, Lachmann, S.M, Thomas, D.P, Hazleman, B.L. (1986). Thermographic diagnosis in athletes with patellofemoral arthralgia. *Journal of Bone and Joint Surgery British*, Vol.68-B(1). pp42-44.

Grappe, F., Candau, R., Barbier, B., Hoffman, M., Belli, A., Rouillon, J.D. (1999). Influence of tyre pressure and vertical load on coefficient of rolling resistance and simulated cycling performance. *Ergonomics*, Vol.10. pp.1361-1371.

Henke, T. (1998). Real-time feedback of pedal forces for the optimisation of pedaling technique in competitive cycling. *In Proceeding of the 16th Symposium of the International Society of Biomechanics in Sports*, University of Konstanz, Germany.

Hildebrandt, C., Raschner, C. and Ammer, K. (2010). An overview of Recent Application of Medical Infrared Thermography in Sports Medicine in Austria, *Sensors*, Vol.10. pp.4700-4715

Holmer, I. (1974). Energy cost of arm stroke leg kick and the whole stroke in competitive swimming styles. *Europ. J. Appl. Physiol.*, Vol.33. pp.105–118.

Hoover, K.C., Burlingame S.E., Lautz C.H. (2004). Opportunities and challenges in concrete with thermal imaging. *Concrete International*, Vol.26(12). pp.23-27

Horvath, SM, Hollander, J L (1949). Intra-articular temperature as a measure of joint reaction. *J Clin Invest*, Vol.73. pp.441-469.

Huttunen, P., Lando, N.G., Meshttsheryakov, V.A., Lyutov, V.A. (2000). Effects of long-distance swimming in cold water on temperature, blood pressure and stress hormones in winter swimmers. *Journal of Thermal Biology*, Vol.25. pp.171–174.

Huygen, FJ, Niehof, S, Klein, J, Zijlstra, FJ. (2004). Computer-assisted skin videothermography is a highly sensitive quality tool in the diagnosis and monitoring of complex regional pain syndrome type I. *Eur J Appl Physiol*, Vol.91. pp.516-540.

Hsu, Y.L., Tai, C.Y., Chen, T.C. (2000). Improving thermal properties of industrial safety helmets. *International Journal of Industrial Ergonomics*, Vol.26 (1). pp.109-117.

Jansky, L., Vavra, V., Jansky, P., Kunc, P., Knzkova, I., Jandova, D., Slovacek, K. (2003). Skin temperature changes in humans induced by local peripheral cooling. *Journal of Thermal Biology*, Vol.28. pp.429–437.

Jiang, L.J.; Ng, E.Y.K.; Yeo, A.C.B.; Wu, S.; Pan, F.; Yau, W.Y.; Chen, J.H.; Yang, Y. (2005). A perspective on medical infrared imaging. *J. Med. Eng. Tech.*, Vol.29. pp.257–267.

Jones, S.M, Passfield L. (1998).Dynamic calibration of bicycle power measuring cranks. In: Haake SJ (ed). The Engineering of sport. *Oxford: Blackwell Science*, pp.265-274.

Kaminski, A., Jouglar, J., Volle, C., Natalizio, S., Vuillermoz, P.L., Laugier, A. (1999). Non-destructive characterization of defects in devices using infrared thermography. *Microelectronics Journal,* Vol.30. pp.1137–1140

Kautz, S., Hull, M. (1995). Dynamic optimisation for equipment setup problems in endurance cycling. *J. Biomech.,* Vol.28. pp.1391-1401.

La thermographie infrarouge: Principes, technologies, applications Gilbert Gaussorgues, (1999)

Lavoie, J.M., Montpetit, R. (1986). Applied physiology of swimming. *Sport Med.,* Vol.3. pp.165–189.

Maly, M.R, Culham, E.G, Costigan, P.A. (1987).Static and dynamic biomechanics of foot osteoarthritis in people with medial compartment knee osteoarthritis. *Clinical Biomechanics,* Vol.17. pp.603-610.

Mangine, RE, Siqueland, KA, Noyes, FR. (1987). The Use of Thermography for the Diagnosis and Management of Patellar Tendinitis. *JOSPT,*Vol.9(4). pp.132-140.

Melnizky, P., Schartelmüller, T., Ammer, K. Prüfung der intra- und interindividuellen Verlässlichkeit der Auswertung von Infrarot-Thermogrammen. *Eur. J. Thermol.,* Vol.7. pp.224–226

Mougios, V., Deligiannis, A. (1993). Effect of water temperature on performance, lactate production and heart rate at swimming of maximal and submaximal intensity. *J. Sports Med. Phys. Fitness,* Vol.33. pp.27–33.

Moussay, S., Bessot, N., Gauthier, A., Larue, J., Sesboue, B., Davenne, D. (2003). Diurnal variations in cycling kinematics. *Chronobiol Int.,* Vol.20. pp. 879-892

Planck, M. (1901). On the law of distribution of energy in the normal spectrum. *Ann. Phys.,* Vol. 4, pp.553.

Robinson, S., Somers, A. (1971). Temperature regulation in swimming. *J. Physiol.,* Vol.63. pp.406-409.

Schmidt, F.W, Henderson, R.E, Wolgemuth, C.H (1993). Introduction to thermal sciences. *New York: Wiley.*

Selfe, J., Sutton, C., Hardaker, N.J., Greenhalgh, S., Karki, A., Dey, P., (2010). Anterior knee pain and cold knees: A possible association in women. *The knee,* Vol.17. pp.319 - 323.

Selfe, J , Sutton, C, Hardaker, N.J, Greenhalgh, S, Karki, A, Dey, P (2010). Anterior knee pain and cold knees: A possible association in women. *The Knee,* Vol.17. pp.319-323.

Steketee, J. (1973). Spectral emissivity of skin and pericardium. *Phys. Med. Biol.,* Vol.18. pp.686–694.

Varju, G, Pieper, C.F, Renner, J.B, Kraus V.B. (2004). Assessment of hand osteoarthritis: correlation between thermographic and radiographic methods. *Rheumatology,* Vol.43. pp.915-924.

Wu, C.L., Yu, K.L., Chuang, H.Y., Huang, M.H., Chen, T.W., and Chen, C.H. (2009). the application of Infrared Thermography in the assessment of patients with Coccygodynia before and after manual therapy combined with Diathermy, *Journal of Manipulative and Physiological Therapeutics,* pp. 281-293 www.flirthermography.com

Yanai, T. (2001). Rotational effect of buoyancy in frontcrawl: does it really cause the legs to sink? *Journal of Biomechanics,* Vol. 34. pp.235–243.

Zaproudina, N, Varmavuo, V, Airaksinen, O, Narhi, M. (2008). Reproducibility of infrared thermography measurements in healthy individuals. *Physiol Meas,* Vol.29. pp.515-539.

Part 3

Engineering Applications

8

Thermography Applications in the Study of Buildings Hygrothermal Behaviour

E. Barreira, V.P. de Freitas, J.M.P.Q. Delgado and N.M.M. Ramos
LFC – Building Physics Laboratory, Civil Engineering Department, Faculty of Engineering, University of Porto Portugal

1. Introduction

Infrared thermography (IRT) can be defined as the science of acquisition and analysis of data from non-contact thermal imaging devices. The process of thermal imaging has simplified over the years with the availability of efficient, high resolution infrared cameras that convert the radiation sensed from the surfaces into thermal images (Rao, 2008). Thermography literally means "writing with heat", just as photography implies "writing with light". The invisible infrared radiation emitted by bodies is converted into temperature and displayed as thermal images, the thermographs.

Infrared thermography is a powerful tool for engineers, architects and consultants for use in evaluating existing buildings and structures. Infrared thermography is a fast and reliable tool to assist in identifying potential problems in existing buildings.

Infrared thermography offers several advantages in condition surveying. Recent developments in thermography and image processing made the technique a valuable addition to the repertoire of nondestructive testing methods. Thermography is a non-contact, non-destructive technique. While the potential exists, thermography has not yet been utilized extensively in the assessment of monuments and ancient structures. Condition surveys by conventional techniques cannot detect the presence and source of moisture readily, require access to the surfaces, and are expensive and time consuming. On the other hand, IRT offers a rapid method for assessing large surfaces without the need of a scaffold to reach the area under investigation.

2. A brief history of infrared thermography

Thermography has a long history, although its use has increased dramatically with the commercial and industrial applications of the past fifty years. Sir William Herschel, an astronomer, discovered infrared in 1800. He built his own telescopes and was therefore very familiar with lenses and mirrors. Knowing that sunlight was made up of all the colours of the spectrum, and that it was also a source of heat, Herschel wanted to find out which colour(s) were responsible for heating objects. He devised an experiment using a prism, paperboard, and thermometers with blackened bulbs where he measured the temperatures of the different colours. Herschel observed an increase in temperature as he moved the thermometer from violet to red in the rainbow created by sunlight passing through the prism. He found that the hottest temperature was actually beyond red light. The radiation causing this heating was not visible; Herschel termed this invisible radiation "calorific rays". Today, we know it as infrared.

Approximately thirty years after the infrared discovery the first detector using this type of radiation was developed. This first infrared detector was based in the same principles of the thermocouples and it was referred as “Thermopiles”. In 1880, the discovery of bolometers (materials whose electrical resistance changes with temperature) allowed a significant improvement in sensitivity for infrared rays detection.

Between 1870 and 1920, the technology advance allowed the development of the first quantum detectors, based on the interaction between radiation and matter. The detection nature was changed, because the electrical signal created by the effect of thermal radiation finished and now there is a direct conversion of radiation into electrical signals. With this type of detector the response time was reduced considerably and measurement accuracy was increased.

Between the decades of 30 and 60 several infrared detectors were developed, essentially for military purposes. The wavelengths range that the infrared detectors were sensitive depended on the materials used in its manufacture: Lead sulphide (PbS) sensitive in the range between 1.5 to 3 μm; Indium antimonide (InSb) sensitive in the range between 3 to 5 μm; and Mercury - Cadmium - Tellurium (HgTeCd) sensitive in the range between 8 to 14 μm. All these detectors were working with optical-mechanical scan systems and requiring cryogenic cooling.

The first commercial infrared cameras appeared in the end of 60's. In the '90s a new generation of equipment with array detectors appeared in the market. These new equipments allowed a simultaneous temperature reading at different points and did not require cryogenic cooling systems.

An infrared camera is a non-contact device that detects infrared energy (heat) and converts it into an electronic signal, which is then processed to produce a thermal image on a video monitor and perform temperature calculations. Heat sensed by an infrared camera can be very precisely quantified, or measured, allowing you to not only monitor thermal performance, but also identify and evaluate the relative severity of heat-related problems. Recent innovations, particularly detector technology, the incorporation of built-in visual imaging, automatic functionality, and infrared software development, deliver more cost-effective thermal analysis solutions than ever before. Digital image storage produces calibrated thermal images that contain over 78000 independent temperature measurements that can be measured at any time.

2.1 Infrared thermography applied to building physics

It is essential that studies be performed to evaluate the performance of building materials, especially nowadays when new materials and techniques with unknown characteristics are often being used. These studies are a step towards improving technical solutions and regulations to ensure the building’s durability and to guarantee user comfort and satisfaction.

Most material pathologies are related with temperature action. Therefore, measuring a material's temperature is crucial for understanding the causes of those defects. The use of non-destructive techniques to test a building material may be very useful by making it possible to evaluate a material behaviour without destroying it and without interfering with the users' life (Avdelidis and Moropoulou, 2004; Grinzatoa *et al.*, 1998 and Haralambopoulos and Paparsenos, 1998).

Infrared thermography is a non-destructive testing technology that can be used to determine the superficial temperature of objects. Cameras collect infrared radiation emitted by the surface, convert it into electrical signals and create a thermal image showing the body's superficial temperature distribution (NEC, 1991). In this process, each shade expresses a specific temperature range (see Figure 1).

Fig. 1. Thermogram and visible image of the Carmo Church's façade covered with "azulejos" (ceramic tile) in Porto.

This technology has been applied to buildings for a couple of decades, to evaluate the building performance. According to Hart (1991) thermography can be used to detect insulation defect, air leakage, heat loss through windows, dampness and "hidden details" (subsurface pipes, flues, ducts, wall ties, etc.). It can also be used for examination of heating systems and preventive maintenance.

Despite thermography's potential uses to buildings, its application to building materials has not been greatly studied yet. The parameters that may affect measurements aren't completely understood and interpreting the results becomes difficult and confusing.

The main objective of this work was to evaluate the applicability of thermography in order to study the behaviour of building materials. To do so, some simple experiments were carried out at the Building Physics Laboratory (LFC) of the Engineering Faculty of Porto University (FEUP).

A sensibility study was performed with LFC's equipment to evaluate how measurements are influenced by emissivity, environmental conditions (temperature and relative humidity), colour and reflectivity. It was also possible to visualise the wetting and drying process of specimens, as water evaporation is an endothermic reaction inducing local surface cooling. And, lastly, the comfort of some interior floor coatings was evaluated by comparing thermal images obtained from the sole of a barefoot after having been in contact with different materials.

3. Fundaments of infrared thermography

3.1 The electromagnetic spectrum and the infrared rays

Infrared energy is part of the electromagnetic spectrum and behaves similarly to visible light. It travels through space at the speed of light and can be reflected, refracted, absorbed, and emitted. The wavelength (λ) of infrared energy is about an order of magnitude longer than visible light, between 0.7 and 1000 μm. Other common forms of electromagnetic radiation include radio, ultraviolet, x-ray and γ-ray (see Figure 2).

The electromagnetic spectrum is the range of all possible frequencies of electromagnetic radiation. It is the characteristic distribution of electromagnetic radiation emitted or absorbed by that particular object. The electromagnetic spectrum extends from low frequencies used for modern radio to gamma radiation at the short-wavelength end, covering wavelengths from thousands of kilometers down to a fraction of the size of an atom. The long wavelength limit is the size of the universe itself, while it is thought that the short wavelength limit is in the vicinity of the Planck length, although in principle the spectrum is infinite and continuous.

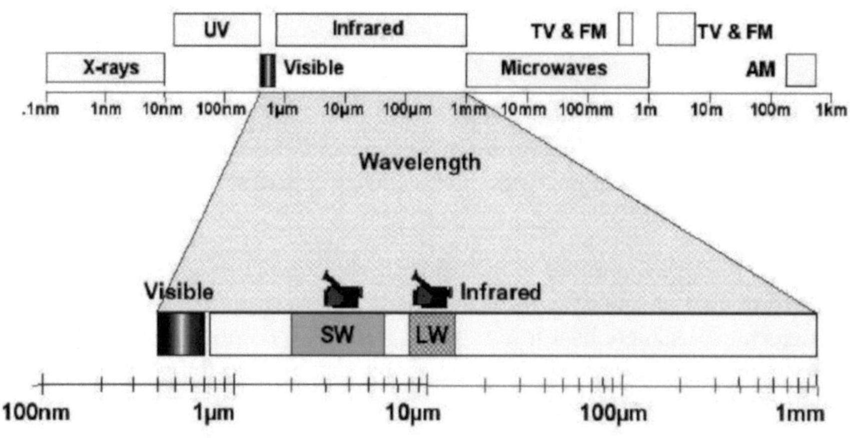

Fig. 2. Electromagnetic Spectrum.

The infrared part of the electromagnetic spectrum covers the range from roughly 1000 μm to 0.75 μm. It can be divided into three parts: far-infrared, from 1000 μm to 20 μm, mid-infrared, from 20 μm to 1.5 μm and near-infrared, from 1.5 μm to 0.75 μm.

3.2 Radiation emitted by a surface

The study of the radiation emitted by real surfaces is related with the concept of blackbody. A blackbody is an ideal surface that absorbs all incident radiation, regardless of wavelength and direction. The blackbody is a diffuse emitter and no other surface can emit more radiation than it, for a certain temperature and wavelength (Incropera and Witt, 2001).

Although there are no perfect absorbers or emitters, in a practical point of view there are two possible ways to guarantee the blackbody behaviour:

- A cavity with a small aperture, whose inner surface is at uniform temperature. If radiation enters the cavity it can be considered almost entirely absorbed as it will experience many reflections before coming out.
- A surface coating with very high absorbance.

The spectral emissive power of the blackbody is ruled by the Plank distribution (Gaussorgues, 1999):

$$\frac{dR(\lambda,T)}{d\lambda} = \frac{2.\pi.h.c^2.\lambda^{-5}}{\exp\left(\frac{h.c}{\lambda.k.T}\right) - 1} \quad (1)$$

where $dR(\lambda,T)/d\lambda$ is the spectral radiance or emissive power per unit area of the blackbody and wavelength, h is the Planck constant (h=6.626176x10^{-34} Js), k is the Boltzmann constant (k=1.380662x10^{-23} J/K), c is the speed of light (c=2.998x10^{8} m/s), λ is the wavelength (m) and T is the temperature (K).

The Planck distribution can be plotted as a family of curves, considering a certain temperature (see Figure 3). From the Planck distribution it is possible to note that the emissive power is zero when the wavelength is zero, it increases continuously with wavelength until a maximum is achieved, corresponding to a wavelength λ_{max}. At any wavelength the emitted radiation increases with temperature (Incropera and Witt, 2001).

The value of the wavelength λ_{max}, corresponding to the maximum emissive power for a certain temperature, is given by the Wien Law that is plotted by the curve in Figure 4.

$$\lambda_{max} = \frac{2898}{T} \quad (2)$$

According to Eq. (2), the maximum spectral emissive power is displaced to shorter wavelengths with increasing temperature. For solar radiation, emitted at a temperature around 6 000 K, the maximum emissive power occurs for a wavelength around 0.50 μm, in the middle of the visible spectrum. For blackbodies emitting at a temperature near the temperature of terrestrial surfaces, around 300K, the peak emission occurs at 10 μm and for the liquid nitrogen (T = 77 K) the peak emission occurs at 40 μm, both in the infrared spectrum.

The Stefan-Boltzmann law results from the integration of the Planck distribution in the domain [$\lambda = 0$; $\lambda = \infty$] and it allows calculating the total amount of radiation emitted by the blackbody at a certain temperature T (in every direction and over all wavelength) (Incropera and Witt, 2001).

$$R_t = \sigma.T^4 \quad (3)$$

where R_t is the total blackbody spectral radiance, σ is the Stefan-Boltzmann constant ($\sigma = \left(2.\pi^5.k^4\right)/\left(15.c^2.h^3\right) = 5.67 \times 10^{-8}\ \mathrm{W/m^2K^4}$) and T is the temperature (K).

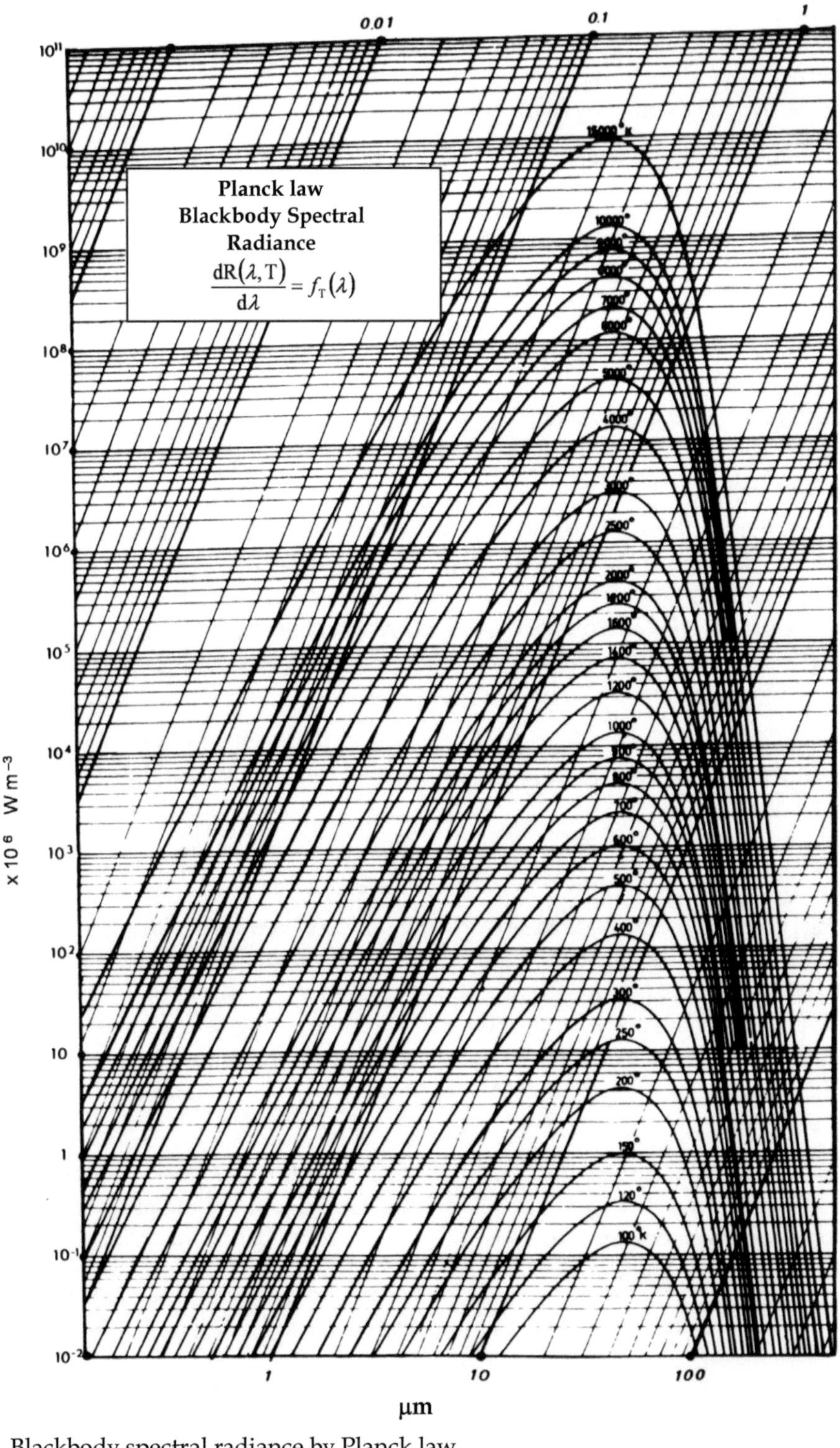

Fig. 3. Blackbody spectral radiance by Planck law.

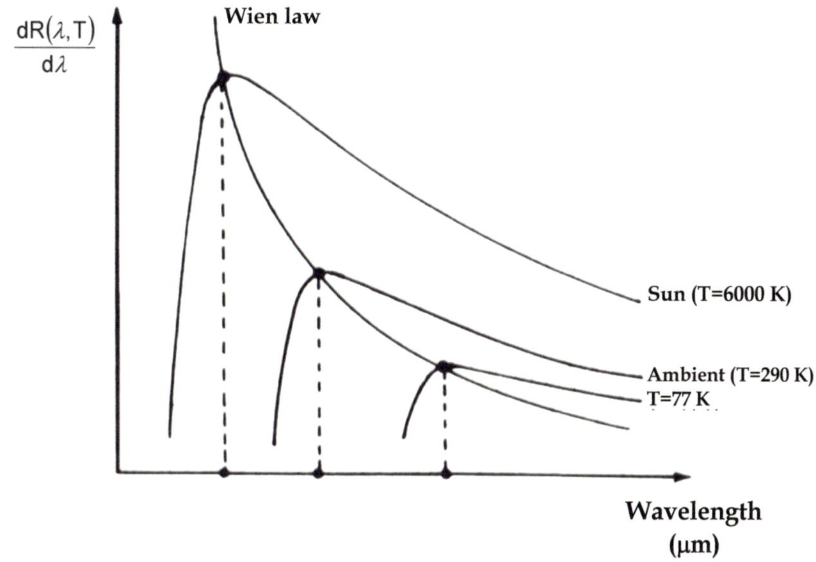

Fig. 4. Wien law (Gaussorgues, 1999).

Normally real surfaces don't behave like a blackbody. The property that traduces the emissive capacity of a real surface is emissivity (ε) that can be defined as the ratio of the radiation emitted by the surface to the radiation emitted by a blackbody at the same temperature (Incropera and Witt, 2001).

$$\varepsilon(\lambda) = \frac{\dfrac{dR(\lambda,T)}{d\lambda}}{\dfrac{dR_{cn}(\lambda,T)}{d\lambda}} \tag{4}$$

When the spectral component of the radiant energy interacts with a semitransparent surface, part of the radiation may be reflected, absorbed and transmitted (see Figure 5). The reflectivity (ρ) is a property that determines the fraction of the incident radiation that is reflected by the surface, the absortivity (α) is a property that determines the fraction of the irradiation absorbed by the surface and the transmissivity (τ) is the property that determines the fraction of the irradiation that is transmitted through the surface. Theses parameters depend on the wavelength, but for the same wavelength they are equal to one (Hagentoft, 2001).

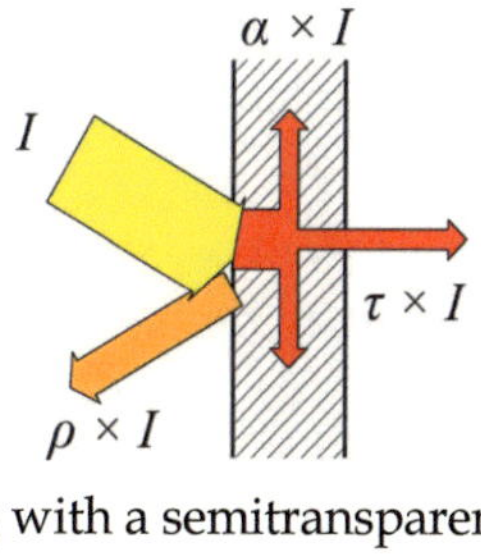

Fig. 5. Radiant energy interacting with a semitransparent surface.

$$\alpha(\lambda)+\rho(\lambda)+\tau(\lambda)=1 \quad (5)$$

where $\alpha(\lambda)$ is the spectral absorption for a given wavelength, $\rho(\lambda)$ is the reflection spectrum for a given wavelength and $\tau(\lambda)$ is the spectral transmittance for a given wavelength.

According to the Kirchhoff Law, at thermal equilibrium, the emissivity of a surface equals its absorptivity (Incropera and Witt, 2001).

$$\alpha(\lambda)=\varepsilon(\lambda) \quad (6)$$

Considering Kirchhoff Law, Eq. (5) can be rewrited as:

$$\varepsilon(\lambda)+\rho(\lambda)+\tau(\lambda)=1 \quad (7)$$

Some materials present specific values for these parameters:

Blackbody	$\varepsilon(\lambda)=1$	$\rho(\lambda)=0$	$\tau(\lambda)=0$
Transparent surface	$\varepsilon(\lambda)=0$	$\rho(\lambda)=0$	$\tau(\lambda)=1$
Perfect mirror	$\varepsilon(\lambda)=0$	$\rho(\lambda)=1$	$\tau(\lambda)=0$
Opaque surface	$\varepsilon(\lambda)+\rho(\lambda)=1$		$\tau(\lambda)=0$
Grey body	$\varepsilon(\lambda)=$ constant		

The non-blackbody emitters for which the emissivity is constant regardless wavelength are called grey bodies. The total amount of radiation emitted by the grey body can be calculated using the Stefan-Boltzmann law and considering the emissivity of the surface, ε.

$$R_t=\varepsilon.\sigma.T^4 \quad (8)$$

3.3 The surface emissivity affecting infrared thermography measurements

Emissivity depends on the wavelength (λ), direction (θ) and temperature (T).

$$\varepsilon=f(\lambda,\theta,T) \quad (9)$$

Generally, for solid materials spectral emissivity don't varies significantly but for liquids and gases the fluctuations are more obvious. For metals, spectral emissivity reduces with wavelength and for non-metals it tends to increase.

The angular dependence of emissivity implies different apparent emissivity's for a non-plane surface. For non-metals the emissivity is high and almost doesn't change for angles between 0º and about 60º from the perpendicular. After 70º its value quickly declines. For metal the emissivity values are lower and practically constant between 0º and 40º, increasing rapidly for higher angles.

Most materials used in buildings are characterized by having emissivity independent from the direction (diffuse surface) and wavelength (a behavior that is similar to the grey body). For that reason, for these materials emissivity can be assumed as constant, for a certain temperature, being quantified considering the normal to the surface and all wavelengths (Hart, 1991):

$$\varepsilon_t = \frac{\int_0^\infty \varepsilon(\lambda)\frac{dR(\lambda,T)}{d\lambda}.d\lambda}{\int_0^\infty \frac{dR(\lambda,T)}{d\lambda}.d\lambda} = \frac{1}{\sigma T^4}\int_0^\infty \varepsilon(\lambda)\frac{dR(\lambda,T)}{d\lambda}.d\lambda \tag{10}$$

The emissivity calculated by Eq. (10) is termed total emissivity and is defined as the ratio of the total energy emitted by the surface to the total energy emitted by a blackbody at the same temperature. Metals present lower values for total emissivity that increases with temperature. Oxide formation on the metal surface changes considerably emissivity. Non-metals have higher total emissivity, normally above 0.80, decreasing with temperature.

Over the temperature range experienced by buildings, from -10° C to 60° C, the temperature dependence of the emissivity can be, however, ignored and there are several databases of emissivity values in the literature. Table 1 presents some examples for metals and non-metals.

Material	Temperature (° C)	Emissivity
Stainless steel	25 - 100	0.79 – 0.80
Steel, non-oxidized, polished	100	0.07 – 0.08
Aluminum, non-oxidized, polished	0 - 100	0.03 – 0.06
Water	0 - 100	0.93 – 0.98
Concrete	-	0.92 – 0.94
Ceramic	21	0.93
White paper	20	0.70 – 0.95
Limestone	38	0.95
Human skin	32	0.98

Table 1. Emissivity of some materials.

Although the availability of emissivity values in the literature, for more accurate results it may be necessary to know the real value of total emissivity for the surface under study. In these cases, some simple methods can be used to define its value.

One of the most common methods consist in measuring the surface temperature using, for example, a thermocouple and adjust the emissivity value of the surface in a manner that the temperature measured by the infrared camera is equal to the one obtained by the thermocouple.

It is also possible to cover on the surface some tape with very high absorption or to paint black a small area of it. After thermal equilibrium is achieved, the temperature on the coated surface is measured, considering emissivity close to 1, as well as the surface that is not coated with the tape or the paint. As the temperature on the two surfaces must be similar, emissivity of the surface must be adjusted until equal values are achieved.

3.4 Other factors affecting infrared thermography measurements

The material reflectivity may cause some problems in infrared radiation measurements. The energy captured by the receptor, resulting from the radiation emitted by the body at temperature T_0, results by the following three effects (see Fig. 6):

- The body reflects a fraction of the energy emitted by the atmosphere, equal to the enegy emitted by a blackbody at temperature T_a.
- If the body is partially transparent, transmits a fraction of the radiation emitted by the background, equal to the energy emitted by a blackbody at temperature T_f.
- The body emits radiation to be at temperature T_0.

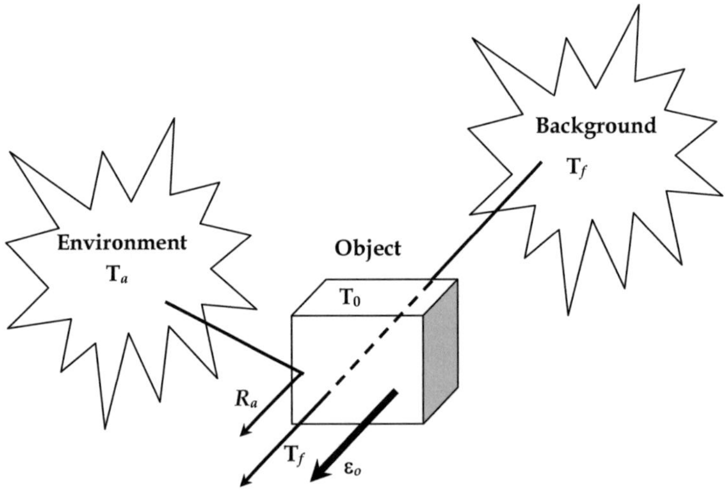

Fig. 6. Factors affecting infrared thermography measurements.

If the purpose of measuring is to determine the object temperature, T_0, from the detected infrared radiation in a spectral range, $\Delta\lambda$, then:

$$R_{\text{det}} = \int_{\Delta\lambda} \rho_a(\lambda)\frac{dR(\lambda,T_a)}{d\lambda}.d\lambda + \int_{\Delta\lambda} \tau_f(\lambda)\frac{dR(\lambda,T_f)}{d\lambda}.d\lambda + \int_{\Delta\lambda} \varepsilon_0(\lambda)\frac{dR(\lambda,T_0)}{d\lambda}.d\lambda \qquad (11)$$

To determine T_0 it is necessary to know the first two terms of Eq. (11) and the body spectral emissivity, $\varepsilon_0(\lambda)$. However, this problem may be simplified if the body is opaque, $\tau_f(\lambda) = 0$, which neglect the second term of Eq. (11). On the other hand, if T_0 is greater than T_a, the first term of Eq. (11) may be neglected, and we only need to know the body emissivity, $\varepsilon_0(\lambda)$.

However, if the body is surrounded by other bodies at different temperatures, greater than the object in study, the term related to its emission, function of its temperature T_0 and emissivity ε_0, is affected by an error due to reflection by this body of the radiation emitted by objects that surround him, with temperature T_a and emissivity ε_a. If ρ_a is the reflection factor of the body in study, the new term is proportional to T_a, ε_a and ρ_a, and assumes a great importance if T_a is greater than T_0 and ρ_a not equal to zero (Gaussorgues, 1999).

Although thermal reflections are common, they can be easily identify during the measurements, as their appearance in the thermal image, as cold or hot areas, will change its relative position depending on the angle of the camera. Once identified, their sources can be eliminated. If surfaces have very low emissivity, like metals, the thermal image will be almost made up of reflected radiation. In this case, the entire counter reflected radiation must be considered in other to estimate the surface temperature. Other solution is to change the surface characteristics, by covering it with paint or tape prior to the measurement (Hart, 1991).

The atmosphere presence between the emission source and the sensor causes disturbances in the measurement. In addition to the attenuation resulting from propagation in the atmosphere, the temperature gradients and turbulence create inhomogeneities in the air refractive index, which cause degradation of image quality. The atmospheric attenuation is the cause of major problems during the measurement because it entails a systematic error, which depends on the spectral range used, the observation distance and weather conditions. For these reasons, the measurements taken at distances great than 10 m should be corrected.

The thermal image of an object depends on the heat transfer between the surface and surroundings. If there are no external heat sources, the surface temperature of a building element is a function of temperature difference between inside and outside and the heat resistance of the various layers.

It is very important to know all the building element layers, the physical phenomenon of heat transfer and ensure that no unexpected thermal influences appeared. On the other hand, if the temperature difference between inside and outside increase the heat flow across the building element increase and the thermal image shows areas with lower thermal resistance. The temperature difference should be taken into account during the results analyse.

The convective heat flux, through or around the buildings, is a result mainly from heating and cooling areas, chimney effect and wind. The upward movement of warm air from radiators or other heat sources and and the cold air down from the windows, can cause distortions in surface temperatures. The chimney effect causes infiltrations on the lower floors and exfiltrations on the upper floors. This phenomenon causes the cooling of the walls on the lower floors and heating the upper floors.

The wind, blowing at a significantly speed, also affects the thermal images. The positive pressure on the façade facing the wind originates air infiltration and negative presures on the remaining façades cause air exfiltration. It is not necessary for the air to completely cross the building envelope to affect surface temperatures, since the air flow on the outer surface of the façade reduces its surface thermal resistance, cooling it down. This effect is differential, since the cooling is higher at the corners.

As the wind or the stack effect, also the mechanical pressurization or depressurization of a building distorts the distribution of surface temperatures of the building element. The pressurization increases the defects when viewed from the outside and smoothes them when viewed from the inside. The opposite occurs when the building is depressurized. The pressure difference in the façade is therefore quite relevant when conducting thermographic testing. In some cases, buildings can even be pressurized or depressurized mechanically, just to highlight the flaws and make them easier to detect.

For a study inside a building, air draughts directed to the area to be measured resulting for example from an open window or vents should be avoided.

The moisture resulting from surface condensation also has a considerable effect on the temperature of the studied element, as it changes the local transmission of heat and causes evaporative cooling. This phenomenon is more severe in areas where there are construction defects, such as thermal bridges or lack of thermal insulation. It is essential to take account of their existence in a later analysis.

A strange heat source and perhaps the most important from the outside is the sun. On clear days, the heat radiated by the sun to the building's façade conceals completely the results of heat transfer through the surrounding. The information obtained inside can also be affected, not only due to heating of the exterior façade, which changes the normal flow of heat from inside to the outside, but also due to the direct effect of sunlight on the surface under study, through the windows. To avoid this interference, the windows must be totally shaded some time before the test begins.

Even on a cloudy day, the diffuse solar radiation can affect the thermal patterns of the exterior of the building. For these reasons, the tests usually are performed outside at night, while the tests conducted indoors take place at night or on overcast days.

The existence of heat sources near the measurement area, such as radiators, artificial lights, vehicles, equipment and people running, may affect the results. The severity of the interference from these sources depends on their radiative power and reflection from the surface under study. The radiation emitted by people is not usually a problem, unlike radiators, lights and machinery, which must be disconnected before the test begins and then have to be taken into consideration when analyzing the results.

The existence of shadows on the building, resulting from the presence of other buildings, trees or other neighboring elements can lead to wrong conclusions if not taken into account when analyzing the results, since highly complex phenomena can provoque change in the heat transfer to the façade.

The distance is another parameter that affects the thermal image, not only because it influences the atmospheric attenuation, but also because the resolution of the thermograms decreases with the distance between the equipment and the object. Each point of the thermogram corresponds to a specific area of the object surface. With increasing distance, each point corresponds to a larger area of the surface and the radiation captured by the equipment becomes a real average of the radiation emitted, lowering the detail level. The ambient temperature can influence the performance of thermal imaging equipment, since when is too low or too high infrared radiation detection systems become less stable.

The transparent bodies or partially transparent in the infrared range, such as glass and some plastics, can also create measurement problems. For example, when the radiation of a transparent surface is captured by the infrared equipment, the image obtained results from the energy emitted and reflected from the surface, plus a portion of energy transmitted through it. The real surface temperature is different of the temperature obtained by the thermographic images, since it contains a portion of the resulting radiation that crosses it.

4. Sensibility case-studies

Several parameters affect thermographic measurements, namely, emissivity, reflectivity, environmental conditions, colour and others (Hart, 1991 and Chown and Burn, 1983). To evaluate the influence of some of these parameters, simple tests were carried out using the LFC's thermography equipment, both in laboratory and "in situ".

One of the laboratory tests consisted of partially immersing two identical specimens of cellular concrete in water followed by a drying period. The tests were performed under steady state conditions, in two climatic chambers with different temperatures and relative humidity. Thermal images were obtained during each test, using four different values of emissivity: 0.62, 0.85, 0.91 and 0.95.

As expected (Hart, 1991; Chown and Burn, 1983; Chew 1998 and Gaussorges, 1999), the results showed that emissivity variation induced changes in the thermal images, during both absorption and drying (see Figure 7). By looking at the thermal images it was possible to say that the images obtained with emissivity 0.62 were quite different from the remaining ones. The differences between the other thermograms (emissivities 0.85, 0.91 and 0.95) were not very significant. However, thermal images obtained with emissivity 0.85 were generally clearer (Barreira and Freitas, 2007 and Barreira, 2004). Thus, if the study aims for a qualitative evaluation of the results, that is, an analysis of superficial temperature differences, the selected emissivity value is not very important (Hart, 1991). Nevertheless, a judiciously selected emissivity value may simplify the interpretation of the thermal image.

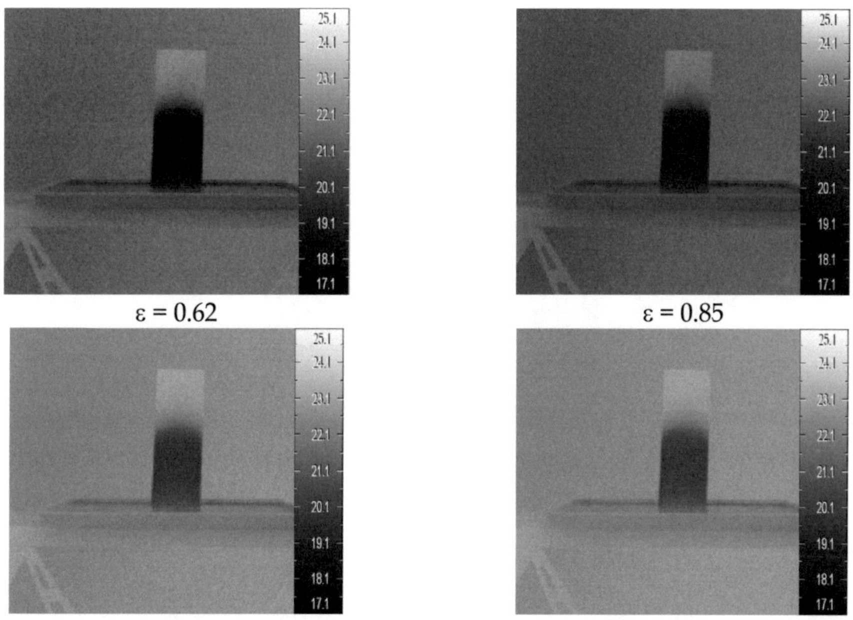

Fig. 7. Thermograms of a cellular concrete specimen after 168 hours of absorption obtained with different emissivities.

Variations in environmental conditions (temperature and relative humidity) also induced changes in the thermal images obtained in the same period (see Figure 8), not only because the environmental conditions may interfere with infrared detection, but especially because superficial evaporation is susceptible to their influence (Barreira and Freitas, 2007 and Barreira, 2004).

At the end of the drying period, the specimen's edges could no longer be distinguished from the image's background (see Figure 9) because their respective temperatures were very similar (Barreira and Freitas, 2007 and Barreira, 2004). Only objects whose temperature

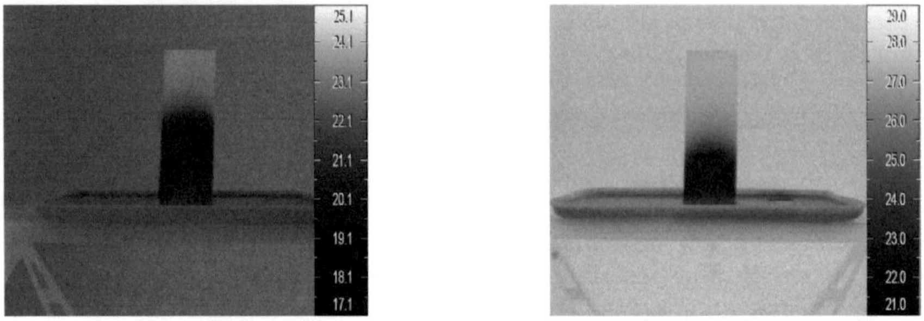

T = 20° C and RH = 60% T = 25° C and RH = 40%

Fig. 8. Thermograms of a cellular concrete specimen after 168 hours of absorption using emissivity 0.85.

varies at least 1° C from the environmental temperature can be detected using thermography. Therefore, this technology cannot be used to study objects in thermal equilibrium or in hygroscopic domain (Hart, 1991 and Santos and Matias, 2002), which may thus restrict thermography to the study of building materials. For example, it is possible to detect thermal resistances from a radiant floor (Barreira and Freitas, 2003), but only if the system is on, inducing temperature variations between the resistances and the background (see Figure 10).

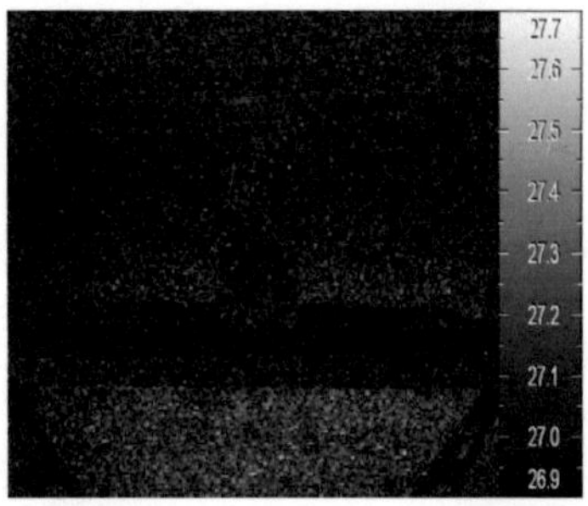

Fig. 9. Thermogram of a cellular concrete specimen after 792 hours (end of the drying period) using emissivity 0.85.

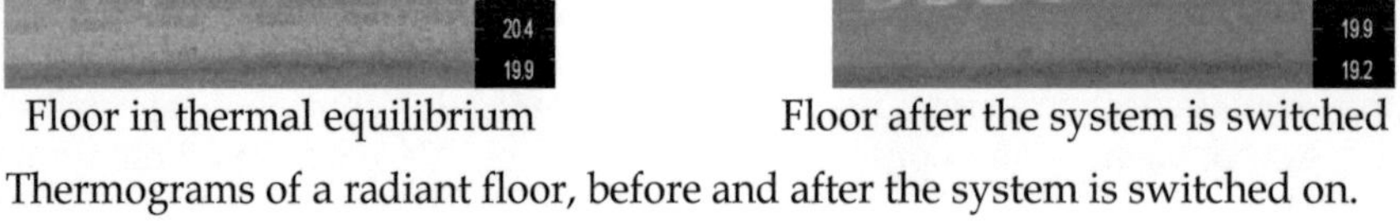

Floor in thermal equilibrium Floor after the system is switched on

Fig. 10. Thermograms of a radiant floor, before and after the system is switched on.

Reflectivity and colour are two important parameters since they may mask defects in building materials or components. Reflectivity is especially important for materials with low emissivity since reflectivity is complementary of emissivity in opaque materials. As such, an object with low emissivity induces a greater superficial temperature variation due to thermal reflection (Hart, 1991). Figure 11 shows the effect of reflectivity on the thermal image of a ceramic floor. The hotter area in the bottom left of the image results from the ceramic surface's thermal reflection (Barreira and Freitas, 2003).

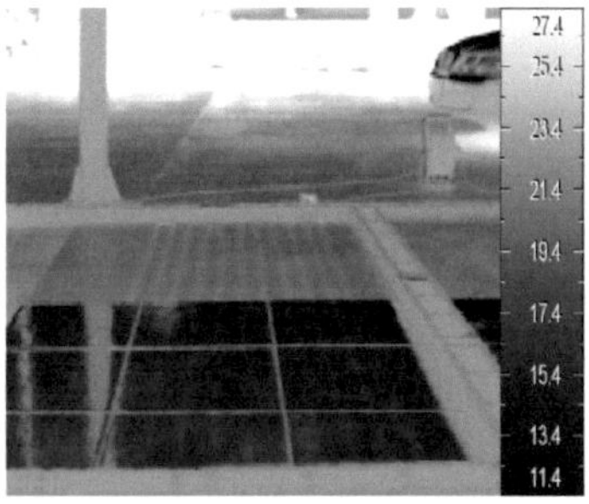

Fig. 11. Thermogram of a ceramic floor coating exhibiting thermal reflection.

Colour's significant influence on thermographic measurements became quite obvious when thermal images were obtained at the Carmo Church in Porto - Portugal. The Carmo Church's east façade is covered with "azulejos" (hand-painted ceramic tiles). The colouring varies between white and several shades of blue (see Figure 1). The thermal images revealed remarkable temperature differences caused by the colour variation. Colour had a greater influence when temperature differences were more pronounced. When temperature differences decreased at the end of the day (see Figure 12), the influence of colour became less important (Barreira and Freitas, 2004a).

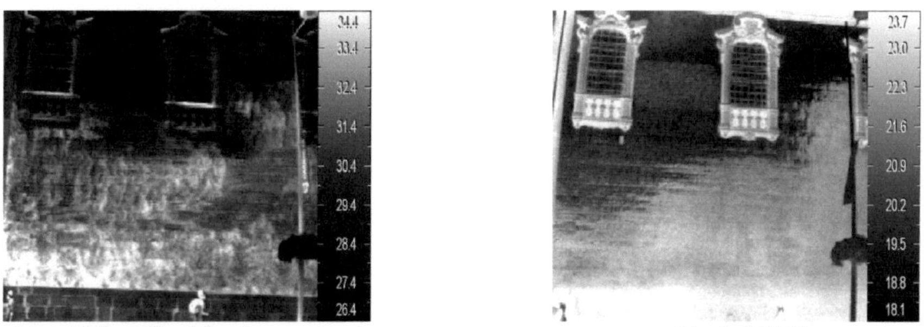

At midday (incident sun rays) At nightfall

Fig. 12. Thermograms of Carmo Church's façade covered with "azulejos" in Porto - Portugal.

4.1 Capillarity absorption and drying of building materials

A cellular concrete specimen was used to visualise capillarity water absorption and subsequent drying. The main properties of cellular concrete can be found in Table 2. The test was performed in a climatic chamber, under steady state conditions of temperature and relative humidity.

During the absorption period, the specimen's water level was visually observed and thermographically detected by its superficial temperature variation (see Figure 13). The temperature varied because of surface evaporation that, being an endothermic reaction, induced local cooling (Santos and Matias, 2002). The visible top water level was shown as

Density	525 kg/m^3
Open porosity [volume]	66%
Critical moisture content	0.24 kg/kg
Capillary saturation moisture content	0.50 kg/kg
Specific heat capacity	1050 J/kg·K
Thermal conductivity	0.3 W/m·K

Table 2. Main properties of cellular concrete measured at Building Physics Laboratory (Freitas, 1992).

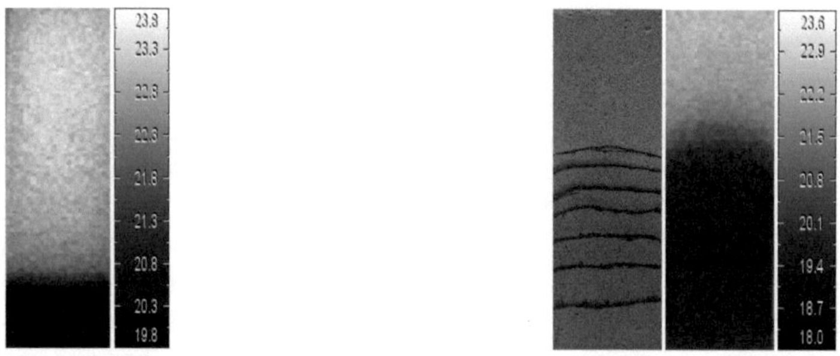

Specimen at 2 hours

Specimen at 408 hours (end of absorption)

Fig. 13. Thermograms obtained during absorption.

the upper limit of the darker shade. The intermediate shades, from dark to light, showed the "wet" to "dry" transition in the specimen's surface.

The drying period began immediately after the specimen was removed from the water. In the first hours of drying, the thermograms still showed significant superficial temperature differences. However, as drying progressed, the colder area decreased and the lighter shades enlarged. This fact revealed faster drying rates along the specimen's top and vertical edges (Barreira and Freitas, 2004a). At the end of the test, superficial temperature was almost uniform and therefore moisture distribution was not perceptible (see Figure 14).

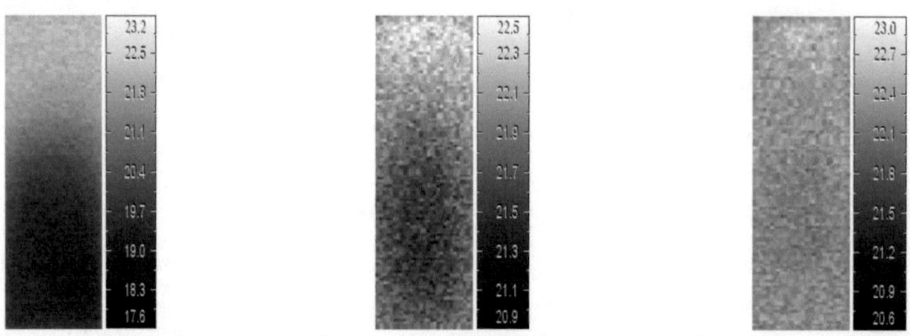

Specimen at 442 hours

Specimen at 578 hours

Specimen at 794 hours (end of drying)

Fig. 14. Thermograms obtained during the drying period.

It may thus be said that thermography allows visualising superficial temperature variations due to rising capillarity. During the drying period, superficial temperature variations may also be detected. Nevertheless, a decrease in moisture induces small superficial temperature variations which are therefore difficult to detect (Barreira and Freitas, 2004a).

4.2 Wetting and drying of building materials

A cellular concrete specimen (see Table 2) was used to visualise wetting and drying. Inside a climatic chamber, water was dropped on the specimen for one hour, at a rate of 11 drops per minute. After the wetting process, the drying period started immediately under the same environmental conditions.

After a one-hour wetting period, a stained area was visually observed and thermographically detected through the superficial temperature variation. The intermediate shades, from dark to light, showed the "wet" to "dry" transition in the specimen's surface (see section 4.1). During the first day of drying the thermograms still showed significant superficial temperature differences. After 48 hours of drying the colder area decreased considerably as the lighter shades enlarged (see Figure 15). This revealed faster drying along the "wet" area's edges (Barreira and Freitas, 2005).

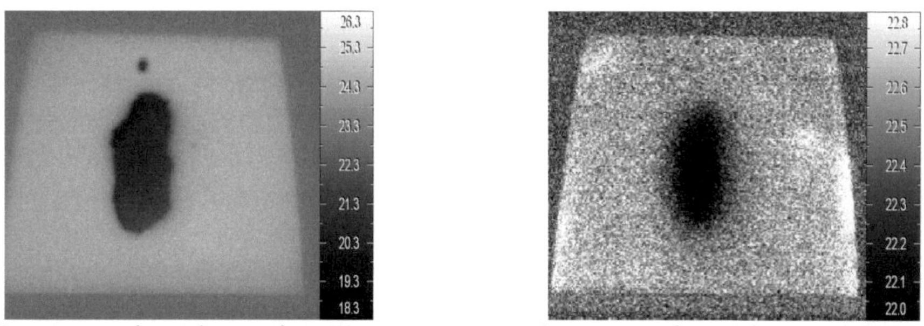

Specimen after 1 hour of wetting　　Specimen after 48 hours of drying

Fig. 15. Thermograms obtained during the drying period.

The test was concluded when moisture detection was hampered by the small superficial temperature differences. However, at the end of the test, the opposite side of the specimen still showed superficial temperature variation, as sketched in Figure 16 (Barreira and Freitas, 2005).

Although thermography is a mean of evaluating the specimen's approximated drying time, it only detects superficial moisture (Barreira and Freitas, 2005).

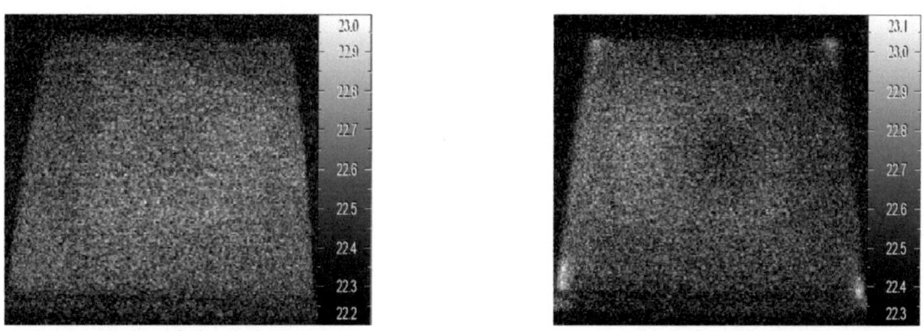

Front side of the specimen　　Opposite side of the specimen

Fig. 16. Thermograms at the end of the drying period.

4.3 Evaluating the thermal comfort of floor coatings

The comfort of interior floor coatings is very important for the well-being of building users, especially in bedrooms and bathrooms where people commonly walk barefoot. Although comfort depends on floor and environmental temperatures, it is also determined by the material's thermal characteristics, namely, thermal diffusivity and effusivity.

Thermography was used to evaluate the comfort of some interior floor coatings by comparing effects from barefoot contact with six different materials. The floor coatings under comparison were carpet, cork, vinyl, wood, ceramic tile and granite (see Table 3). The tests were performed inside a climatic chamber, with temperature and relative humidity similar to a dwelling's environmental conditions. Two different contact periods were adopted: 1 minute to assess initial discomfort and 2 minutes to study discomfort progress over time (Barreira and Freitas, 2004b).

Material	Density [kg/m³]	Specific heat capacity [J/kg·K]	Thermal conductivity [W/m·K]	Thermal diffusivity (a) (*) [m²/s]	Thermal effusivity (b) (**) [W·s½/m²·K]
Cork	150	2008	0.04	$0.13 \cdot 10^{-6}$	110
Wood	500	1500	0.14	$0.19 \cdot 10^{-6}$	324
Ceramic tile	2000	753	1.10	$0.73 \cdot 10^{-6}$	1 287
Granite	2650	837	3.97	$1.79 \cdot 10^{-6}$	2 967

(*) Obtained using Eq. (12); (**) Obtained using Eq. (13)

Table 3. Thermal properties of some materials (Hagentoft, 2001 and SQUARE ONE, 2005).

Before starting the tests, a thermogram was obtained from the sole of a barefoot to show its superficial temperature when the shoe was taken off. This thermogram indicated the foot's standard temperature (see Figure 17). Several thermal images of the sole of the barefoot were collected after the foot was placed in contact with the respective materials.

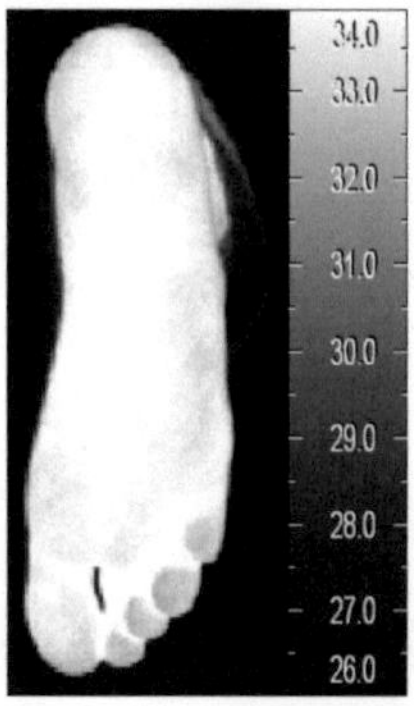

Fig. 17. Thermogram of the foot's standard temperature.

After being in contact with the different materials for 1 minute, the foot's superficial temperature was always less than its standard temperature and decreased according to the respective materials (see Figure 18). Contact with the carpet induced a higher superficial temperature of the foot sole, followed by cork, wood, vinyl, ceramic tile and, lastly, granite (see Table 4).

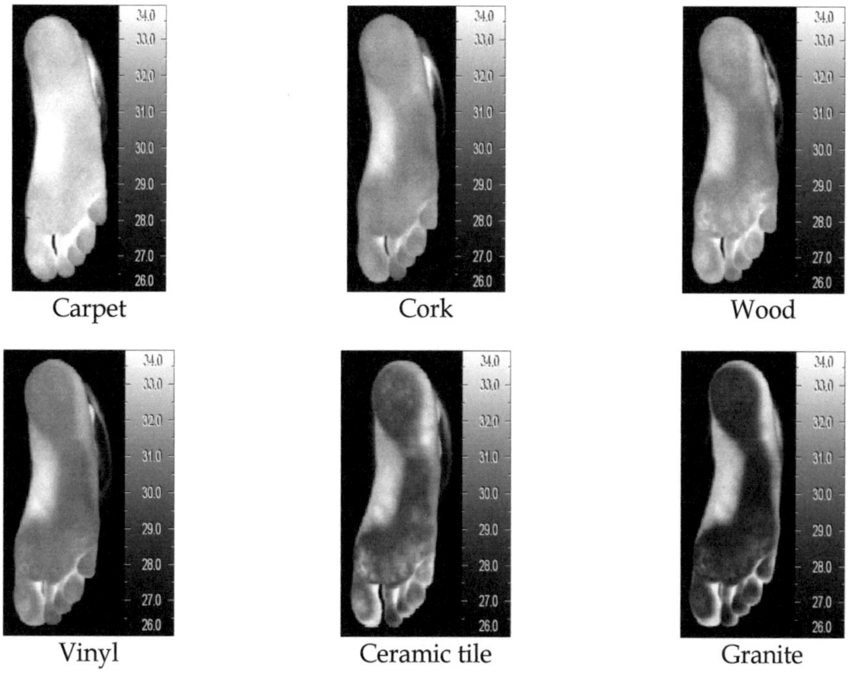

Fig. 18. Thermograms after 1 minute of contact.

Material	Average temperature (° C)	
	1 minute of contact	2 minutes of contact
Standard	*33.5*	
Carpet	32.6	32.7
Cork	31.6	32.6
Wood	31.5	32.5
Vinyl	31.0	32.3
Ceramic tile	30.5	30.6
Granite	29.6	30.1

Table 4. Average superficial foot temperature after 1 and 2 minutes of contact.

After being in contact with the materials for 2 minutes, the foot's superficial temperature was inferior to its standard temperature. The foot's temperature variations resulting from contact with carpet, cork, wood and vinyl were very similar. The ceramic tile and granite induced lower superficial temperatures (see Figure 19). However, the foot's superficial temperature was always higher than its temperature after having been in contact with each material for 1 minute (see Table 4).

Variations in the superficial foot temperature induced by contact with the various floor coatings are caused by the respective material's heat transference characteristics. Contact between the barefoot and the material induces an exchange of heat from the foot's higher temperature to the floor coating's lower temperature, until reaching a thermal equilibrium. This heat transference depends essentially on thermal diffusivity and effusivity (see Table 3).

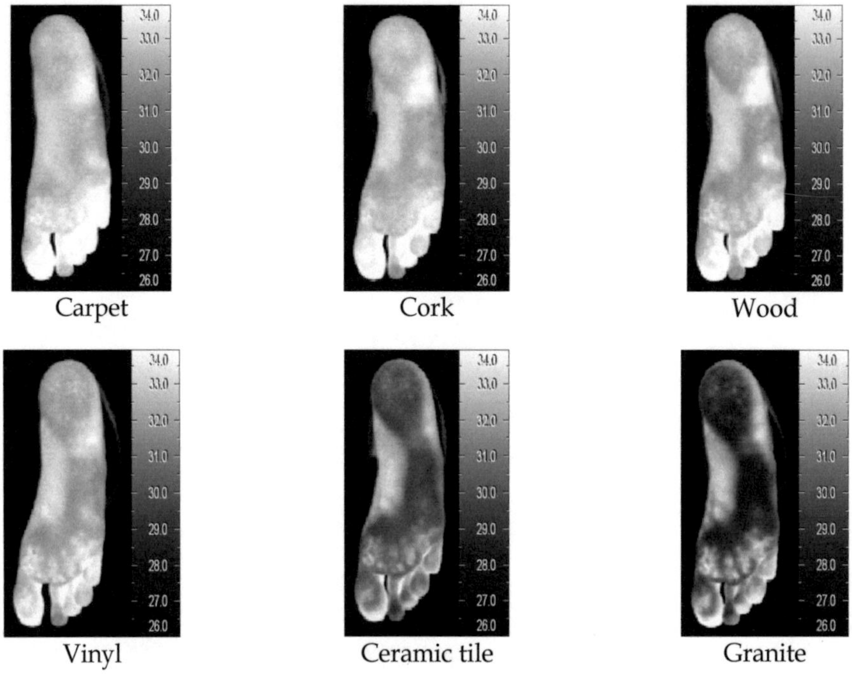

Fig. 19. Thermograms after 2 minutes of contact.

Thermal diffusivity (a) expresses heat transfer inside the material (Hagentoft, 2001). The higher its value, the quicker the heat is exchanged between the foot and the material, which causes more discomfort in the initial contact and induces lower superficial foot temperature.

$$a = \frac{\lambda}{\rho . c} \tag{12}$$

where λ is the thermal conductivity [W/m·K], ρ is the density [kg/m^3] and c is the specific heat capacity [J/kg·K].

Carpet has a lower diffusivity, and thus its contact with the foot induced higher superficial temperature and less initial discomfort. On the contrary, granite has the highest diffusivity and was the most uncomfortable floor coating (Barreira and Freitas, 2004b).

Thermal effusivity (b) expresses the material's capacity to absorb and store heat (Hagentoft, 2001). The higher its value, the greater the heat storage capacity and the longer it takes to reach the thermal equilibrium. The initial discomfort therefore lasts longer.

$$b = \frac{\lambda}{\sqrt{a}} \tag{13}$$

Carpet, cork, wood and vinyl have low effusivity. After 2 minutes of contact, they induced similar superficial foot temperatures that were the closest to the thermal equilibrium temperature. Ceramic tile and granite have higher effusivity and therefore induced lower superficial temperatures and caused discomfort for a longer period (Barreira and Freitas, 2004b).

5. Conclusions

Thermography is a non-destructive testing technology with much potential, but its application to building materials has not been greatly studied yet. Research studies developed at the Building Physics Laboratory (LFC) revealed that emissivity is an essential parameter, since it greatly influences thermographic measurements and may restrict application of this technology to buildings. However, if the study aims for a qualitative analysis of the results, the selected emissivity value is not very important. It was also possible to confirm that thermography cannot be used to study objects in thermal or hygroscopic equilibrium, as temperature differences between the object and the environment must be significant. Colour and thermal reflection must also be considered during thermographic testing, as they may mask the results and cause misinterpretation.

Thermography allowed studying the wetting and drying process of building materials. Temperature differences due to superficial water evaporation provided a means of recognising "wet" and "dry" areas. It was also possible to evaluate a material's approximate drying time since small superficial temperature variations indicate that moisture is rather significant. Thermography, however, detects only superficial moisture.

Thermography was also used to evaluate the comfort of some interior floor coatings. Thermal images were obtained from the sole of a barefoot after having been in contact with different floor materials. Since the superficial temperature variation, after contact between the foot and the material, is related with discomfort, a comparison of thermograms revealed the various material comfort levels.

6. Acknowledgment

The authors acknowledge the financial support provided by Fundação para a Ciência e a Tecnologia (FCT) and FEDER through project PTDC/ECM/114189/2009.

7. References

Avdelidis, N.P.; Moropoulou, A. (2004). Applications of infrared thermography for the investigation of historic structures. *Journal of Cultural Heritage*, Vol. 5, N.º 1, pp. 119-127.

Barreira, E. & Freitas, V.P. (2003). Thermal Images Obtained from Different Solutions of Radiant Floor. Internal Report LFC/IC 134.2003, LFC, FEUP, Porto, Portugal (in Portuguese).

Barreira, E. & Freitas, V.P. (2004a). Infrared Thermography Applications in the Study of Building Hygrothermal Behaviour. *CIB W40 Meeting*, Caledonian University, Glasgow.

Barreira, E. & Freitas, V.P. (2004b). Thermal Comfort Evaluation of Floor Coatings Using Infrared Thermography . *Construlink International Journal*, Vol. 3, No. 8, pp. 30-38 (in Portuguese).

Barreira, E. & Freitas, V.P. (2005). Barreira, E. & P. de Freitas, V. Importance of Thermography in the Study of ETICS Finishing Coatings Degradation Due to Algae and Mildew Growth. *Proceedings of the 10th International Conference On Durability of Building Materials and Components* (10DBMC), CSTB, Lyon.

Barreira, E. & Freitas, V.P. (2007). Evaluation of building materials using infrared thermography. *Construction and Building Materials*, Vol. 21, N.º 1, pp. 218-224.
Barreira, E. (2004). Thermography Applications in the Study of Buildings Higrothermal Behaviour. MSc Theses, Porto, FEUP, Portugal (in Portuguese).
Chew, M.Y.L. (1998). Assessing building facades using infra-red thermography. *Structural Survey*, Vol. 16, N.º 2, pp. 81-86.
Chown, G.A. & Burn, K. N. (1983). Thermographic identification of buildings enclosure defects and deficiencies. *Canadian Building Digest* 229, Canada, NRC-IRC.
Freitas, V.P. (1992). The Moisture Migration in Buildings Walls – Analysis of the Interface Phenomenon. PhD Theses, Porto, FEUP, Portugal (in Portuguese).
Gaussorgues, G. (1999). La thermographie infrarouge – Principes, Technologies, Applications. Fourth Édition, Edition TEC & DOC, Paris, France.
Grinzatoa, E.; Vavilovb, V.; Kauppinen, T. (1998). Quantitative infrared thermography in buildings. *Energy and Buildings*, Vol. 29, N.º 1, pp. 1-9.
Hagentoft, C. (2001). Introduction to Building Physics. Student Litteratur, Sweden.
Haralambopoulos, D.A.; Paparsenos, G.F. (1998). Assessing the thermal insulation of old buildings - The need for in situ spot measurements of thermal resistance and planar infrared thermography. *Energy Conversion and Management*, Vol. 39, N.º 1-2, pp. 65-79.
Hart, J.M. (1991). A practical guide to infra-red thermography for building surveys. Garston, Watford, BRE.
Incropera, F.P.; Witt, D. P. (2001). Fundamentals of Heat and Mass Transfer. John Wiley & Sons.
NEC San-ei Instruments, Ltd (1991). TH1 101 Thermo Tracer – Operation Manual, Japan.
Rao, P. (2008). Infrared thermography and its applications in civil engineering. *The Indian Concrete Journal*, Vol. 82, N.º 5 , pp. 41-50.
Santos, C.P. & Matias, L. (2002). Application of Thermography for Moisture detection - A laboratory research study. *Proceedings of the XXX IAHS World Congress on Housing*, Coimbra, Portugal.
SQUARE ONE (2005). Research PTY LTD Properties of materials (http://www.squ1.com/index.php).

9

Nondestructive Evaluation of FRP Strengthening Systems Bonded on RC Structures Using Pulsed Stimulated Infrared Thermography

Frédéric Taillade, Marc Quiertant, Karim Benzarti, Jean Dumoulin and Christophe Aubagnac
Université Paris-Est, IFSTTAR, F-75015 Paris France

1. Introduction

In civil engineering, strengthening or retrofitting of reinforced concrete (RC) structures by externally bonded Fiber-Reinforced Polymer (FRP) systems is now a commonly accepted and widespread technique (Hollaway, 2010; Quiertant, 2011). However, the use of bonding techniques always implies following rigorous installation procedures (440.2R-08 Committee ACI, 2008; AFGC, 2011; FIB, 2001) and application personnel have to be trained in conformity with installation procedures to ensure both durability and long-term performances of FRP reinforcements. The presence of bonding defects can significantly affect the structural performance and durability of the strengthening systems. Defects have then to be detected, located and evaluated in order to estimate if injection or replacement is needed. In these conditions, conformance checking of the bonded overlays through *in situ* nondestructive evaluation (NDE) techniques is highly suitable. The quality-control program should involve a set of adequate inspections and tests.

Visual inspection and acoustic sounding (hammer tapping) are commonly used to detect delaminations (disbonds) (Fig.1). However, these current practices are unable to provide relevant information about the depth (in the case of multilayered FRP systems) and width of debonded areas and they are not capable of evaluating the level of adhesion between the FRP and the substrate (partial delamination, damage or poor mechanical properties of the polymer adhesive). Adherence properties of FRP systems installed on concrete substrates can be evaluated by conducting on site pull-off adhesion tests on witness panels specifically bonded on test zones (Fig.2).

Consequently, different authors have developed nondestructive methods to assess the quality of the FRP/concrete adhesive bond, based on microwave (Akuthota et al., 2004), acousto-ultrasonic (Ekenel & Myers, 2007), impact-echo (Maerz & Galecki, 2008), shearography (Hung, 2001; Taillade et al., 2011; 2006), infrared thermography (Galietti et al., 2007; Valluzzi et al., 2009) or a coupling of these two latter techniques (Lai et al., 2009; Taillade et al., 2010).

This chapter is devoted to the pulsed stimulated infrared thermography technique applied to the detection and the characterization of the depth and width of adhesion defects (delaminations or adhesive disbonds) of FRP externally bonded on RC structures.

Fig. 1. Inspection with acoustic sounding (hammer tapping).

Fig. 2. Pull-off method.

In a first part, the principle of pulsed stimulated infrared thermography is recalled ; laboratory investigations are then presented in a second part. The laboratory samples contain different defects (with calibrated size and depth) inserted between the concrete substrate and the carbon FRP laminate bonded to its surface. Experiments were conducted in laboratory on the dedicated samples and complementary 2D numerical simulations were also carried-out. Analysis methods of thermograms are presented. Thermal signatures of different geometries of defects are studied in the cases of pulse and square heating thermal excitations. Main advantages of each stimulated technique are discussed in relation to the targeted application. Results from experiments based on long pulse approach are also discussed in details.

In the second part of this chapter, a case study of field application is presented for the proposed method. Inspection is carried-out using a hand-held heating device and an infrared camera. Such a simple technology enables real time NDE in the field with a high efficiency.

2. Principle of pulsed stimulated infrared thermography

For many years, the Pulsed Stimulated Infrared Thermography technique has been used to control aerospace structures, in particular to detect and characterize delaminations in carbon/epoxy composites (Maldague, 2001).

Pulse heating principle consists in heating the surface of the composite during a period τ and measuring the temperature distribution on the sample surface with an infrared camera (Fig.3).

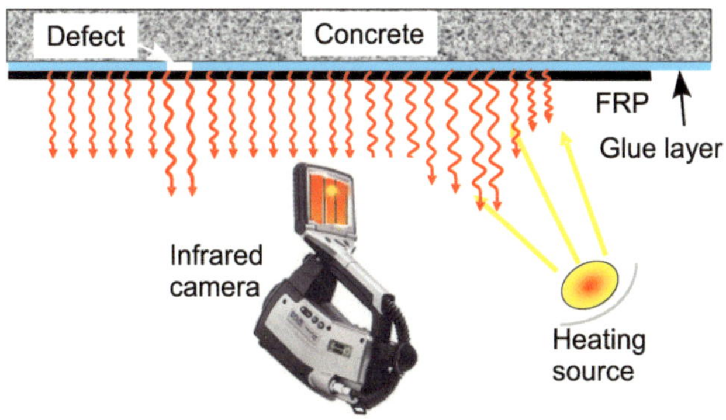

Fig. 3. Principle of stimulated infrared thermography.

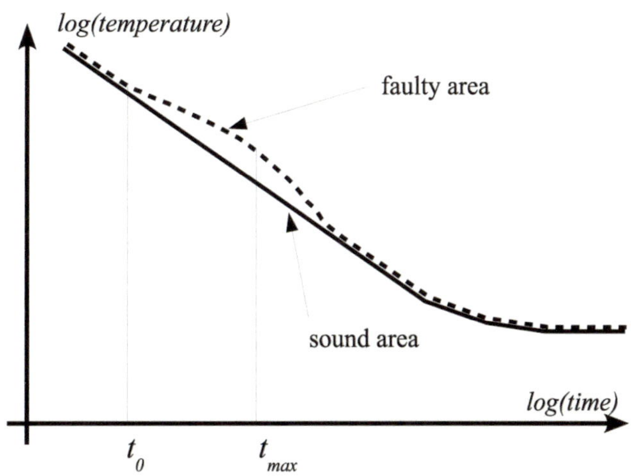

Fig. 4. Thermograms of sound and faulty regions.

Detection and localization of the subsurface defects can then be performed using adequate image analysis approaches (Ibarra-Castanedo et al., 2004). Characterization of the resistive subsurface defects can be achieved by monitoring the emergence of a thermal contrast (Balageas et al., 1987) between sound and faulty areas (Fig.4) after the pulse illumination (thermal relaxation phase). The thermal contrast C_T could be expressed by:

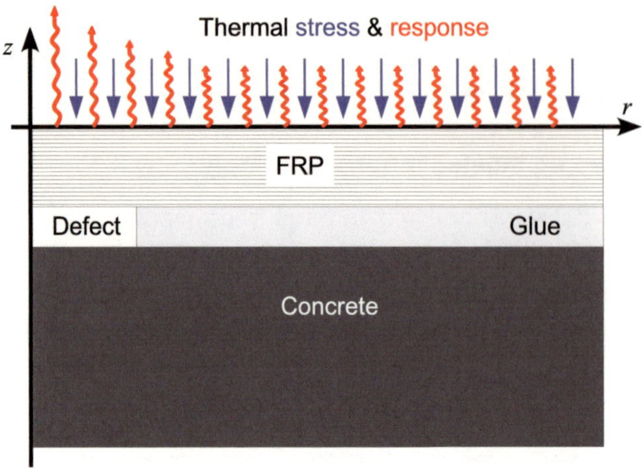

Fig. 5. Axisymmetric finite element model.

$$C_T = \frac{T}{T^{sound}} - 1 \quad (1)$$

where T^{sound} and T are respectively the temperature above sound and faulty regions.

Using the thermal diffusion time concept, hypothesis of heat diffusion in a semi infinite body and assuming the period τ is infinitely short (Dirac pulse), the depth d of the defect can be deduced from the time t_{max} associated to the maximum thermal contrast using the expression:

$$d = \sqrt{\alpha t_{max}} \quad (2)$$

where $\alpha = \lambda/\rho c$ is the thermal diffusivity of the material through the thickness direction with ρ, c and λ are respectively the density, heat capacity and thermal conductivity of the material.

It follows that for defects of same nature but localized at different depths, their localization is based on the detection of thermal contrast appearing at different time on thermal image sequences. For a same thermal solicitation, thermal contrast fades while defect depth increase. So, localization of defects requires to analyze the whole sequence of thermal images acquired during and after thermal solicitation.

3. Finite element simulation for test calibration

In this section, it is proposed to calculate the thermal time response of a sample (carbon FRP / polymer adhesive / concrete) with a bonding defect and subjected to an external heat pulse of finite duration. It is assumed that thermal stress is applied uniformly over the composite surface. To simplify the simulations, an orthotropic behavior is assumed for the FRP material while concrete and polymer adhesive are considered as isotropic. The bonding defects are assumed to be circular areas of finite diameter characterized by a lack of glue (Fig.5). The finite element software enables one to solve the Fourier heat transfer equation:

$$\rho c \frac{\partial T}{\partial t} = div\,(\lambda\, \mathbf{grad}\, T) \quad (3)$$

The FRP reinforcement, which is based on bonded carbon fabrics in this case, is simulated by a definite thickness of carbon/epoxy laminate exhibiting equivalent properties. Thicknesses of the FRP laminate, glue layer and concrete substrate are respectively 2 *mm*, 0.2 *mm* and 20 *mm*. Since a delamination of surface area 6.5 cm^2 is considered as the threshold above which repair should be undertaken (Maerz & Galecki, 2008), various diameters ranging from 10 to 40 *mm* have been chosen for the bonding defects.

Thermal properties considered in the numerical calculations are given in Table 1 for the different materials.

Table 1. Thermal properties of the materials.

Material	ρ $(kg.m^{-3})$	c $(J.K^{-1}.kg^{-1})$	λ $(W.m^{-1}.K^{-1})$
Epoxy	1200	1200	0.2
Concrete	2300	900	1.8
Composite	1500	850	4.2 along fiber 0.7 perpendicular to the fiber

Figure 6 shows the computed thermal response of the composite surface heated with a thermal flux equal to 1000 $W.m^{-1}$ for 1 s. Time evolutions of the temperature near sound and faulty areas are depicted for defect diameters equal to 20, 30 and 40 *mm*. Maxima of the thermal contrast (Fig. 7) are respectively observed 9.0 s, 12.5 s and 16.4 s after the end of the heating period, according to the diameter of the defect. Using the composite thermal diffusivity in the transverse direction (perpendicular to the fibers) and equation 2, these characteristic times enable one to estimate an average value of the defect depth, as well as an expanded uncertainty ($k = 2$): $d = 2.7 \pm 0.8$ *mm*. Although the accuracy is low, it remains in the same range as uncertainties on the thermal properties of materials (typically 20%) and it should be underlined that the thickness of the adhesive layer is not known precisely in most practical cases (typical uncertainty of 30%). We prefer this technique, very simple to implement, rather than the early detection method. Nevertheless, it is to note that the early time detection related to different diameters of the defect is merged at the same short times $t_0 \approx 2.2$ s but it is necessary to apply a threshold of detection factor (Krapez et al., 1994) depending on the noise level of the experiment, in order to assess the depth of the defect with a good accuracy.

To increase the measured thermal contrast, it is possible to apply the thermal flux for a longer period and/or to use a high sensitivity infrared camera (NETD of the order of 25 *mK*). Moreover, in practice, the second solution is almost unrealistic due to the prohibitive cost of this type of IR camera which is not suitable to field inspections. By increasing pulse duration, the contrast can be enhanced but the maxima of the thermal contrast is delayed (Fig.8) and Equation 2 is not applicable directly any more.

4. Analysis method

Different analysis tools (Balageas et al., 1987; Ibarra-Castanedo et al., 2004; Maldague, 2001) can be used. They are based on techniques of contrast enhancement (increase in the defect signature), thermal images sequence decomposition on basis (data compression) and image segmentation (localization of defects on thermal images).

A first approach to reduce the number of thermal images to be analyzed in a sequence (Ibarra-Castanedo et al., 2004) consists in using frequency analysis tools. The Fourier

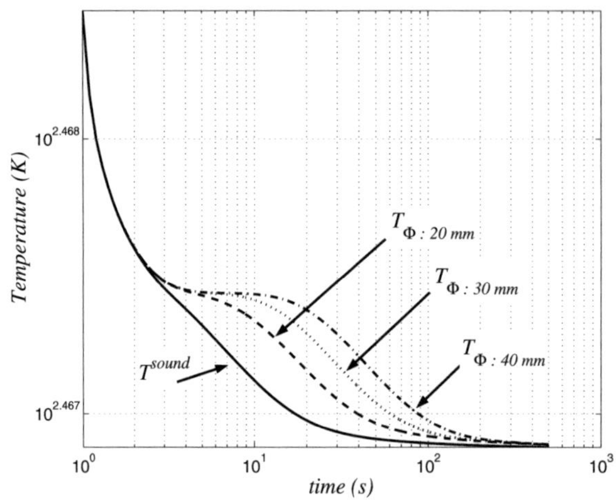

Fig. 6. Thermograms of sound and faulty regions for three defect diameters.

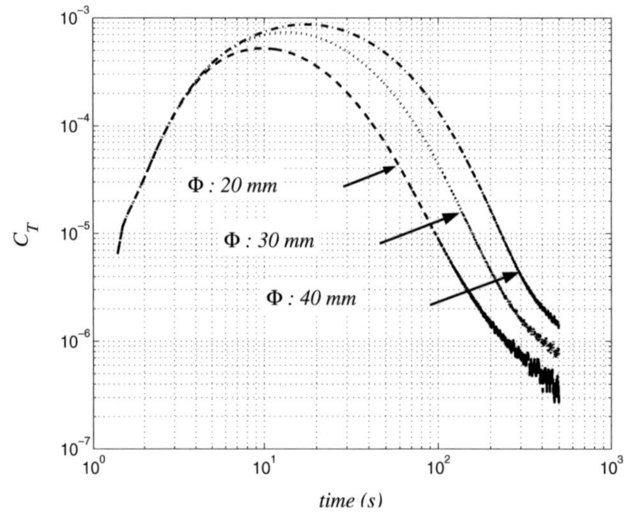

Fig. 7. Thermal contrast simulation for three defect diameters.

transform (Equ.4) is applied to temporal evolution of each pixel of the thermal image ($T(t)$):

$$F_n = \Delta t \sum_{m=0}^{N-1} T(m\Delta t) exp(-j2\pi n/N) \tag{4}$$

where Δt is the sampling time, F_n is the complex image of the nth frequency and N the maximum number of the frequencies.

Magnitude and phase maps calculated are then analyzed to locate defects.

Another approach is based on Singular Value Decomposition (SVD) which is an interesting tool for the extraction of the spatial and temporal information from a thermographic matrix

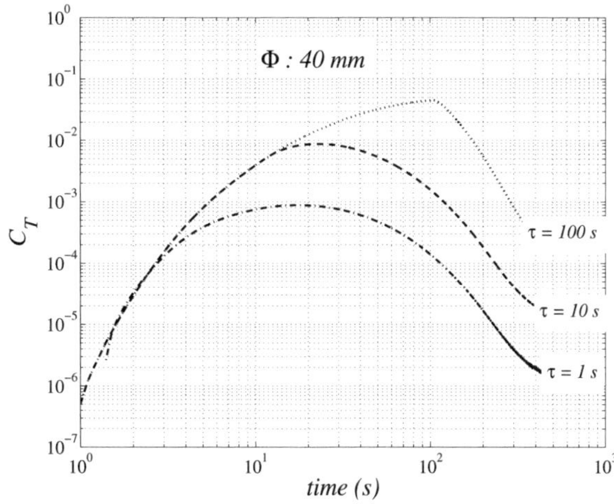

Fig. 8. Thermal contrast simulation for three thermal flux durations τ and 40 *mm* defect diameter.

in a compact or simplified manner (Rajic, 2002). The SVD of a $M \times N$ matrix A ($M > N$) can be calculated as follows:

$$A = U \sum V^T \tag{5}$$

where U is a $M \times N$ orthogonal matrix, $\sum$ is a diagonal $N \times N$ matrix (with the singular values of A in the diagonal), and V^T is the transpose of a $N \times N$ orthogonal matrix (characteristic time).

Hence, to apply the SVD to thermographic data, the 3D thermogram matrix representing time and spatial variations has to be reorganized as a 2D $M \times N$ matrix A. This can be done by rearranging the thermograms for every time as columns in A, in such a way that time variations will occur column-wise while spatial variations will occur row-wise.

Under this configuration, the columns of U represent a set of orthogonal statistical modes known as Empirical Orthogonal Functions (EOF) that describe the spatial variations of data. On the other hand, the Principal Components (PC), which represent time variations, are arranged row-wise in matrix V^T. The first EOF will represent the most characteristic variability of the data; the second EOF will contain the second most important variability, and so on. Usually, original data can be adequately represented with only a few EOF. Typically, a 1,000 thermal images sequence can be replaced by 5 to 10 EOF and analyzed to locate defects.

When the defect is located (spatially) in the image sequence using one of the previous methods, the method can be refined in order to improve the determination of the defect depth, i.e. (i) to be insensitive to the material anisotropy in terms of thermal diffusivity (Krapez et al., 1994) if we determine the early detection time t_0 (Fig.4) and (ii) to take the non uniformity of the heat flux into account (Krapez et al., 1992). As shown in figure 8, in the case of a finite pulse duration τ, equation 2 must be modified.

The first order correction consists in moving the time scale origin toward the pulse barycenter (Degiovanni, 1987). Moreover, in the case of an infinitely extended defect located in a homogeneous medium, Krapez (Krapez, 1991) has proposed an abacus to apply a correction and take the pulse duration into account (Fig.9).

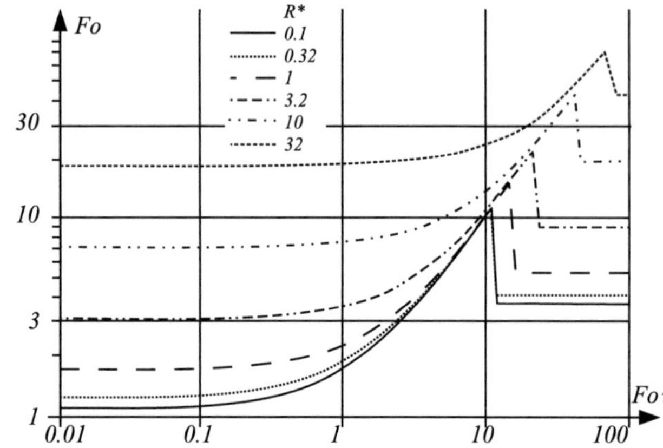

Fig. 9. Variation of the Fourier number F_0 *vs.* the pulse duration for different thermal resistances of the defect.

This abacus (Fig.9) gives the variation of the Fourier number F_0 ($F_0 = \alpha t_{max}/d^2$) as a function of the pulse Fourier number $F_{0\tau}$ ($F_{0\tau} = \alpha\tau/d^2$) for various thermal resistances of the defect R^*, where R^* is the ratio between the discontinuity resistance of the defect R and the resistance of the front layer ($R^* = R/\ (d/\lambda)$).

Although the composite can not be considered as an isotropic material and defects have a finite size, it is proposed to use this abacus in our case in order to improve the estimation of the defect's depth.

Equation 2 is then used in a first approximation to estimate the depth of the defect d. This depth enables one to compute the different parameters R^* and $F_{0\tau}$ used in the abacus (Fig.9). Finally, a value of F_0 is determined and we use it to improve the depth estimation and so on (Fig.16):

$$d' = \sqrt{\frac{\alpha t_{max}}{F_0}} \tag{6}$$

where d' is the new value of depth.

5. Experimentations

Laboratory tests have been carried out to evaluate the performance of the proposed NDE method. A concrete slab ($400 \times 300 \times 15\ mm^3$) has been manufactured and externally reinforced by three superimposed layers of pultruded FRP plates (thickness of 1.2 *mm*) with intermediate glue layers of thickness 1 *mm*, as shown in figure 10. Bonding defects were simulated by locally replacing the adhesive by polytetrafluoroethylene (PTFE) discs (0.5 *mm* thick), placed either between the concrete surface and the lower FRP plate, or between two adjacent FRP layers. The final specimen contains discs of three different diameters (10, 20 and 30 *mm*), located at three different depths (1.2, 3.4 and 5.6 *mm*).

The surface of the specimen was heated during 50 *s* using a flexible electric cover (electric power is about 1000 *W* and the surface is $1 \times 0.9\ m^2$). To visualize the temperature of the sample surface during the cooling phase after external heating, we used an infrared camera which produces images of 320×240 pixels and composed of uncooled microbolometer

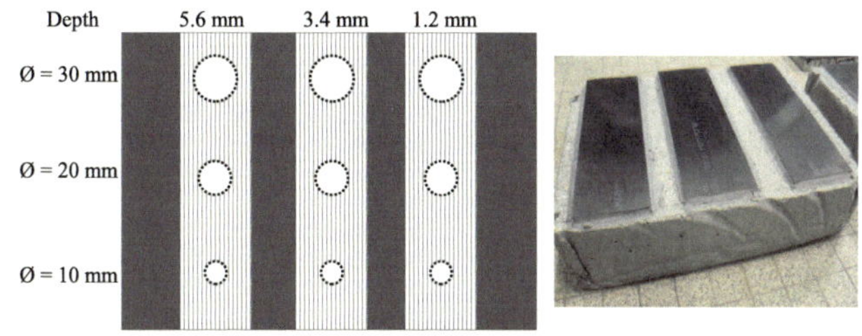

Fig. 10. Concrete slab reinforced with bonded FRP plates (3 superimposed layers) and containing calibrated defects.

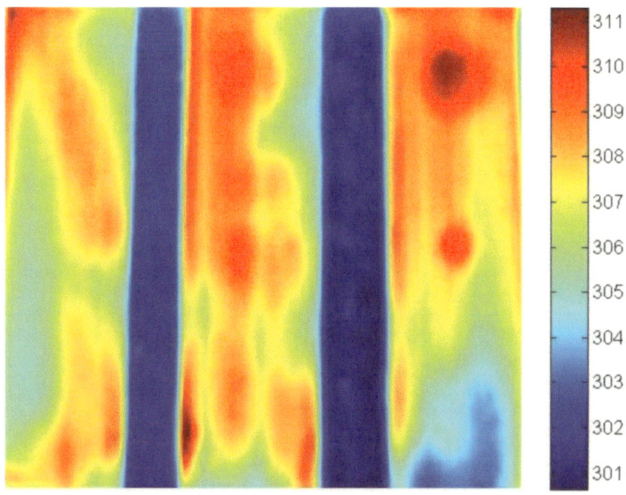

Fig. 11. Thermal image at the beginning of thermal relaxation.

detectors allowing to see temperature differences as low as 80 mK in the range from $-40°C$ to $+2,000°C$. The spectral response is comprised between 7.5 and 13 μm.

Figures 11 and 12 show thermal images of the sample at the beginning of the thermal relaxation and 52 s after the end of the heating stage. On these figures, we notice the non homogeneity of the heating.

Using SVD method to analyze the sequence of acquired thermal images, one can select only few images to localize the defects (Fig.13). Furthermore, SVD method in that case partly corrects effects of the non homogeneity of the previous heating.

The thermograms (Fig.14) and the thermal contrast (Fig.15) are computed above the larger defect (diameter = 30 mm). The maximum contrast appears 7 s, 33 s and 120 s after the end of the heating respectively to the depth. Using equations (2) and (6) iteratively, it is possible to retrieve the defect's depth with a good accuracy in the three cases considered here.

Taking the thermal diffusivity perpendicular to the FRP into account, and after some iterations (Fig.16), the depth of the defects and its expanded uncertainty ($k = 2$) were estimated to 1.2 ± 0.2 mm, 3.3 ± 0.3 mm and 6.3 ± 0.3 mm which can be compared to the actual depth values of 1.2, 3.4 and 5.6 mm. Globally, a fairly good agreement was obtained. However, it is

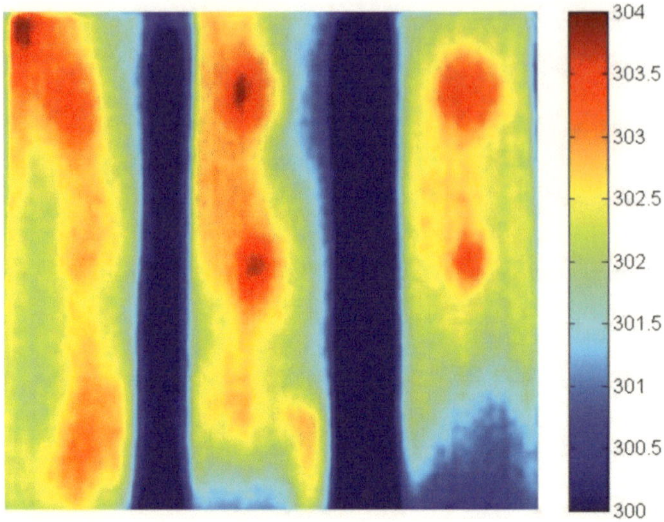

Fig. 12. Thermal image 52 *s* after the end of the heating stage.

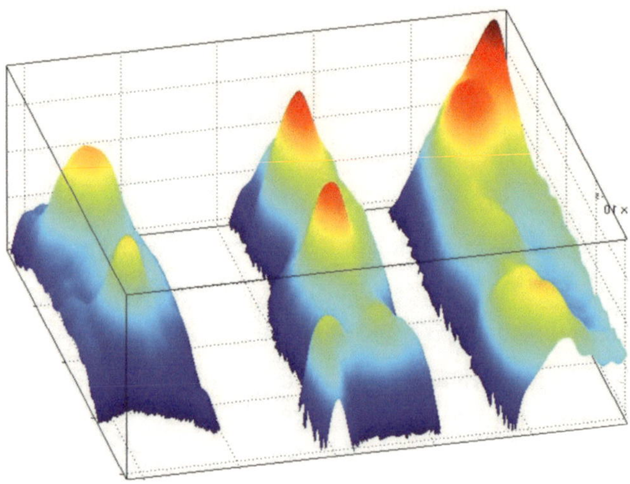

Fig. 13. 3D view of EOF map obtained with SVD method.

to note that an increased deviation was observed for the defect depth of 5.6 *mm*, since in this case measured temperature values were very close to the ambient noise.

6. Field inspection - a case study

In this part, the feasibility of the thermographic method for routine inspection of strengthened concrete structures is illustrated through a case study conducted on an existing RC structure. The field test presented in this section only focuses on the detection of bonding defects. Evaluation of the depths of localized defects was not performed here.

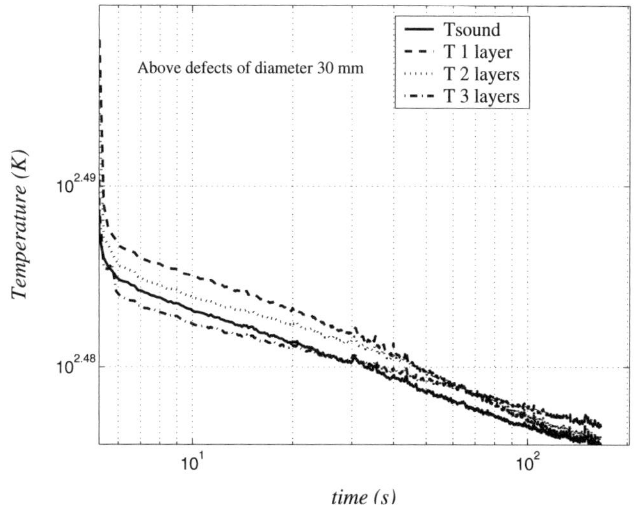

Fig. 14. Experimental thermograms of sound (solid line) and faulty regions above the larger defects (dashed line for a depth of 1.2 *mm*, dotted line for a depth of 3.4 *mm* and dash-dot line for a depth of 5.6 *mm*).

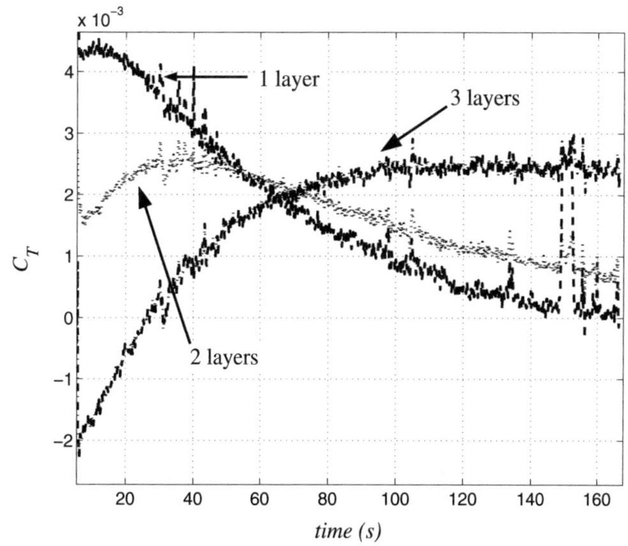

Fig. 15. Thermal contrast *vs.* computed time for regions above the larger defects (dashed line for a depth of 1.2 *mm*, dotted line for a depth of 3.4 *mm* and dash-dot line for a depth of 5.6 *mm*).

6.1 Description of the bridge and repair works

The bridge under study is located near Besançon in France, over the Doubs river. It was built in the 60ies. The bridge consists in three distinct and independent sections, i.e, 2 access spans and a main central structure. The latter is divided itself into three spans, respectively 29, 54 and 29 m long, and composed of two box-girders made of prestressed concrete.

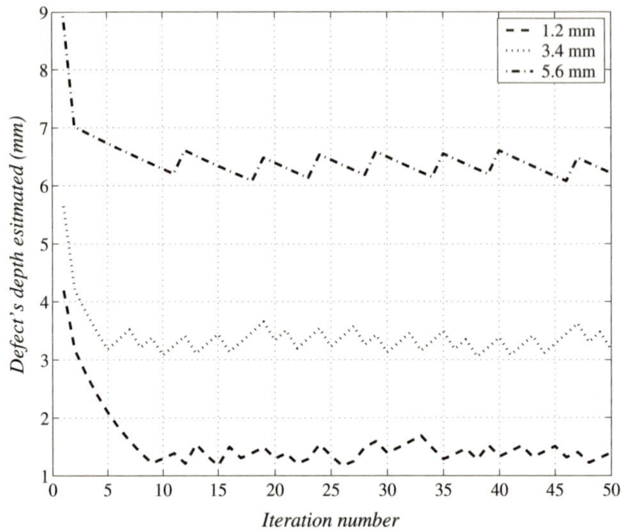

Fig. 16. Iterative estimation of the defect's depth (dashed line for a depth of 1.2 *mm*, dotted line for a depth of 3.4 *mm* and dash-dot line for a depth of 5.6 *mm*).

Fig. 17. View of the bridge under consideration; FRP repaired zones correspond to the white parts on the girders.

A visual inspection conducted in the 90ies revealed extensive transverse cracking of lower slabs of box-girders at mid-span. Such a deterioration was mainly attributed to an inadequate of the thermal gradients consideration in the initial design and to a lack of the inter-element continuity of longitudinal prestressing in lower slabs.

In order to prevent brittle failure at mid-span, it was decided to repair the cracked box-girders by bonding carbon fibre sheets according to the wet lay-up process (onsite impregnation). A recalculation of the structure was performed in order to optimize the repair design with respect to the shear stress distribution. Finally, composite reinforcements were installed at the outer side of the web of girders as shown in figure 17.

Fig. 18. Inspection operations.

6.2 Thermographic inspection of the FRP repairs

The first operational evaluation of the innovative thermographic method was accomplished during the inspection of the CFRP installation. The *in situ* inspection procedure is based on the use of an uncooled infrared camera coupled with a hand-held thermal excitation device consisting of an infrared lamp or an electric cover.

Such a simple set-up offers a fully portable real-time assessment system. The main difficulty of the inspection was the accessibility to the FRP bonded areas, which was resolved by using a truck mounted lift-platform (Fig. 18).

Figure 19 shows the geometrical configuration of the controlled area. The thermal solicitation is imposed by heating the FRP surface with an infrared lamp. Two examples of detected defects are presented in figure 20. The top image shows a wrapping defect and the bottom

Fig. 19. Shematic representation of the survey area by active infrared thermography.

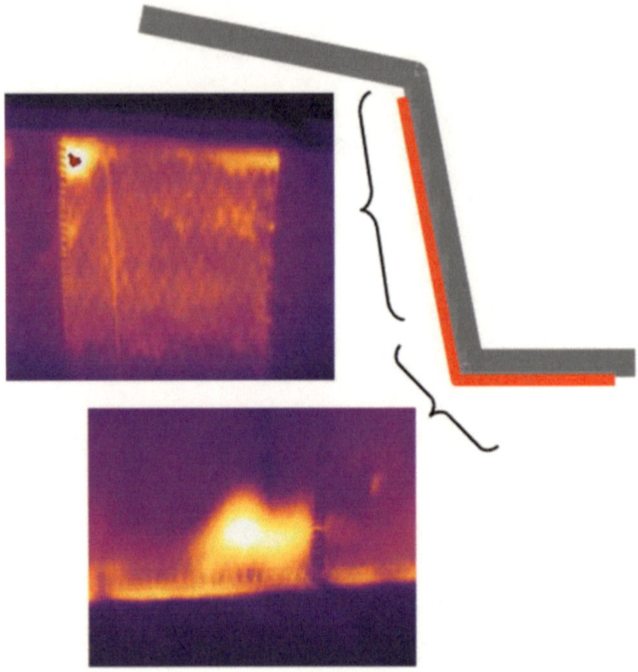

Fig. 20. Thermal images showing defect on bonded CFRP wrap.

image a gluing defect. These two small debonds were confirmed afterwards by hammer tapping.

7. Conclusion

In this paper, basic principles of the pulsed stimulated infrared thermography technique used for NDE of bonded overlays are briefly recalled. A finite element simulation of the thermal time response of bonding defects on a concrete sample reinforced by externally bonded FRP makes it possible to calibrate this NDE technique for this particular application. Moreover, a theoretical analysis of the thermograms is developed in order to quantify the defect depth. Besides, a laboratory evaluation, performed on FRP-strengthened concrete sample containing calibrated defects, has demonstrated the effectiveness of the method for detecting and assessing the depth of the bond defects. Defects were located between concrete and external FRP reinforcements or between two layers of FRP.

Thermography offers a simple method with real time and full field imaging capabilities. Moreover, hand portability of the thermal imaging equipment, including the heating source, is well adapted to field application. Furthermore, feasibility of the thermographic inspection method into the field was demonstrated during the inspection of a recently CFRP strengthened bridge. In this last validation test, only qualitative evaluation of the adhesive bond was performed (detection of the bonding defects). Based on satisfactory laboratory and field results, it is the author's point of view that a coupling of the two methods (pulsed stimulated infrared thermography and analysis of the thermograms) will offer an effective NDE tool for the evaluation of FRP strengthening systems bonded on concrete structures.

8. References

440.2R-08 Committee ACI (2008). Guide for the design and construction of externally bonded FRP systems for strengthening concrete structures, *Technical report*, ACI, Michigan (US).

AFGC (2011). Réparation et renforcement des structures en béton au moyen des matériaux composites, *Technical report*, Bulletin scientifique et technique de l'AFGC. in French.

Akuthota, B., Hughes, D., Zoughi, R., Myers, J. & Nanni, A. (2004). Near-field microwave detection of disbond in carbon fiber reinforced polymer composites used for strengthening cement-based structures and disbond repair verification, *Journal of Materials in Civil Engineering* 16(6): 540–546.

Balageas, D., Déom, A. & Boscher, D. (1987). Characterization and nondestructive testing of carbon-epoxy composites by a pulsed photothermal method, *Materials Evaluation* 45(4): 461.

Degiovanni, A. (1987). Correction de longueur d'impulsion pour la mesure de la diffusivité thermique par méthode flash, *International Journal of Heat and Mass Transfert* 30(10): 2199–2200.

Ekenel, M. & Myers, J. (2007). Nondestructive evaluation of RC structures strengthened with FRP laminates containing near-surface defects in the form of delaminations., *Science and Engineering of Composite Materials.* 14(4): 299–315.

FIB, T. G. . (2001). Externally bonded FRP reinforcement for RC structures, *Technical Report 14*, Fib bulletin 14, Lausanne, Switzerland.

Galietti, U., Luprano, V., Nenna, S., Spagnolo, L. & Tundo, A. (2007). Non-destructive defect characterization of concrete structures reinforced by means of FRP, *Infrared Physics & Technology* 49: 218–223.

Hollaway, L. (2010). A review of the present and future utilisation of FRP composites in the civil infrastructure with reference to their important in-service properties., *Construction and Building Materials* 24(12): 2419–2445.

Hung, M. Y. Y. (2001). Shearography and applications in nondestructive evaluation of structures, *Proceedings of the International Conference on FRP Composites in Civil Engineering (CICE 2001)*, pp. 1723–1730.

Ibarra-Castanedo, C., González, D., Klein, M., Pilla, M., Vallerand, S. & Maldague, X. (2004). Infrared image processing and data analysis, *Infrared Physics* 46: 75–83.

Krapez, J.-C. (1991). *Contribution à la caractérisation des défauts de type délaminage ou cavité par thermographie stimulée*, PhD thesis, Ecole Centrale de Paris.

Krapez, J.-C., Boscher, D., Delpech, P., Déom, A., Gardette, G. & Balageas, D. (1992). Time-resolved pulsed stimulated infrared thermography applied to carbon-epoxy non destructive evaluation, *Quantitative Infrared Thermography (QIRT 92)*.

Krapez, J.-C., Lepoutre, F. & D. Balageas, . (1994). Early detection of thermal contrast in pulsed stimulated thermography, *8th International Topical Meeting on Photoacoustic and Photothermal Phenomena*.

Lai, W. L., Poon, S. C. K. C. S., Tsang, W. F., Ng, S. P. & Hung, Y. Y. (2009). Characterization of flaws embedded in externally bonded CFRP on concrete beams by infrared thermography and shearography, *Journal of Nondestructive Evaluation* 28(1): 27–35.

Maerz, N. H. & Galecki, G. (2008). Preservation of missouri transportation infrastructures: Validation of FRP composite technology, *Technical Report Volume 4 of 5 Non-Destructive Testing of FRP Materials and Installation, Gold Bridge*, Prepared by Missouri S&T and Missouri Department of Transportation.

Maldague, X. P. V. (ed.) (2001). *Theory and practice of infrared technology for non-destructive testing*, John Wiley & sons Inc.

Quiertant, M. (2011). *Strengthening concrete structures by externally bonded composite materials*, ISTE-Wiley, chapter Chapter 23. of Organic Materials for Sustainable Construction, pp. 503–525.

Rajic, N. (2002). Principal component thermography for flaw contrast enhancement and flaw depth characterisation in composite structures, *Composite Structures* 58: 521–528.

Taillade, F., Quiertant, M., Benzarti, K. & Aubagnac, C. (2010). Shearography and pulsed stimulated infrared thermography applied to a nondestructive evaluation of FRP strengthening systems bonded on concrete structures, *Construction and Building Materials* 25(2): 568–574.

Taillade, F., Quiertant, M., Benzarti, K., Aubagnac, C. & Moser, E. (2011). Shearography applied to the non destructive evaluation of bonded interfaces between concrete and CFRP overlays, *European Journal of Environmental and Civil Engineering* 15(4): 545–556.

Taillade, F., Quiertant, M. & Tourneur, C. (2006). Nondestructive evaluation of FRP bonding by shearography, *Proceeding of the Third International Conference on FRP Composites in Civil Engineering (CICE 2006)*, Miami, Florida, US, pp. 327–330.

Valluzzi, M. R., Grinzato, E., Pellegrino, C. & Modena, C. (2009). IR thermography for interface analysis of FRP laminates externally bonded to RC beams, *Materials and Structures* 42(1): 25–34.

10

Thermal Imaging for Enhancing Inspection Reliability: Detection and Characterization

Soib Taib[1], Mohd Shawal Jadin[2] and Shahid Kabir[3]
[1]*School of Electrical and Electronic Engineering*
USM Engineering Campus, Nibong Tebal
[2]*Faculty of Electrical and Electronic Engineering*
Universiti Malaysia Pahang, Pekan, Pahang
[3]*Sustainable Materials and Infrastructure Cluster*
Collaborative μ-Electronic Design Excellence Centre
Universiti Sains Malaysia, Nibong Tebal, P. Pinang
Malaysia

1. Introduction

Reliable performance of an equipment or structure depends on pre-service quality and in-service degradation of the equipment or structure under operating conditions. The role of non-destructive testing (NDT) is to ensure integrity, and in turn, reliability of equipment or structure. Besides, NDT can also monitor in-service degradation and to avoid premature failure of the equipments/structures and prevent accidents as well as save human life. Up to now, NDT has been used in various fields of applications such as the inspection of electrical power plant, substation, storage tanks, bridges, aircraft, pressure vessel, rail, pipeline and so on. Efficient and reliable NDT evaluation techniques are necessary to ensure the safe operation of complex parts and construction in an industrial environment for assessing service life, acceptability, and risk, as well as for reducing or even eliminating human error. Hence, making the inspection process to be fully automated could produce a more reliable, reproducible, faster evaluation and also sustainability.

Previously, due to the lack of effective computational and analytical tools, the data interpretation depends strongly on the experienced and expert of NDT personnel. Nonetheless, since the advancements in computer engineering, modern electronic systems, material science and other related fields made a major impact on all or many of the NDT methods. Data acquisition, analysis and interpretation were automated to increase the reliability and thus reduce the effect of human errors and wrong diagnosis. As NDT is not a direct measurement method, the nature and size of defects must be obtained through analysis of the signals obtained from inspection. Signal and image processing have provided powerful techniques to extract information on material characterization, size, defect detection, and so on. For instance, in the case of images, the major processing and analysis methods include image restoration and enhancement, morphological operators, wavelet transforms, image segmentation, as well as object and pattern recognition, facilitating extraction of special information from the original images, which would not, otherwise, be available. Therefore, this chapter will emphasize the application of thermal image processing in assessing the reliability of concrete structure and diagnosing the condition of electrical equipments.

2. Infrared thermography

Infrared radiation was discovered in 1800 by William Hershel, who used a prism to refract sunlight onto thermometers placed just beyond the red end of the visible spectrum generated by the prism. He found that this area had the highest temperature of all, contained the most heat, and therefore contained a form of light beyond red light. Herschel's experiment was important, not only because it led to the discovery of infrared light, but because it was the first experiment that showed there were forms of light not visible to the human eye (Hellier, 2001).

2.1 IRT principles

Human eyes can only see light in the visible spectrum, ranging from about 400 nm to a little over 700 nm. The electromagnetic spectrum is a band of all electromagnetic waves arranged according to frequency and wavelength. As shown in Fig. 1, the wavelength spectrum of infrared light ranges from about 1 mm down to 750 nm. All objects emit energy proportional to its surface temperature. However, the energy radiated can only be detected by an infrared detector that depends on the emissivity coefficient of the surface under measurement.

The core of the camera is the infrared detector, which absorbs the IR energy emitted by the object (whose surface temperature is to be measured) and converts it into electrical voltage or current. Any object emits energy proportional to its surface temperature. However, the energy really detected (by the infrared detector) depends on the emissivity coefficient of the surface under measurement. The emissivity tells us how much of the thermal radiation from an object that is emitted due to the temperature of the object. All objects above absolute zero (0 Kelvin) emit infrared radiation. The Stefan-Boltzmann law describes the total maximum radiation that can be released from a surface. Since thermal imaging systems only respond to a small portion of the spectrum, it is necessary to introduce Planck's blackbody law.

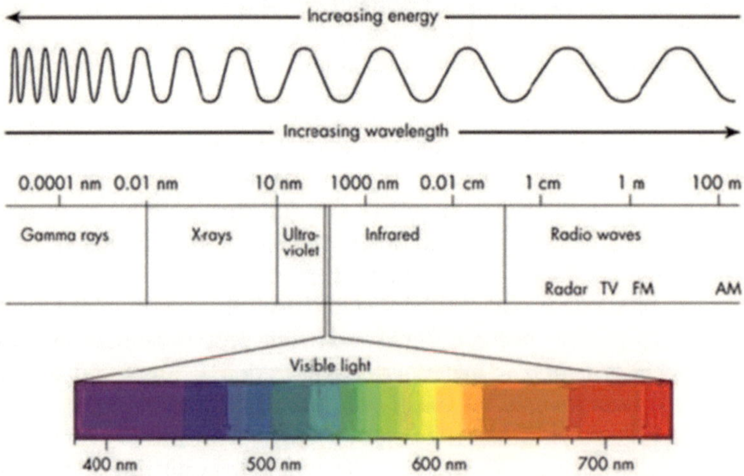

Fig. 1. Electromagnetic wavelength

Planck derived the law as in equation (1), which describes the spectral distribution of the radiation intensity from a black body where the emissivity of the surface, ε is equal to 1 (Holst, 2000).

$$\varepsilon_{\lambda b} = \frac{C_1}{\lambda^5\left(e^{C_2/\lambda T} - 1\right)} \quad \frac{W}{m^2 - \mu m} \tag{1}$$

where $\varepsilon_{\lambda b}$ is the black body monochromatic radiation intensity, C_1 (3.7411 x10^8 W-$\mu m^4/m^2$) and C_2 (1.4388 x10^4 μm-K) are the first and second radiation constants respectively; λ is the wavelength of the radiation being considered and T is the absolute temperature of the blackbody. By integrating Planck's law over the entire spectrum (λ = 0 to ∞), the total hemispherical radiation intensity is obtained.

$$\varepsilon_b = \sigma T^4 \tag{2}$$

where σ is the Stefan–Boltzmann constant (5.67051 x 10-8 W/m2K). It has to be pointed out that equation (2) describes the radiation emitted from a black body which is the maximum value radiated by a body at a given temperature. Real objects almost never comply with this law although they may approach the behaviour of a black body in certain spectral intervals. A real object generally emits only a part ε_λ of the radiation emitted by a black body at the same temperature and at the same wavelength. By introducing the quantity,

$$\varepsilon = \frac{\varepsilon_\lambda}{\varepsilon_{\lambda b}} \tag{3}$$

which is called the spectral emissivity coefficient, equation (2) can be rewritten for real bodies by simply multiplying its second term by ε_λ. When averaged over all wavelengths, the total power density for a non-black body object is

$$emissivity = \varepsilon\sigma T^4 \tag{4}$$

As infrared energy functions outside the dynamic range of the human eye, special equipment is needed to transform the infrared energy to another signal, which can be seen. For this purpose, infrared imagers were developed to see and measure this heat. There are two general types of infrared instruments that can be used for condition monitoring: infrared thermometers and infrared focal plane area (FPA) cameras (Braunovic, 2007). Infrared thermometers only provide a temperature reading at a single and relatively small point on a surface area. Another type of instrument that can provide a one-dimensional scan, or line of comparative radiation, is the line scanner. This type of instrument provides a somewhat larger field of view in predictive maintenance applications compared to the infrared thermometer (Mobley, 2002). Further advancements in infrared thermographic technology started with the development of the FPA. Based on FPA technology, nowadays various types of IR imagers with more advanced and sophisticated features have been developed (Epperly et al 1997). Although FPA technology is more expensive than infrared thermometers and line scanners, it provides more flexible and accurate measurements.

The basic concepts of the IR imager, commonly known as the thermographic camera, is that it can capture an image of the thermal pattern and measure the emissive power of a surface in an area at various temperature ranges. The digital output image of IRT is called a thermogram. Each pixel of a thermogram has the specific temperature value, and the image's contrast is derived from the differences in surface temperature. An infrared imaging system detects radiation in the infrared part of the electromagnetic spectrum and produces images from that radiation. All objects emit infrared radiation and the amount of emitted radiation increases with temperature. Therefore, infrared imaging allows us to see variations in temperature. In an infrared imaging system, there are two types of IR detectors i.e. mid wave (MW) and long wave (LW) which are operated in the range of 2–5 μm and 8–14 μm of the electromagnetic spectrum band, respectively (Minkina & Dudzik, 2009). These bands do not cover the full infrared spectrum because not all parts of the spectrum are suitable for infrared imaging. The reason for this is that the atmospheric transmission of infrared radiation is low in some ranges of the spectrum. This means that the atmosphere will block infrared radiation in these ranges, thus making these wavelengths unsuitable for infrared imaging. Most infrared cameras today work in the MW or LW ranges (Wretman, 2006). The energy detected depends not only on the emissivity coefficient of the surface under measurement but also on the environment. In fact, a fraction may be either absorbed by the atmosphere between the object and the camera, or added as reflected by the surface from the surroundings. This part will be discussed further in the following parts of this chapter.

2.2 Type of IRT

Infrared thermography is generally classified in two types, passive and active thermography (Kumar et al, 2009). In passive thermography, the temperature gradients are present in the materials and structures under tests naturally. In active thermography, the relevant thermal contrasts are induced by an external stimulus (Santos, 2008). The passive method has been widely applied in diverse areas such as production, predictive maintenance, medicine, detection of forest fire, thermal efficiency survey of buildings, road traffic monitoring, agriculture and biology, detection of gas and in NDT. In all these applications, abnormal temperature profiles indicate a potential problem to take care of.

In active infrared thermography, the sample is heated by an external controlled heat source and its surface temperature is monitored as a function of time through changes of emitted infrared radiation. The specific thermal properties of the material under test influence transport of heat thus causing surface temperature to change with respect to areas with different thermal properties. Active thermography is a very popular method in NDT applications such as for detecting crack in structure. There are many methods that have been used in active thermography. Table 1 shows the summary of active thermography methods and its characteristics (Maldague, 2000).

Contrary to active thermography, passive thermography approach does not require external heat source. This is because the heat flow necessary for the evaluation already exists naturally. Passive thermography only to pinpoints anomalies since the heating energy source is difficult to measure. Therefore the accuracy and reliability of passive thermography is not a major concern. In many applications, passive thermography applies relative temperature from the similar object or surrounding temperature. This is well known as qualitative measurement that will be discussed later. One of the applications of passive

Methods	Characteristic
Pulse Thermography (PT)	Fast inspection relying on a thermal stimulation pulse, with duration going from a few milliseconds for high thermal conductivity material inspection (such as metal parts) to a few seconds for low thermal conductivity specimens (such as plastics, graphite epoxy components).
Step Heating (SH)	Contrary to PT scheme for which the temperature decay is of interest (after the heat pulse), the increase of surface temperature is monitored during the application of a step heating pulse ('long pulse'). Variations of surface temperature with time are related to specimen features.
Lock-in thermography	Based on thermal waves generated inside the specimen under study in the permanent regime. Here, at a frequency, the specimen is submitted to a sine modulation heating, which introduces highly attenuated a dispersive thermal waves of frequency inside the material (in close to the surface region).
Vibrothermography	A mechanical vibration induced externally to the structure direct conversion from mechanical to thermal energy occurs and the heat is released by friction precisely at locations where defects such as cracks and desalinations are located.

Table 1. Method of active thermography

thermography is for preventive and predictive maintenance. In construction for example the passive thermography can be used in the search of hidden defects or damages in the road or bridge pavement structure, together with information on the degradation mechanism, serves as an early diagnostic tool, which completes the methodologies utilised for the survey of the state of the paving (Stimolo, 2003)

2.3 A Review of IRT applications

In electrical power systems, the developed IRT plays a vital role in inspecting and diagnosing the integrity of electrical power equipments. It has become one of the preferred methods for assessing equipment conditions online especially in electrical transmission and distribution systems (Lindquist & Bertling, 2008). IRT can be used to monitor the thermal behaviour of the power equipment, as well as the structure of a system. It senses the emission of infrared energy (i.e. temperature) to detect thermal anomalies, which are hotter or colder than they should be. Through this the inspector can then locate and identify the incipient problems within the system. While heat is not a perfect indicator of all problems in electrical systems, heat produced by abnormally high electrical resistance often precedes electrical failures (Hellier, 2001). Although the technique for inspecting electrical systems is quite straightforward, there are several things that need to be considered. Some of the factors, such as environmental effects and equipment conditions, will normally affect the analysis results, especially during an outdoor inspection of a power substation, for example. Direct inspection without considering these factors definitely will result in inaccurate measurements. A good electrical thermographer must contend with several problems related to the electrical equipments, the infrared instrument, and the interpretation of data.

2.3.1 Measurement and analysis methods

There are two ways for temperature measurement. The first is known as quantitative, which is to take the exact temperature values of the objects. The second type is qualitative, which takes the relative temperature values of a hotspot with respect to other parts of the equipment with similar conditions. Infrared thermography can be used as both a qualitative and a quantitative tool. Some applications do not require obtaining exact surface temperatures. In such cases, it is sufficient to acquire thermal signatures, which are characteristic patterns of relative temperatures of phenomena or objects. This method of qualitative visual inspection is expedient for collecting a large number of detailed data and conveying them in a fashion that can be easily interpreted. In contrast, accurate quantitative thermography demands a more rigorous procedure to extract valid temperature maps from raw thermal images (Griffit et al, 2001).

A widely used method of using thermography in electrical equipment inspection is by employing the ΔT criteria (Chou & Yao, 2009)(Lindquist et al., 2005). Qualitative measurements are sometimes called comparative thermography. When the comparative technique is used appropriately and correctly, the differences between the two (or more) samples will often be indicative of their condition (Hellier, 2001). The severity or the level of overheating of the electrical equipments will refer to the temperature-rating table. This table is usually divided into three or four different categories to indicate the maintenance priority based on the equipment's temperature rise with respect to other similar component (Lindquist et al., 2005). Table 2 shows the maintenance testing specifications for electrical equipment published by the InterNational Electrical Testing Association (NETA) ("Standard for Infrared Inspection of Electrical Systems & Rotating Equipment," 2008). NETA provides guidelines for thermal inspections of electrical equipment. These guidelines are based on differences in temperature from one phase conductor or component to another. Recommended action is dependent on the difference in the temperatures.

Priority	ΔT between similar components under similar load (°C)	ΔT over ambient temperature (°C)	Recommended Action
4	1 - 3	1 - 10	Possible deficiency, warrants investigation
3	4 – 15	11 - 20	Indicates probable deficiency; repair as time permits
2	---	21 – 40	Monitor until corrective measures can be accomplished
1	> 15	> 40	Major discrepancy; repair immediately

Table 2. Maintenance testing specifications for electrical equipment

Fig. 2 shows an example of a hotspot and its reference point. A hot area is the suspected component and the reference must be another similar component with the same condition. It could be similar components in other phases. In the ΔT method, the temperature differences between suspected and normal component is calculated as:

$$\Delta T = T_{hot} - T_{ref} \quad (5)$$

where T_{hot} is the warm or hot temperature value of the suspected components while T_{ref} is the reference temperature value from the normal operating component. The severity of the hot spot is then checked using Table 3 under the column 'ΔT between similar components under similar load (°C)'. Action should be taken according to the level of priority. The advantage of this practical method is to establish "failure" or "no failure" condition and the emissivity has only a minor impact on the result (Chou & Yao, 2009). A drawback is that the temperature tables are usually only found in handbooks and guidelines; which is not a standard benchmark. Moreover, the ΔT criterion does not say anything about whether the equipment temperature limits are actually exceeded and also will not expose systematic failures affecting all three phases' connection (Lindquist et al., 2005).

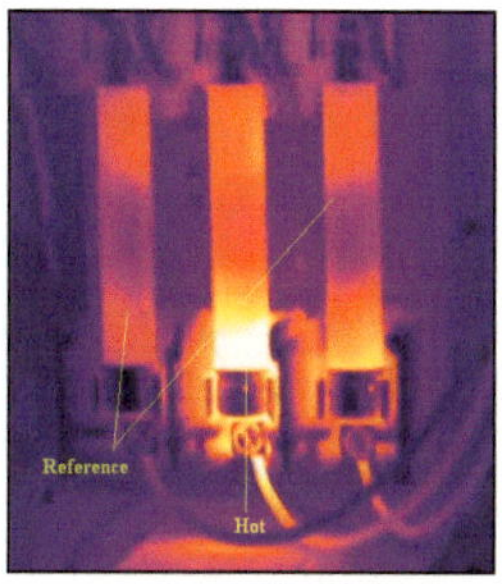

Fig. 2. Example of a hot fuse; the other fuses are used as the reference.

Environmental factor	Effect on IRT measurement
Ambient air temperature	An increase in air temperature will result in an increase in the measured temperature component. At a very high or very low-temperature, the IR system becomes less stable.
Precipitation/humidity (snow, rain, fog, etc.)	It can result in evaporative cooling. The temperature differences (either phase to phase or rise over ambient) can be dramatically reduced, leading to a misinterpretation of the data. Condition that is only slightly warm may be cooled below a point where they can be detected.
Wind or other convection	Wind speeds less than 5 mph can contribute to a significant cooling effect on a high resistance fitting. Wind speeds above 5 mph can reduce the temperature difference between the components and ambient to a few degrees above the ambient.
Sun or solar radiation	Solar heating of components, especially those with a high absorption of the sun's energy (such as aged conductors), will mask small thermal differences.

Table 3. Environmental factors effect on IRT inspection

In quantitative measurement, the reference is the ambient temperature. The observation is established by measuring the absolute temperature of electrical equipment under the same ambient conditions. As the reference temperature has to be measured, it requires an even greater understanding of the variables influencing radiometric measurement, as well as a grasp of its limitations. The temperature rise is calculated as:

$$\Delta T = T_{hot} - T_{amb} \tag{6}$$

where T_{hot} is the hot-spot temperature of the measured component and T_{amb} is the ambient temperature at the time of measuring. Again, the column 'ΔT over ambient temperature' in Table 3 is used for testing specifications.

2.3.2 Improving inspection techniques

Newer developments in modern IRT equipment have improved the quality of measurements. Most modern IR imagers can resolve surface temperature differences of 0.1°C or less (Hellier, 2001)(Griffith et al., 2001). An infrared thermographic system is essentially imaging IR radiometers that can provide IR images continuously and in real-time, just like the images provided by a normal video camera. Complete thermographic systems also integrate an advanced image processing and display system. Despite the advantage of modern designs of IRT cameras, there are still several factors that need to be considered when doing an inspection. Even if temperatures can be measured accurately, several other factors must be taken into account if the real influence of the abnormal temperature difference is to be accurately accounted for (Snell & Spring, 2003). This is a very critical aspect, especially for an outdoor inspection. The inspection of electrical power systems using IRT can be divided into three main areas; substation, underground distribution, and aerial distribution (Azmat & Turner, 2005). When accurate measurements are required, all the influence factors had to be identified during the IRT image is captured.

Generally, the factors that affect the accuracy of IRT measurements can be categorized as procedural, technical and environmental/ambient conditions (Santos et al., 2008)(Hellier, 2001). The procedural factor concerns the thermographer itself. This factor can be minimized if certified or qualified personnel are employed. For technical factors, the issue is normally relates to the emissivity of the equipment under inspection, load current variation, distance of the object being inspected, and the IRT camera specifications. For an outdoor or uncovered inspection, such as at a power substation, environmental effect is a critical issue. The data relates to the environmental factors are considered crucial and should be collected prior to inspection. Table 3 summarizes the environmental factors effects that need to be considered when doing an IRT inspection (Snell & Renowden, 2000).

A proper observation should be made before starting any IRT inspection. In most of the cases, essential information of the target location is provided. In this case, the history of the target location and electrical power equipments had to be considered. Among the important data needed for an inspection are the load variations, type of equipment and the materials used in building the. For an accurate measurement, the right and suitable tool should be selected. It is recommended that for an extensive outdoor inspection, especially during sunny periods, long-wave (generally 8 μm -14 μm) sensing of IRT systems should be used. It had been proven that within this wave band, the thermal detectors provide greater sensitivity to ambient temperature objects and insensitive to atmospheric attenuation (Balaras & Argiriou, 2002), (Epperly et al., 1997). Short-wave systems should be used only on a limited basis depend on the condition of loads and time (Hellier, 2001). In summary, Table 5 shows all the factors that can affect the IRT measurement related to the target equipment and the inspection tools.

3. Significance of imaging for inspection reliability

Digital signal and image processing is a widely used engineering term that, in a broad sense, can be described as a transformation converting signal data into useful information using digital computers. Although digital signal and image processing methods have been commonly employed in various scientific and engineering fields, their application in NDT is very recent. Back to the previous technique, NDT was implemented in a manual or semi-automatic method where the operator will accept or reject the decision. This approach is subject to error and wrong interpretation. Furthermore, most data recording and analysis methods are primitive. Therefore, utilizing the digital signal and image processing for the inspection can minimize the operator dependence because it uses automated data analysis, thus improving NDT inspection reliability. Both techniques can be employed to improve the signal-to-noise ratio (SNR) therefore increased the detection capabilities of the defect or fault. Detailed definitions of the detected problem such as type of defect, shape, size and severity of defects has acquired a great significance lately. This is because of the need for this information for implementation of such methodologies as retirement-for-cause and remaining-life analysis. In summary, applying automated imaging inspection can greatly improve the NDT inspection reliability including the use of non NDT technologies (Zu et all, 2011). Digital signal processing has played and will continue to play a very significant role in NDT. To provide a more complete understanding about this subject, the remainder of this chapter presents a deep review of how the IRT image processing technique can enhance and practically implemented for detecting and characterizing the inspection reliability of concrete structure and electrical equipment.

Since the early 1960s, infrared thermography (IRT) has been used in many fields of application, such as military, industrial, civil engineering and medical, as well as electrical engineering. IRT is a non-contact, non-destructive, visualizing technique, which is becoming an important means for quality control in production and in service inspection (Junyan et al., 2008). Due to its advantages in terms of being non-contact, free from electromagnetic interference, safe, reliable, providing large inspection coverage and fast data interpretation, IRT has taken a very important role in condition monitoring especially in predictive and preventive maintenance techniques (Azmat & Turner, 2005). Furthermore, inspections can be done without shutting down operation of the system.

4. Problem concerning faults and damage detection of power system equipment and concrete infrastructure

4.1 Fault diagnosis in power equipment

Electrical devices are usually rated in power, which indicates the maximum amount of energy the device can consume without being damaged. If the device operated above its specifications, the excess power causes the atoms present in the device's material to resonate and resist the flow of electricity. This resistance to the flow of electricity will generates heat, which in turn, overheats the device and reduces its life cycle and efficiency. Another major problem that usually created within the utility equipment is the change of resistance due to loose connections. The loose connection causes the electricity to occupy a smaller area of the defective connection that is required for proper flow. This phenomenon will and therefore increases the resistance and temperature of the connection. Any change in resistance will cause the equipment to consume more power than the intended load (Azmat & Turner, 2005). Fig. 3

shows an example of infrared image and the visual light image of a circuit breaker. The red colour region in the infrared image indicates the hot spot and possible anomaly.

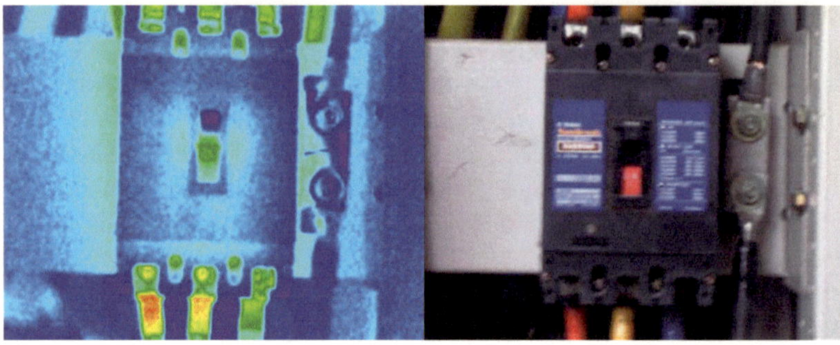

Fig. 3. Infrared image and visible light image

Faults in electrical power systems can be classified into a few categories, such as poor connection, short or open circuit, overload, load imbalance and improper component installation (Kregg, 2004),(Cao et al., 2008). In most cases, poor connections are among the more common problems in transmission and distribution lines of electrical power systems (Azmat & Turner, 2005). According to a thermographic survey conducted during the period of 1999-2005 (Martínez & Lagioia, 2007), it was found that 48% of the problems were found in conductor connection accessories and bolted connections. This is mainly due to loose connections, corrosion, rust, and non-adequate use of inhibitory grease. On the other hand, 45% of the thermal anomalies appear in disconnector contacts. Most of the anomalies are due to deformations, deficient pressure of contact, incorrect alignment of arms, and accumulation of dirt. Only 7% of the problems were found in electrical equipment. In diagnosing faults at power substation using IRT, it was found that transformers have taken priority over other equipment. This is due to the fact that transformers are the most costly equipment in a power substation. The common causes of failure in transformers are oil leakage and inferiority in internal insulation, which can lead to catastrophic destruction and power outage (Utami et al., 2009). For normal transformer installation, operating temperatures rise over ambient 65 °C for oil-filled and 150 °C air-cooled transformers respectively (Balaras & Argiriou, 2002). Temperatures above these operating points will cause breakdown in the insulation winding and therefore causing an electrical short-circuit.

By utilizing IRT, the thermal image will clearly indicate problem areas. The suspected areas then can be easily located and the problems indentified. Nevertheless, in some cases, the interpretation of thermographic images cannot be done directly except by an experienced and qualified thermographers because most of the thermographic characteristics had to be understood. According to (Hou, 1998), faults in electrical equipment can be divided into two kinds, external or internal, depend on location. However, internal faults are difficult to identify because they are much more complex. Table 4 summarizes the faults that commonly occur in electrical power equipments.

4.2 Automated diagnostic system

In industrial application, the sophisticated diagnosis system is the choice in order to get fast and accurate data with minimum maintenance cost. Most of the IRT cameras that are

Equipments	Common problem
Lightning arrestor, circuit breaker, disconnector, splices, transmission lines, compression clamps	Poor electrical connection: Loose, corroded or improper connection or splices. Poor breaker connection.
	Inoperative capacitor
	Fail lightning arrestor
	overloading
	Broken conductor strands
Bus duct	Unbalance load
	High resistance in joints
	Bus plug-ins
	Fuse connections
Switches, bus bars, capacitor bank, fuses, load centres, motor control centres	Poor electrical connection: Loose, corroded or improper connection and contacts.
	Unbalance load
	Harmonics, eddy currents and hysteresis
	Overloading
Transformer	Poor electrical connection: Loose, corroded or deteriorated connection.
	Unbalance load
	Overloading
	Low fluid level
	Overheated bushing
	Blocked cooling tubes
Motor and generators	Unbalance load
	Blocked cooling passage
	Shorted or open winding
	Overheating of brushed, slip rings and commutators.
	Overloading
Lighting	Poor electrical connection
	Overheating ballast

Table 4. Faults and their thermographic image characteristics

available today come with analysis software and have the capability to provide the inspection report. Furthermore, there is also stand-alone analysis software that can be used for any type of thermographic image. Digital images are uploaded into the computer directly from the IRT camera for further analysis. Most of the software may have various analysis functions, such as spot, area, isotherms, and line thermal measurements, as well as size measurements. Analysis can be extended beyond the image by displaying the numerical data in a spreadsheet or in various standard graphical forms, such as a histogram (Hellier, 2001). However, despite the power and ease of use of the software, the analysis process still needs qualified or experienced personnel. Also, most of the conventional analyses are time consuming in preparing the final report. Therefore, applying an intelligent system in thermographic image analysis can overcome this limitation. In recent years, rapid development in computer vision based on image processing techniques and the integration of artificial intelligence has provided many advantages in monitoring and diagnosing problems. In electrical power systems, the application of IRT for automatic diagnosis using intelligent systems is still in the early stages.

This is due to the complex analysis and various factors that need to be considered in developing such system. Most of the research in automatic diagnosis of electrical power equipment and machine condition monitoring started in early 2000 and it become more complicated with the used of advanced materials. Table 5 shows the factors related to the target equipment and the inspection tool.

Equipments factor	**Characteristic**
Electrical loads	Temperature of the connection will increase as the load increases. For light load problems in the early stages of failure will be less thermally. It is recommended that during the inspection, the load on the line should be at least 40%.
Equipment emissivity	Most of the conductors have quite low emissivity, typically 0.1-0.3. While greasy, black, overheated and aged conductors can have emissivity values as high as 0.97, it is often difficult to assess this visually from a distance.
Thermal gradient	The heat of high resistance is usually being generated at some internal point to the surface. There exists a thermal gradient between the hottest spot inside the equipment and the surface being viewed.
IRT device (camera)	Factors that must be considered are resolution, both spatial and measurement, detected waveband, sensitivity as well as the signal processing speed.
Distance and angle	The resolution of the IRT image decreases with distance. Acute angles present less information than images taken at right angles.

Table 5. Factors related to the target equipment and the inspection tool

In order to automatically analyze the condition of electrical equipment, image processing techniques can be used to extract the thermal profiles within the electrical equipments. Image processing techniques generally consist of several steps: pre-processing, segmentation, feature extraction, classification and decision making. The straightforward approach is to follow these steps one by one in bottom-up order. There are three techniques that can be used to determine the thermal severity of electrical equipments through thermal image analysis. The first one is a direct interpretation by identifying the real maximum temperature for each of electrical equipment and evaluating their condition based on the ΔT criteria. The maximum temperature is determined by finding the highest pixel value within the selected region. Calculating the histogram or histogram distance is another method that can be used for finding the similarity between two objects. In this case, the histogram for each region is computed and compared with other regions in order to get the ΔT. Another approach is to analyze the gradient of the segmented region. One of the advantages of utilizing the gradient analysis technique is that the source of the hotspot in electrical equipment can be identified.

Based on previous research, the simplest method of identifying hotspot regions within a thermal image of electrical equipment is to use thresholding techniques (Chou & Yao,

2009),(de Oliveira & Lages, 2010). The hot spot area is detected by filtering the image using a certain threshold value. Then, the hotspot region is extracted using morphological segmentation where the maximum gray pixel value determines the maximum temperature of the hotspot region. The reference temperature is derived from the average gray values of the equipment outside of the hotspot region. The level of severity of the electrical equipment is calculated by comparing the hotspot temperature and the reference temperature (Chou & Yao, 2009)(de Oliveira & Lages, 2010) (Baoshu et al., 2006). In another approach, the watershed transformation algorithm is used for segmenting the hotspot regions in the thermal image of electrical equipment (Almeida et al., 2009).

For diagnosing the thermal fault within electrical equipment, certain feature descriptions are created for the regions of interest. For the classification process, various intelligent techniques, such as neuro-fuzzy (Almeida et al., 2009), artificial neural network (Shafi'i & Hamzah, 2010) and support vector machine (SVM) algorithm (Li et al., 2006)(Rahmani et al., 2010) are used to determine the condition of the electrical equipment. The thermal profiles of electrical equipment can also be extracted by analyzing their real temperature values. The real temperature values for each pixel in the image can be extracted directly from its RGB data. This method is quite straightforward but has a problem with high processing time due to the large feature vectors to be computed by an artificial neural network (ANN) algorithm (Shafi'i & Hamzah, 2010). The previous research with various hotspot detection techniques and fault classification method is summarized in Table 6.

Reference	Automatic hotspot detection technique	Fault classification method
(Ying-Chieh Chou & Yao, 2009), (de Oliveira & Lages, 2010)	Thresholding	Calculating and comparing the real temperature values
(Laurentys Almeida et al., 2009)	Watershed segmentation	Neuro-fuzzy
(Shafi'i & Hamzah, 2010)	RGB image data	ANN
(Baoshu Li et al., 2006), (Rahmani et al., 2010)	Thresholding	SVM
(Wretman, 2006),(Smedberg, 2006)	Finding repeated pattern and smooth but steep image gradient	ANN
(Korendo & Florkowski, 2001)	Manually find region	Invariant coefficient method
(Younus & Bo-Suk Yang, 2010)	discrete wavelet decomposition	Bio-orthogonal wavelet algorithm
(Wong, Tan, Loo, & Lim, 2009)	RGB	RGB value comparison

Table 6. Automatic Diagnosing System of Electrical Equipment

Based on previous research except (Almeida et al., 2009), most of the diagnoses systems are only analyze the captured thermal image without considering other important variables. In this research, the only thing that distinguishes this research from others is the input variables. Besides considering environmental factors, this study also includes the identification variables of the electrical equipment, such as pollution index, rated voltage, material and manufacturer of the equipment. The whole system diagram and the input variables are depicted in Fig. 4(a) and Fig. 4 (b) respectively.

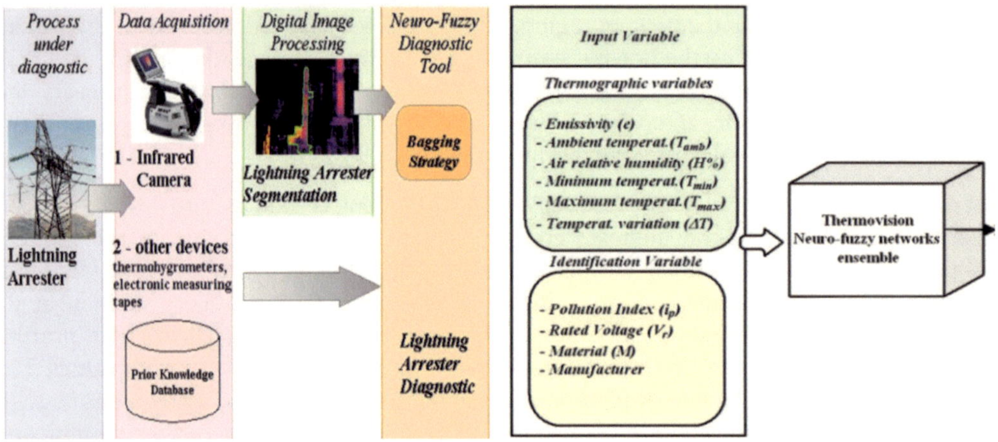

(a) (b)

Fig. 4. Intelligent surge arrester diagnosis system (a) system block diagram (b) input variables for neuro-fuzzy classifier

Instead of using classical bottom-up approach, Wretman (Wretman, 2006) and Smedberg (Smedberg, 2006) have successfully segmented the IRT image of electrical installation by using the top-down approach of image processing method. In other words, identifying the interesting region will be done first by detecting and grouping a regular repetitive structure. The tasks for finding repeated objects in the image can be broken down into two separate steps: (i) finding interesting features in the image and describing these using pre-specified descriptors, and (ii) comparing all the features and look for matches. In this research, all the interesting features were detected by modifying the scale-invariant feature transform (SIFT) algorithm (Lowe, 2004).

4.3 Damage detection of concrete infrastructure

The deterioration of concrete infrastructure is a growing problem worldwide; many structures are approaching the end of their service lives and need maintenance or rehabilitation in order to remain functional. In spite of recent increases in public infrastructure investments, infrastructure is deteriorating faster than it is being renewed. Various factors can contribute to the deterioration of concrete infrastructure; mechanical stress and fatigue, and chemical and environmental conditions are among the major causes concrete distress (Scott et al., 2003). Damage such as cracks, may exist in concrete even before the structure is subjected to any external loading. An excessive water-cement ratio, improper curing, and creation of high temperatures during the hardening process may result in shrinkage, which is the direct cause of cracking. These cracks later expand and widen during service due to freeze and thaw cycles and the intrusion of moisture. This process is especially critical for large concrete structures, such as dams, due to placement of massive amounts of concrete during construction. Even an initially sound concrete dam can develop cracks during its service life. Since a concrete dam is always in contact with water, relatively small size cracks will eventually become wider and develop into holes or delaminations, and decompression joints or bedding in the shallow bedrock. Assessing the safety of concrete gravity dams against sliding requires a detailed investigation of the cracks and other discontinuities in the concrete structure and the rock foundation underneath. This is achieved through characterization of the mechanical properties of the materials (concrete and rocks), and especially the shear strength of the different types of discontinuities found throughout the structure and the foundation. Traditionally, a log is kept of the discontinuities found in cores drilled from the investigated structure. This method has the advantage of providing specimens for petrographic examinations and allows the testing of specific properties, such as compressive strength, Young modulus or permeability. However, information on the condition of the discontinuities is sometimes altered or lost due to drilling operations, even if a triple-tube coring system is used.

For instance, cracks might be created during drilling or transportation of the samples. Also, planes of cohesive weakness can separate after drilling, which modify the evaluation of the shear properties of the structure. The orientation of the core is another parameter that can be lost during drilling, if the procedure is not properly done. Since drilled cores are usually collected from dams for testing concrete and rocks, the borehole itself can be used to perform a detailed investigation and collect additional information on the surrounding materials. Borehole geophysical logs have been used for more than 50 years, mainly for oil mining. These methods provide continuous quantitative and statistical measurement of the depth, thickness, and orientation of features such as fractures and joints. Borehole imaging can actually provide better data than core samples, since the equipment used (the televiewers) depict in-situ conditions, and are not subjected to incomplete core recovery. Furthermore, tools are magnetically referenced to true north, thus eliminating the need for oriented cores.

Map-like surface cracking may indicate an adverse reaction between cementitious alkalis and aggregates. This reaction, known as the alkali-aggregate reaction (AAR) is a potentially harmful process in concrete containing reactive aggregates, and can lead to varying degrees of cracking in structures, as well as differential movement and misalignment of concrete elements and mechanical installations (Bérubé et al., 2000). AAR has been recognized in more than 50 countries around the world; it is likely that the problems associated with AAR exist in a larger number of countries, but concrete distress in several instances may have been attributed to other causes.

4.4 Advances in thermography imaging for sub-surface damage detection

Visual colour and greyscale imagery of concrete greatly extend natural vision capabilities in terms of colour and greyscale perception. Human vision is relatively poor at differentiating the brightness and colour features in the scene being viewed, whereas greyscale digital imagery can provide hundreds of levels of grey and colour digital imaging allows the quantitative differentiation of millions of different colours. Such a range of image perception is unattainable by the human eye, but is extremely useful for quantitative image analysis. There is, however, a need for the development of effective image analysis techniques in order to derive the information needed from the concrete imagery. Surface damage, such as cracks, are usually treated as objects, and are thus quantified through techniques that first segment the objects from the background to extract shape or object features, and then classify the images based on those features.

Visual colour and greyscale imagery of concrete greatly extend natural vision capabilities in terms of colour and greyscale perception. Human vision is relatively poor at differentiating the brightness and colour features in the scene being viewed,whereas greyscale digital imagery can provide hundreds of levels of grey and colour digital imaging allows the quantitative differentiation of millions of different colours. Such a range of image perception is unattainable by the human eye, but is extremely useful for quantitative image analysis. There is, however, a need for the development of effective image analysis techniques in order to derive the information needed from the concrete imagery. Surface damage, such as cracks, are usually treated as objects, and are thus quantified through techniques that first segment the objects from the background to extract shape or object features, and then classify the images based on those features.

A variety of image processing techniques can be used to characterize the damage in concrete data; among these methods are edge-detection algorithms (Abdel-Qader et al., 2003). Edges are considered to be areas with strong intensity contrasts in an image, causing a jump in intensity from 1 pixel to the next. In image data of damaged concrete, these edges would characterize boundaries between areas of sound concrete and deterioration, such as cracks. However, in their study on the classification of pits and cracks in corrosion images, Livens et al. (Livens et al., 1996) found that segmentation approaches worked well on individual images, but proved unsatisfactory when applied to a large set of samples due to the variability in the background. So they adopted a method based on the analysis of the textured appearance of the pits and cracks in the images, which was successfully employed to discriminate between the two types of damage. Furthermore, according to He and Wang (He & Wang, 1991), a good understanding or a more satisfactory interpretation of an image should include the description of both spectral and textural aspects of the image.

Other approaches used for damage characterization include transform-based techniques. Wavelet transforms are powerful tools often employed in image processing applications. The main advantage of this transform remains in its ability to locally describe signal frequency content. Through the wavelet transform (Ksantini, 2003), an image is decomposed into several high-frequency images containing wavelet coefficients representing details with increasing scale and different orientations (Foucher et al., 2001). More specialized methods that may be used to detect deterioration in concrete images are statistical-based approaches. These techniques allow for the analysis of the textural content in an image. Statistical texture methods analyse the spatial distribution of grey values by computing local features at each point in the image, and deriving a set of statistics from the distributions of the local features (Haralick, 1979).

Different types of concrete damage each have a specific texture typical of the type of deterioration, which should permit their discrimination through texture analysis methods. There are very few studies that have applied image processing techniques, such as texture analysis, to extract textural features in order to obtain concrete deterioration information from optical imagery. The analysis of concrete structure can be done by extracting the texture information through the grey level co-occurrence matrix texture analysis approach and the deterioration features in the concrete imagery is detected through the artificial neural network classifier in order to obtain more accurate damage characterization and assessment. These methods are applied to three types of concrete imagery, thermographic, colour and greyscale digital imagery as shown in Fig. 5, in order to evaluate their effectiveness in providing surface damage information.

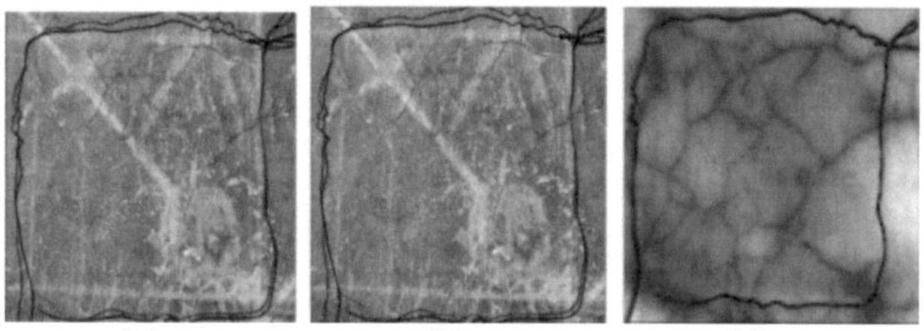

(a) (b) (c)

Fig. 5. Greyscale, thermographic, and colour images of GRAI-3 slab: (a) Greyscale, (b) Colour and (c) Thermographic.

Images were taken of concrete specimens exhibiting various levels of surface cracking associated with the alkali-aggregate reaction (AAR). This reaction occurs between some reactive aggregates and alkali hydroxides in the concrete pore solution. AAR leads to the swelling and cracking of concrete. The amount of cracking is closely related to the expansion level, and other indicators of concrete deterioration, such as loss of rigidity, decreasing mechanical properties, etc. (Rivard & Ballivy, 2005). Experiments were conducted on two sets of concrete specimens. The first set consists of three concrete blocks measuring 40×40×70 cm3 each, which were exposed outdoors to the elements for over ten years at the CANMET site in Ottawa (Canada); CAN-1, CAN-2, and CAN-3 present low, medium and high amounts of damage, respectively. The second set consists of three concrete slabs, 100×100×25 cm in size each, batched and kept at the GRAI laboratory (University of Sherbrooke). These slabs were wrapped with damp terry cloth and stored at ambient air (20±2 °C). As with the CANMET blocks, GRAI-1, GRAI-2, and GRAI-3 present low, medium and high amounts of damage, respectively. Table 7 and 8 provides more details on the concrete mixtures for the CANMET and GRAI specimens.

Concrete mixtures	**GRAI**			**CANMET**		
	GRAI-1	**GRAI-2**	**GRAI-3**	**CAN-1**	**CAN-2**	**CAN-3**
Cement content (kg/m^3)	210	390	390	423	423	425
Density (kg/m^3)	2223	2326	2340	2303	2303	2317
Total Na_2O_{eq} (kg/m^3)	3.81	3.25	5.25	1.69	3.81	5.31
W/C	0.75	0.66	0.66	0.42	0.42	0.42

Table 7. Mixture proportions for CANMET and GRAI specimens.

Average measurement	GRAI			CANMET		
	GRAI-1	**GRAI-2**	**GRAI-3**	**CAN-1**	**CAN-2**	**CAN-3**
P-wave velocities (m s^{-1})[a]	3810	3590	3440	4909	4513	4402
Expansion (%)[b]	0.000	0.060	0.100	0.025	0.283	0.340

[a] Based on 11 measurements.

[b] Based on side and surface measurements.

Table 8. Average measurements of P-wave velocities and expansion levels

4.5 Statistical texture analysis using GLCM

Statistical texture methods analyze the spatial distribution of grey values in an image by computing local features at each point in the image, and deriving a set of statistics from the distributions of the local features. Depending on the number of pixels defining the local feature, statistical methods can be classified into first-order (one pixel), second-order (two pixels) and higher-order (three or more pixels) statistics. The outputs of the derived features are images in which the pixel values have been changed to reflect a particular feature, or texture; therefore, the resulting feature images are also known as texture features (Schowengerdt, 1997). A second-order histogram is an array that is formed based on the probabilities that pairs of pixels, separated by a certain distance and a specific direction, will have co-occurring grey levels. This array, or second-order histogram, is also known as the grey level co-occurrence matrix (GLCM). Since the co-occurrence matrix expresses the two-dimensional distribution of pairs of grey level occurrences, it can be considered a summary of the spatial and spectral frequencies of the image. A large number of texture features have been proposed; as many as fourteen different features that can be derived from these matrices are described by Haralick et al. (Haralick et all, 1973). However, only some of these are widely used. This is because many of the features are redundant, due to their high correlation. Thus they are not all useful for describing a particular texture. Some of the texture features that can be extracted from the GLCM are image contrast, correlation, dissimilarity, mean, variance, standard deviation, second moment, energy and entropy.

The most effective features are selected through a process consisting of visual analysis, histogram analysis, and analysis of correlation matrices. In this study, the thermographic image of the GRAI-3 slab, which exhibits a fair amount of deterioration associated with AAR and presents quite a bit of textural variation, was used in the feature selection process, since features found appropriate for this image will be suitable for the other images as well. For the first step in the feature selection process of the second-order statistics, visual analysis of the feature images revealed that the visual quality of the contrast and correlation features was not adequate; the contrast and correlation features were thus initially considered for discarding due to their poor quality in terms of visual information. This is shown in Fig 6.

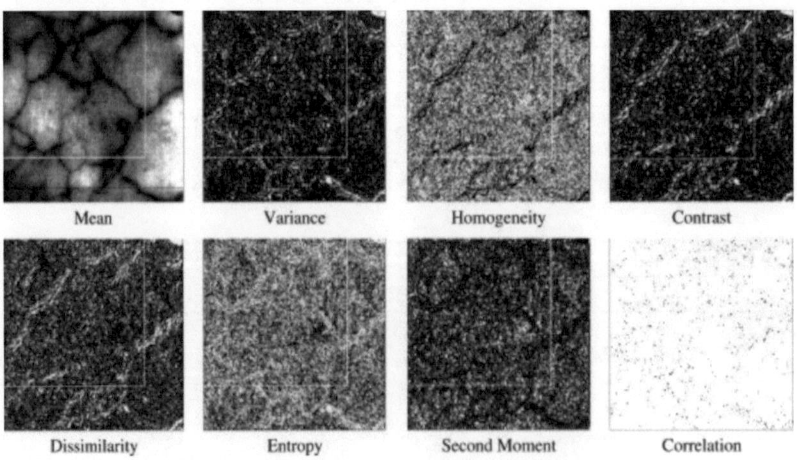

Fig. 6. Second-order texture features from green band of TIR G3 image.

Analysis of the histograms of the feature images confirmed that the two features, contrast and correlation, should be eliminated because of the narrow peaks they presented in their respective histograms, which signify a lack of textural information. The histogram of the variance feature also demonstrated a lack of texture information, so this feature was also considered for elimination (see Fig. 7).

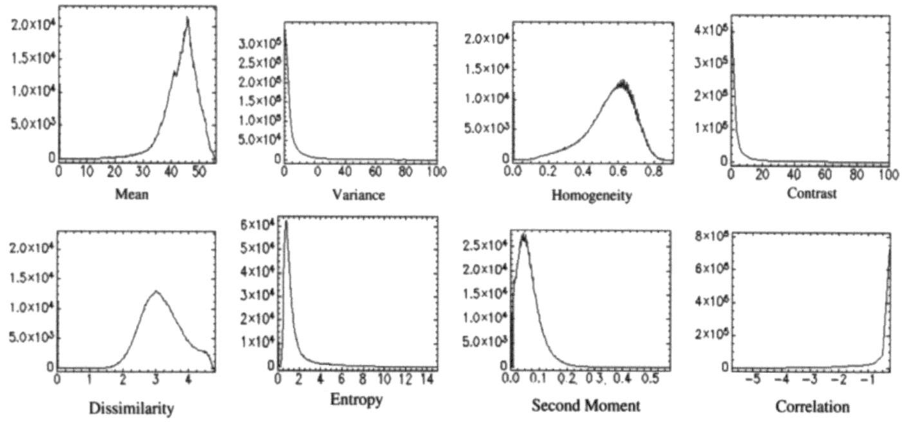

Fig. 7. Histograms of second-order features for greyscale map-crack image

Finally, the correlation matrix of the feature images was calculated; analysis of the correlation matrix further confirmed the removal of the contrast, correlation and variance features, due to their relatively high correlation with the other features. This analysis also indicated the need to discard the entropy and second moment features as well, due to the same reason (see Table 9). As a result, the mean, homogeneity and dissimilarity second-order features were selected for use in this study.

Features	**Mean**	**Var**	**Homo**	**Cont**	**Diss**	**Ent**	**SM**	**Corr**
Mean	1	−0.312	0.684	−0.240	−0.229	0.174	0.425	0.228
Var	−0.312	1	−0.515	0.833	0.857	0.500	−0.360	−0.376
Homo	0.684	−0.515	1	−0.512	−0.617	−0.324	0.797	0.450
Con	−0.240	0.833	−0.512	1	0.942	0.470	−0.338	−0.391
Diss	−0.229	0.857	−0.617	0.942	1	0.679	−0.486	−0.525
Ent	0.174	0.500	−0.324	0.470	0.679	1	−0.571	−0.454
SM	0.425	−0.360	0.797	−0.338	−0.486	−0.571	1	0.335
Corr	0.228	−0.376	0.450	−0.391	−0.525	−0.454	0.335	1

Table 9. Correlation matrix of second-order texture features for GRAI-3 thermographic image

The success of the GLCM method depends on the choice of the distance and the direction between the pixels, and the window size. The appropriate distance between pixels depends on how fine or coarse the texture of interest is. Small distances are usually used for fine textures since pixels close to each other will present enough variation in their grey values to characterize these textures, whereas greater pixel distances are generally used for more coarse textures because variations in the grey values occur in pixels farther away from each other. It has been found that small distances produce the best results (Karathanassi, Iossifidis, & Rokos, 2000), since they are appropriate for textures that are fine, as well as for those that are coarse. A distance of 1 pixel means that the pixels in the pair are located right next to each other. Concerning the direction between pixels, four directions can be used: 0° (horizontal), 45° (diagonal), 90° (vertical) and 135° (diagonal); however, selecting which direction to use can be difficult. One method consists of calculating the GLCM features for all four directions and then taking their averages. The most common choice for the direction between pixels found in the literature, however, is 0°; this means that the pixels in the pair are located horizontally with respect to each other.

After extraction of the textural information from the images, the next step consists of classifying and quantifying the different classes of texture. Artificial neural networks (ANNs) classification approach based on MLP architecture was used to extract the crack patterns from the concrete imagery. Four input nodes were used to represent the following four input features: the original input image and the three selected second-order features. Three output nodes were used to correspond to the following three target classes: wide crack, narrow crack and no crack. For the training algorithm, the popular error back-propagation method was employed. In order to avoid poor classifications or inaccurate estimates of the elements, efforts were made to choose a sufficient number of training pixels for each class, in order to ensure adequate representation. After regions of the image were selected as training data and separate areas were selected as testing data, the software implemented the MLP, which then performed a classification using the four input features. This was done for each of the three image types, thermographic, colour and greyscale, of the CANMET and GRAI specimens. The results of the classification can be presented in two forms: a classified image (also known as a thematic map) which shows the spatial distribution of the various classes in which each pixel is assigned a symbol or colour that relates it to a specific class, and a table that summarizes the number of pixels in the whole image that belongs to each class. The Kappa coefficient was adopted to assess the accuracy of the results obtained.

Results of the classifications show that the greyscale imagery performed fairly well, with an overall classification accuracy range of 72.3–76.5% for the CANMET blocks, and 68.7–75.3% for the GRAI slabs. Classifications using the colour imagery were slightly better than the greyscale imagery, with accuracies ranging from 71.4% to 75.2% for CANMET blocks and 70.9–72.0% for the GRAI slabs. The thermographic imagery, however, produced the highest overall classification accuracies, which range from 73.1% to 76.3% for the CANMET blocks and 74.2-76.9% for the GRAI slabs. [Table 10] and [Table 11] show the classification accuracies obtained for each class, as well as the Kappa coefficients and overall accuracies for each specimen of the CANMET blocks and GRAI slabs, respectively. Fig. 8 presents the classified images for the greyscale, colour and thermographic imagery of the GRAI-3 slab.

Since the infrared thermography performed better than the other two types of imagery, only the results obtained from the thermographic image classifications were used to determine the various levels of AAR damage. The tabular results of the classifications performed on the thermographic imagery of the specimens are presented in Table 12. Among the CANMET blocks, specimen 1 presented the least amount of surface deterioration at 3.9% in the form of narrow cracks, specimen 2 had a moderate amount of narrow cracks (8.2%) and wide cracks (3.6%) for a total of 11.8% surface deterioration, and specimen 3 revealed the

	Thermographic			Colour			Greyscale		
	1	2	3	1	2	3	1	2	3
Kappa coefficient	0.73	0.73	0.74	0.69	0.71	0.74	0.74	0.73	0.76
Overall accuracy (%)	74.5	73.1	76.3	71.4	75.2	74.1	72.3	76.5	75.4
Classes	Accuracy (%)								
Wide crack	81.3	79.9	83.5	78.6	78.1	78.4	76.8	78.1	78.4
Narrow crack	79.7	78.6	81.5	77.7	70.9	77.5	76.3	70.9	77.5
No crack	82.4	76.7	80.6	75.3	74.5	74.2	73.6	71.4	74.2

Table 10. Classification accuracies for CANMET blocks

	Thermographic			Colour			Greyscale		
	1	2	3	1	2	3	1	2	3
Kappa coefficient	0.75	0.74	0.76	0.72	0.74	0.74	0.69	0.72	0.74
Overall accuracy (%)	75.6	76.9	74.2	70.9	71.6	72.0	68.7	74.1	75.3
Classes	Accuracy (%)								
Wide crack	76.7	78.1	80.0	73.4	76.1	79.9	70.4	77.1	79.7
Narrow crack	74.7	75.2	73.4	70.1	72.9	73.4	73.7	72.9	73.4
No crack	76.6	74.9	79.1	72.6	76.4	78.2	73.9	71.4	77.0

Table 11. Classification accuracies for GRAI slabs

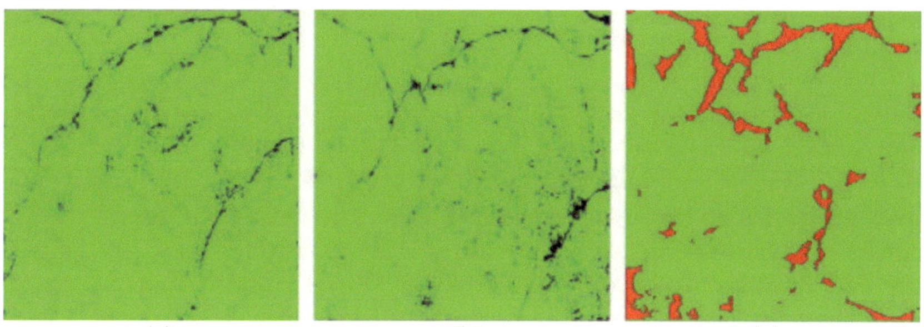

(a) (b) (c)

Fig. 8. Classified images of greyscale, colour and TIR images of GRAI-3 specimen: (a) Greyscale classified, (b) Colour classified and (c) TIR classified.

greatest amount of narrow cracks at 19.7%, as well as a number of wide cracks at 14.3%, for a total surface deterioration of 34.0%. For the GRAI slabs, specimen 1 had a total surface damage of 1.14% comprised of narrow cracks, specimen 2 presented 14.1% total damage made up of 8.8% narrow cracks and 5.4% wide cracks, and specimen 3 showed 14.0% narrow cracks and 9.1% wide cracks for a total surface deterioration of 23.1%.

Further analysis of the surface damage was performed after converting the classified images into binary images. This process simplifies the image by assigning the pixels that represent a damage value of 1 (black) and the background pixels a value of 0 (white). Manual or automated methods are then used to count or sum the pixels to calculate total crack length, as well as average crack width. In order to quantify the total length of wide cracks, pixels

Specimens	Classes	CANMET		GRAI	
		Image pixels	Image (%)	Image pixels	Image (%)
3	Wide crack	37 356	14.3	23 907	9.1
	Narrow crack	51 852	19.7	36 674	14.0
	No crack	172 936	66.00	201 563	90.3
	Total pixels	262 144	100.00	262 144	100.00
2	Wide crack	9 489	3.6	14 078	5.4
	Narrow crack	21 365	8.2	22 937	8.8
	No crack	231 290	88.2	225 129	85.8
	Total pixels	262 144	100.00	262 144	100.00
1	Wide crack	0	0.00	0	0.00
	Narrow crack	10 119	3.9	5 943	1.1
	No crack	252 025	96.1	256 201	98.9
	Total pixels	262 144	100.00	262 144	100.00

Table 12. Tabular representation of thermographic classifications

along the length of each branch of the cracks were summed and the total multiplied by the pixel resolution of 0.26 mm. For the CANMET blocks, a total length of 237.4 mm of wide cracks was calculated for specimen 3. For specimen 2, the total length was found to be 97.6 mm, and for specimen 1, the total length was 0 mm. Determination of average crack width was done by measuring the width of the wide cracks at several points. Each square represents one pixel at a resolution of 0.26 mm. As a result, the average width of cracks in the CANMET blocks was found to be 1.6 mm for specimen 3, 0.8 mm for specimen 2, and 0 mm for specimen 1.

These findings are supported by in-situ data recorded for the CANMET blocks and the GRAI slabs. CAN-3 was prepared with the highest alkali content, and showed the highest expansion level. On the other hand, CAN-1 showed the lowest expansion level as the concrete was mixed with a low level of alkali content. The highest values for the total length of wide cracks as well as for the average width of cracks found for the CAN-3 specimen also relate well to its having the lowest P-wave velocities (Table 8), indicating the highest deterioration level. The absence of wide cracks in the CAN-1 sample, which had a value of 0 mm for the average width of cracks, as well as for the total length of cracks, corresponds well to the lower percentage for average expansion levels measured on the blocks, where the CAN-2 sample showed a higher percentage, and with the CAN-3 sample having the highest percentage of expansion.

As for the GRAI slabs, the absence of wide cracks in the GRAI-1 specimen, which had a value of 0 mm for the average width of cracks, as well as for the total length of cracks, is corroborated by its having the lowest expansion level, indicating very little damage. A higher level of expansion was measured on the GRAI-2 specimen, with the GRAI-3 specimen having the highest measurement for expansion level among the slabs. Fig. 9 demonstrates the relationship between the test measurements and the damage quantities obtained for the three CANMET blocks and the three GRAI slabs. Fig. 9(a) presents a comparison of the total amount of crack damage and expansion levels, Fig. 9(b) is a comparison of the total crack length and expansion levels, and Fig. 9(c) is a comparison of total crack damage and P-wave velocities.

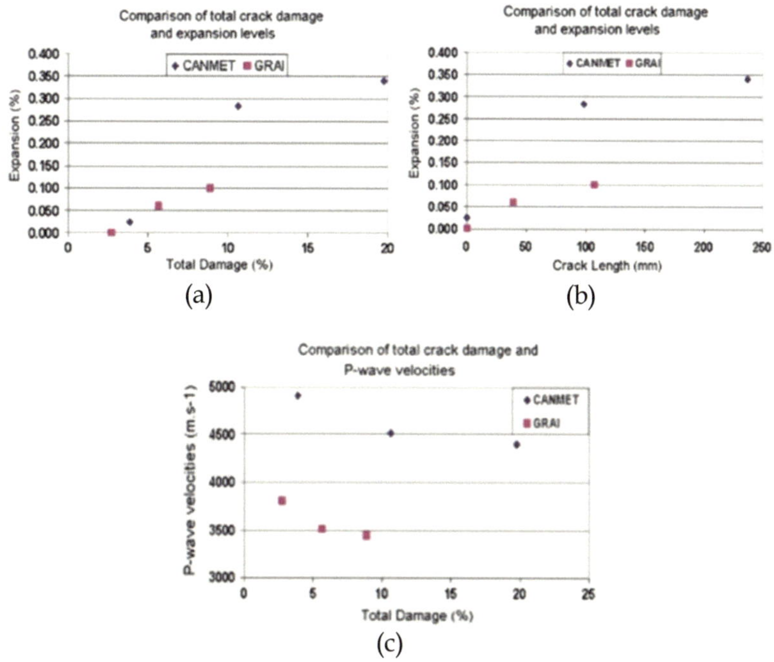

Fig. 9. Comparison of test measurements and damage information: (a) Crack damage vs expansion levels, (b) Crack length vs expansion levels and (c) Crack damage vs P-wave velocities

5. Recommendations for future research and development

There are many things that need to be done in improving the quality of NDT inspections using IRT. This includes the technology of IRT equipment, method of inspection, advanced methods of fault diagnosis and so on. Of course, this will involve various fields of study. Since the demand for NDT inspection reliability and condition monitoring is increasing, a robust and rapid analytical tool is required to do inspections. This part will highlight some recommendations for future research in order to improve the reliability inspection. The main factor that mostly affects the quality of inspection is the IRT equipment itself. Advances in manufacturing processes of thermographic detectors have dramatically increased both yield and quality while reducing production costs. However, the quality of inspection is related to the image resolution. Poor resolution will produce bad interpretation of inspection images. Therefore, for more accurate and correct data interpretation, it is recommended to use the latest technology for IRT cameras. Besides the resolution, the modern IRT cameras have very high thermal sensitivity. Some cameras even have the capability to adjust data measurements on screen, like object emissivity, temperature, etc.

In monitoring the condition of electrical equipments, the adoption of continuous thermal imaging can deliver increased benefits over periodic thermal inspection, especially in respect of mission critical electrical equipment. There is a big advantage to continuous thermal monitoring, since faults can occur at any time. In addition, it is not operator dependent, nor is it dependent on the time of inspection, which is usually when the equipment is running at load. Another benefit of real-time continuous monitoring is that the system can give warning signals or alarms if anomalies occur at any time. Therefore, action can be taken immediately. Perhaps one of the most important advantages of continuous monitoring is the ability to integrate continuous monitoring into existing Supervisory Control and Data Acquisition (SCADA) systems, enabling real-time remote monitoring without the need for separate systems or reports, which is something that cannot be achieved with periodic thermal inspections. Real-time imaging systems are not only rapid, compact and frequency agile, but they also have greater resolution, which improves imaging analysis and recording capabilities. Due to the high demand for preventive maintenance in electrical power systems, there is a need to have more reliable and robust intelligent systems. To date, most of the developed intelligent systems could not be used for all types of electrical equipments. This is due to the different features of the equipments. Therefore, a new intelligent system model has to be developed. Another issue in vision technology is the image quality. For inspections being done outdoors, the captured image will normally be affected by noise. Therefore, further study is needed on more advanced image processing techniques and developing new algorithms that could solve these problems.

Specifically in analyzing the civil concrete structure, studies can be conducted in order to determine which bands of RGB are more suitable for these two types of images in an effort to reduce the number of features that need to be computed, and results compared with those of greyscale imagery to see if there is any significant difference. Other topics for future studies can also be considered. One topic concerns the application of the statistical analysis. This research dealt with only first-order and second-order statistics; higher-order statistics were not commonly employed with remote-sensing imagery previously due to the computational costs involved when working with large image dimensions. Since concrete imagery has relatively much smaller image dimensions, and computer efficiency has steadily increased, the use of third-and higher-order statistics for the texture analysis of concrete imagery can also be further experimented.

Another subject is the development of a standard set-up for data acquisition, which would control the resolution and uniformity of large-scale data. Additional studies can comprise the development of a model for incorporating concrete image data from various NDT imaging techniques, such as optical images, which present image data of the surface, infrared thermography and acoustics, which are used for subsurface conditions, and ground penetrating radar, which is employed to obtain below-surface information of a structure. Furthermore, the image analysis model employed in this research has the potential to be developed as a component for automated damage assessment, which can be incorporated into a structural health monitoring system for concrete infrastructure. Automation of the system would allow for the assessment of a large volume of data, which could be used to establish a database of monitoring imagery, inspection results, etc. Since the imaging and inspection data can be stored in a digital format, image and data retrieval using metadata and content-based methods can be employed in order to compare the damage characteristics with previous inspection results and information. Data concerning a particular structure can be put together to form a three-dimensional representation of the condition using GIS techniques. This can aid in monitoring the condition of a structure; a history of inspection results can thus be examined and compared in order to quantitatively establish changes that occur with time.

6. Conclusion

In this chapter we covered briefly fundamentals of IRT imaging in NDT for enhancing the inspection reliability for both applications in monitoring the condition of electrical power equipment and damage detection in concrete structure. Inspection by utilizing thermal imaging especially for analyzing electrical installations and concrete structures presents many challenges due to the fact that these equipments and materials are non-homogeneous images. Some improvements in analysis methods need to be considered in order to avoid misinterpretation or inaccurate analysis of IRT data. Recent trends in IRT inspection showed that there is a need to apply an automatic intelligent system. A more advanced system could improve the quality of inspections. Therefore, some effort must be made to design a new approach of IRT inspection and to develop a better model for an intelligent system. For more complex image analysis, more robust and effective image processing techniques must be applied. Further development could enhance automatic processing capabilities in the form of automatic recognition of the measured objects and their critical parts.

7. Acknowledgment

This research was supported by Fundamental Research Grant Scheme (FRGS), Universiti Sains Malaysia (USM) and Universiti Malaysia Pahang (UMP).

8. References

Abdel-Qader, I., Abudayyeh, O., & Kelly, M. E. (2003). Analysis of Edge-Detection Techniques for Crack Identification in Bridges. *Journal of Computing in Civil Engineering*, Vol.17, No.4, pp.255.

Azmat, Z., & Turner, D. J. (2005). Infrared thermography and its role in rural utility environment. *Proceedings of Rural Electric Power Conference*, pp. B2/1-B2/4.

Balaras, C. A., & Argiriou, A. A. (2002). Infrared thermography for building diagnostics. *Energy and Buildings*, Vol.34, No.2, pp. 171-183.

Baoshu Li, Xiaohui Zhu, Shutao Zhao, & Wendong Niu. (2006). HV Power Equipment Diagnosis Based on Infrared Imaging Analyzing. *Proceedings of International Conference on Power System Technology*, pp. 1-4.

Bérubé, M.-A., Durand, B., Vézina, D., & Fournier, B. (2000). Alkali-aggregate reactivity in Québec (Canada). *Canadian Journal of Civil Engineering*, Vol.27, No.2, pp. 226-245.

Braunovic, M. (2007). *Electrical contacts fundamentals, applications and technology*. CRC Press, ISBN 9781574447279, Boca Raton.

Cao, Y., Gu, X.-ming, & Jin, Q. (2008). Infrared technology in the fault diagnosis of substation equipment. *Proceedings of 2008 China International Conference on Electricity Distribution*, pp. 1-6, Guangzhou, China.

Epperly, R. A., Heberlein, G. E., & Eads, L. G. (1997). A tool for reliability and safety: predict and prevent equipment failures with thermography. *Petroleum and Chemical Industry Conference, 1997. Record of Conference Papers. The Institute of Electrical and Electronics Engineers Incorporated Industry Applications Society 44th Annual*, pp. 59-68.

Foucher, S., Benie, G. B., & Boucher, J.-M. (2001). Multiscale MAP filtering of SAR images. *IEEE Transactions on Image Processing*, Vol.10, No.1, pp. 49-60.

Griffith, B., Türler, D., & Goudey, H. (2001). *Infrared thermographic systems: A review of IR imagers and their use*. Berkeley CA: John Wiley and Sons.

Haralick, R.M. (1979). Statistical and structural approaches to texture. *Proceedings of the IEEE*, Vol.67, No.5, pp. 786-804.

Haralick, Robert M., Shanmugam, K., & Dinstein, I. (1973). Textural Features for Image Classification. *IEEE Transactions on Systems, Man, and Cybernetics*, Vol.3, No.6, pp. 610-621.

He, D.-C., & Wang, L. (1991). Texture features based on texture spectrum. *Pattern Recognition*, Vol.24, No.5, pp. 391-399.

Hellier, C. (2001). *Handbook of Nondestructive Evaluation* (1st ed.). McGraw-Hill Professional. ISBN 0070281211

Holst, G. (2000). *Common sense approach to thermal imaging*. SPIE Optical Engineering Press. ISBN 9780819437228, Bellingham Wash.

Junyan, L., Jingmin, D., Yang, W., Hui, L., & Zijun, W. (2008). An IR lock-in thermography nondestructive test system Based on the image sequence processing. *Proceedings of 17th World Conference on Nondestructive Testing*. Shanghai, China.

Karathanassi, V., Iossifidis, C., & Rokos, D. (2000). A texture-based classification method for classifying built areas according to their density. *International Journal of Remote Sensing*, Vol.21, No.9, pp. 1807-1823.

Korendo, Z., & Florkowski, M. (2001). Thermography based diagnostics of power equipment. *Power Engineering Journal*, Vol.15, No.1, pp. 33-42.

Kregg, M. A. (2004). Benefits of using infrared thermography in utility substations. *Thermosense XXVI*, pp. 249-257,Orlando, FL, USA.

Ksantini, R. (2003). *Analyse multirésolution et recherche d'images.* M.Sc. Thesis, University of Sherbrooke, Sherbrooke, Québec, Canada.

Laurentys Almeida, C. A., Braga, A. P., Nascimento, S., Paiva, V., Martins, H. J. A., Torres, R., & Caminhas, W. M. (2009). Intelligent Thermographic Diagnostic Applied to Surge Arresters: A New Approach. *IEEE Transactions on Power Delivery*, Vol.24, No.2, pp. 751-757.

Lindquist, T. M., Bertling, L., & Eriksson, R. (2005). Estimation of disconnector contact condition for modelling the effect of maintenance and ageing. *in Power Tech, 2005 IEEE Russia*, pp. 1-7.

Lindquist, T. M., & Bertling, L. (2008). Hazard rate estimation for high-voltage contacts using infrared thermography. *Proceedings of Reliability and Maintainability Symposium*, pp. 231-237.

Livens, S., Scheunders, P., Van de Wouwer, G., Van Dyck, D., Smets, H., Winkelmans, J., & Bogaerts, W. (1996). A Texture Analysis Approach to Corrosion Image Classification. *Microscopy, microanalysis, microstructures*, Vol.7, pp. 143-152.

Lowe, D. G. (2004). Distinctive Image Features from Scale-Invariant Keypoints. *International Journal of Computer Vision*, Vol.60, No.2, pp. 91-110.

Maldague, X. (2000). Applications Of Infrared Thermography In Nondestructive Evaluation. *Trends in Optical Nondestructive Testing (invited chapter)*. Available from http://citeseerx.ist.psu.edu/viewdoc/summary?doi=10.1.1.33.2908

Martínez, J., & Lagioia, R. (2007). Experience performing infrared thermography in the maintenance of a distribution utility. *Proceedings of 19th International Conference on Electricity Distribution*, Vienna: CIRED.

Minkina, W., & Dudzik, S. (2009). *Infrared Thermography*. John Wiley & Sons, Ltd. ISBN 9780470682234, Chichester, UK

Mobley, R. (2002). *An introduction to predictive maintenance*. Butterworth-Heinemann, ISBN 9780750675314, Amsterdam;New York.

Niancang Hou. (1998). The infrared thermography diagnostic technique of high-voltage electrical equipments with internal faults. *Proceedings of 1998 International Conference on Power System Technology*, Vol.1, pp. 110-115.

de Oliveira, J. H. E., & Lages, W. F. (2010). Robotized inspection of power lines with infrared vision. *Proceedings of 2010 1st International Conference on Applied Robotics for the Power Industry (CARPI 2010)*, pp. 1-6, Montreal, QC, Canada.

Rahmani, A., Haddadnia, J., & Seryasat, O. (2010). Intelligent fault detection of electrical equipment in ground substations using thermo vision technique. *Proceedings of 2010 2nd International Conference on Mechanical and Electronics Engineering*, Vol. 2, pp. V2-150-V2-154.

Rivard, P., & Ballivy, G. (2005). Assessment of the expansion related to alkali-silica reaction by the Damage Rating Index method. *Construction and Building Materials*, Vol.19, No.2, pp. 83-90.

dos Santos, L., Bortoni, E. C., Souza, L. E., Bastos, G. S., & Craveiro, M. A. C. (2008). Infrared thermography applied for outdoor power substations. *Thermosense XXX*, Vol.6939, pp. 69390R-11). Orlando, FL, USA.

Schowengerdt, R. A. (1997). *Remote Sensing, Second Edition: Models and Methods for Image Processing*. Academic Press. ISBN 0126289816.

Scott, M., Rezaizadeh, A., Delahaza, A., Santos, C. G., Moore, M., Graybeal, B., & Washer, G. (2003). A comparison of nondestructive evaluation methods for bridge deck assessment. *NDT & E International*, Vol.36, No.4, pp. 245-255.

Shafi'i, M. A., & Hamzah, N. (2010). Internal fault classification using Artificial Neural Network. *Proceedings of 2010 4th International Power Engineering and Optimization Conference*, pp. 352-357.

Smedberg, M. (2006). *Thermographic Decision Support – Detecting and Classifying Faults in Infrared Images*. M.Sc Thesis. Royal Institute of Technology, Stockholm, Sweden.

Snell, J., & Renowden, J. (2000). Improving results of thermographic inspections of electrical transmission and distribution lines. *Proceedings of IEEE 9th International Conference on Transmission and Distribution Construction, Operation and Live-Line Maintenance Proceedings*. pp. 135-144.

Snell, Jr., & Spring, R. W. (2003). The new approach to prioritizing anomalies found during thermographic electrical inspections. *Thermosense XXV*, Vol.5073, pp. 222-230, Orlando, FL, USA.

Standard for Infrared Inspection of Electrical Systems & Rotating Equipment. (2008). Infraspection Institute. Available from http://www.armco-inspections.com/ files/ir/-Electrical%-20Rotating%-20Std.pdf

Stimolo M (2003), Passive infrared thermography as inspection and observation in bridge and road construction, International Symposium NDT-CE, Available from www.ndt/articleindice03/papers/v083.htm

Utami, N. Y., Tamsir, Y., Pharmatrisanti, A., Gumilang, H., Cahyono, B., & Siregar, R. (2009). Evaluation condition of transformer based on infrared thermography results.

Proceedings of 2009 IEEE 9th International Conference on the Properties and Applications of Dielectric Materials. pp. 1055-1058, Harbin, China.

Verma Anuj Kumar, Ray, A. K., Singh, S. P., Banerjee, D., & Schabel Samuel. (2009). A Review of Recent Advances in the Use of Thermography in Pulp and Paper Industry. *Indian Pulp and Paper Technical Association (IPPTA)*, Vol.21, No.2, pp. 55-58.

Wong, W. K., Tan, P. N., Loo, C. K., & Lim, W. S. (2009). An Effective Surveillance System Using Thermal Camera. *Proceedings of 2009 International Conference on Signal Acquisition and Processing*, pp. 13-17, Kuala Lumpur, Maylaysia.

Wretman, D. (2006). *Finding Regions of Interest in a Decision Support System for Analysis of Infrared Images*. M.Sc Thesis). Royal Institute of Technology, Stockholm, Sweden.

Ying-Chieh Chou, & Yao, L. (2009). Automatic Diagnostic System of Electrical Equipment Using Infrared Thermography. *Proceedings of International Conference of Soft Computing and Pattern Recognition*, pp. 155-160.

Younus, A. M., & Bo-Suk Yang. (2010). Wavelet co-efficient of thermal image analysis for machine fault diagnosis. *Proceedings of Prognostics and Health Management Conference*, pp. 1-6.

Zu, Y. K., Tian G. Y., Lu R. S. & Zhang H. (2011). A Review of Optical NDT Technologies. *Sensors, 11*, pp 7773-7798, www.mdpi.com/journal/sensors